Biodegradable Polymers as Drug Delivery Systems

DRUGS AND THE PHARMACEUTICAL SCIENCES

A Series of Textbooks and Monographs

Edited by
James Swarbrick
School of Pharmacy
University of North Carolina
Chapel Hill, North Carolina

Volume 1. PHARMACOKINETICS, *Milo Gibaldi and Donald Perrier*
(out of print)

Volume 2. GOOD MANUFACTURING PRACTICES FOR
PHARMACEUTICALS: A PLAN FOR TOTAL QUALITY
CONTROL, *Sidney H. Willig, Murray M. Tuckerman, and
William S. Hitchings IV* (out of print)

Volume 3. MICROENCAPSULATION, *edited by J. R. Nixon*

Volume 4. DRUG METABOLISM: CHEMICAL AND BIOCHEMICAL
ASPECTS, *Bernard Testa and Peter Jenner*

Volume 5. NEW DRUGS: DISCOVERY AND DEVELOPMENT,
edited by Alan A. Rubin

Volume 6. SUSTAINED AND CONTROLLED RELEASE DRUG DELIVERY
SYSTEMS, *edited by Joseph R. Robinson*

Volume 7. MODERN PHARMACEUTICS, *edited by Gilbert S.
Banker and Christopher T. Rhodes*

Volume 8. PRESCRIPTION DRUGS IN SHORT SUPPLY: CASE
HISTORIES, *Michael A. Schwartz*

Volume 9. ACTIVATED CHARCOAL: ANTIDOTAL AND OTHER
MEDICAL USES, *David O. Cooney*

Volume 10. CONCEPTS IN DRUG METABOLISM (in two parts), *edited
by Peter Jenner and Bernard Testa*

Volume 11. PHARMACEUTICAL ANALYSIS: MODERN METHODS
(in two parts), *edited by James W. Munson*

Volume 12. TECHNIQUES OF SOLUBILIZATION OF DRUGS,
edited by Samuel H. Yalkowsky

Volume 13. ORPHAN DRUGS, *edited by Fred E. Karch*

Volume 14. NOVEL DRUG DELIVERY SYSTEMS: FUNDAMENTALS, DEVELOPMENTAL CONCEPTS, BIOMEDICAL ASSESSMENTS, *Yie W. Chien*

Volume 15. PHARMACOKINETICS, Second Edition, Revised and Expanded, *Milo Gibaldi and Donald Perrier*

Volume 16. GOOD MANUFACTURING PRACTICES FOR PHARMACEUTICALS: A PLAN FOR TOTAL QUALITY CONTROL, Second Edition, Revised and Expanded, *Sidney H. Willig, Murray M. Tuckerman, and William S. Hitchings IV*

Volume 17. FORMULATION OF VETERINARY DOSAGE FORMS, *edited by Jack Blodinger*

Volume 18. DERMATOLOGICAL FORMULATIONS: PERCUTANEOUS ABSORPTION, *Brian W. Barry*

Volume 19. THE CLINICAL RESEARCH PROCESS IN THE PHARMACEUTICAL INDUSTRY, *edited by Gary M. Matoren*

Volume 20. MICROENCAPSULATION AND RELATED DRUG PROCESSES, *Patrick B. Deasy*

Volume 21. DRUGS AND NUTRIENTS: THE INTERACTIVE EFFECTS, *edited by Daphne A. Roe and T. Colin Campbell*

Volume 22. BIOTECHNOLOGY OF INDUSTRIAL ANTIBIOTICS, *Erick J. Vandamme*

Volume 23. PHARMACEUTICAL PROCESS VALIDATION, *edited by Bernard T. Loftus and Robert A. Nash*

Volume 24. ANTICANCER AND INTERFERON AGENTS: SYNTHESIS AND PROPERTIES, *edited by Raphael M. Ottenbrite and George B. Butler*

Volume 25. PHARMACEUTICAL STATISTICS: PRACTICAL AND CLINICAL APPLICATIONS, *Sanford Bolton*

Volume 26. DRUG DYNAMICS FOR ANALYTICAL, CLINICAL, AND BIOLOGICAL CHEMISTS, *Benjamin J. Gudzinowicz, Burrows T. Younkin, Jr., and Michael J. Gudzinowicz*

Volume 27. MODERN ANALYSIS OF ANTIBIOTICS, *edited by Adorjan Aszalos*

Volume 28. SOLUBILITY AND RELATED PROPERTIES, *Kenneth C. James*

Volume 29. CONTROLLED DRUG DELIVERY: FUNDAMENTALS AND APPLICATIONS, Second Edition, Revised and Expanded, *edited by Joseph R. Robinson and Vincent H. L. Lee*

Volume 30. NEW DRUG APPROVAL PROCESS: CLINICAL AND REGULATORY MANAGEMENT, *edited by Richard A. Guarino*

Volume 31. TRANSDERMAL CONTROLLED SYSTEMIC MEDICATIONS, *edited by Yie W. Chien*

Volume 32. DRUG DELIVERY DEVICES: FUNDAMENTALS AND APPLICATIONS, *edited by Praveen Tyle*

Volume 33. PHARMACOKINETICS: REGULATORY · INDUSTRIAL · ACADEMIC PERSPECTIVES, *edited by Peter G. Welling and Francis L. S. Tse*

Volume 34. CLINICAL DRUG TRIALS AND TRIBULATIONS, *edited by Allen E. Cato*

Volume 35. TRANSDERMAL DRUG DELIVERY: DEVELOPMENTAL ISSUES AND RESEARCH INITIATIVES, *edited by Jonathan Hadgraft and Richard H. Guy*

Volume 36. AQUEOUS POLYMERIC COATINGS FOR PHARMACEUTICAL DOSAGE FORMS, *edited by James W. McGinity*

Volume 37. PHARMACEUTICAL PELLETIZATION TECHNOLOGY, *edited by Isaac Ghebre-Sellassie*

Volume 38. GOOD LABORATORY PRACTICE REGULATIONS, *edited by Allen F. Hirsch*

Volume 39. NASAL SYSTEMIC DRUG DELIVERY, *Yie W. Chien, Kenneth S. E. Su, and Shyi-Feu Chang*

Volume 40. MODERN PHARMACEUTICS, Second Edition, Revised and Expanded, *edited by Gilbert S. Banker and Christopher T. Rhodes*

Volume 41. SPECIALIZED DRUG DELIVERY SYSTEMS: MANUFACTURING AND PRODUCTION TECHNOLOGY, *edited by Praveen Tyle*

Volume 42. TOPICAL DRUG DELIVERY FORMULATIONS, *edited by David W. Osborne and Anton H. Amann*

Volume 43. DRUG STABILITY: PRINCIPLES AND PRACTICES,
    Jens T. Carstensen
Volume 44. PHARMACEUTICAL STATISTICS: PRACTICAL AND
    CLINICAL APPLICATIONS, Second Edition, Revised
    and Expanded, *Sanford Bolton*
Volume 45. BIODEGRADABLE POLYMERS AS DRUG DELIVERY
    SYSTEMS, *edited by Mark Chasin and Robert Langer*

Additional Volumes in Preparation

Biodegradable Polymers as Drug Delivery Systems

edited by

Mark Chasin
Nova Pharmaceutical Corporation
Baltimore, Maryland

Robert Langer
Massachusetts Institute of Technology
Cambridge, Massachusetts

Marcel Dekker, Inc.   New York • Basel • Hong Kong

Library of Congress Cataloging-in-Publication Data

Biodegradable polymers as drug delivery systems/edited by Mark
    Chasin, Robert Langer.
        p.   cm. -- (Drugs and the pharmaceutical sciences; v. 45)
        Includes bibliographical references.
        Includes index.
        ISBN 0-8247-8344-1 (alk. paper) O9582/ .
        1. Drug delivery systems.   2. Polymers--Metabolism.   3. Drugs-
    -Vehicles--Biodegradation.   I. Chasin, Mark.   II. Langer, Robert S.
    III. Series
        [DNLM: 1. Biodegradation.   2. Drugs--administration & dosage.
    3. Infusion Pumps, Implantable.   4. Polymers--therapeutic use.   W1
    DR893B v. 45/QV 800 B615]
    RS199.5.B56   1990
    615'.19--dc20
    DNLM/DLC
    for Library of Congress   90-3908
                                                                  CIP

This book is printed on acid-free paper.

MARCEL DEKKER, INC.
270 Madison Avenue, New York, New York 10016

Current printing (last digit):
10 9 8 7 6 5 4 3 2 1

PRINTED IN THE UNITED STATES OF AMERICA

Preface

Over the past decade, the use of polymers for the administration of pharmaceutical and agricultural agents has increased dramatically. This field of <u>controlled release technology</u> has changed from being merely useful in research to having a significant clinical impact. As we look ahead toward advances in the next decade, one of the areas of greatest practical consequence in medical therapeutics of controlled release technology will be the development of biodegradable polymer systems. Such polymers offer the great advantage of enabling either site-specific or systemic administration of pharmaceutical agents without the need for subsequent retrieval of the delivery system. These polymers offer many other advantages as well.

While this book is focused on drug delivery, the value of biodegradable polymers is not limited to this field. Biodegradable polymers will be useful in other areas of medical therapeutics, such as sutures and bone plates and other types of prostheses. The polymers will also be useful in nonmedical fields, for disposable plastics, bottles, diapers and many other entities.

It was our intention in formulating this book to take selected polymers that have been widely studied and provide a comprehensive review of their properties, synthesis, and formulations. We hope that this will be useful to individuals who have been in the field for a long time and who would like to have all the information together in one place, as well as to individuals who are new to the field and would like to understand more about the various properties of biodegradable polymers.

Mark Chasin

Robert Langer

Contents

Preface   iii

Contributors   vii

1. Controlled Release of Bioactive Agents from Lactide/
   Glycolide Polymers   1
   Danny H. Lewis

2. Polyanhydrides as Drug Delivery Systems   43
   Mark Chasin, Abraham Domb, Eyal Ron, Edith Mathiowitz,
   Kam Leong, Cato Laurencin, Henry Brem, Stuart Grossman,
   and Robert Langer

3. Poly-ε-Caprolactone and Its Copolymers   71
   Colin G. Pitt

4. Poly(ortho esters)   121
   Jorge Heller, Randall V. Sparer, and Gaylen M. Zentner

5. Polyphosphazenes as New Biomedical and Bioactive
   Materials   163
   Harry R. Allcock

6. Pseudopoly(amino acids)   195
   Joachim Kohn

7. Natural Polymers as Drug Delivery Systems   231
    Simon Bogdansky

8. Liposomes   261
    *Ulla K. Nässander, Gert Storm, Pierre A. M. Peeters,
    and Daan J. A. Crommelin*

Index   *339*

Contributors

Harry R. Allcock Department of Chemistry, The Pennsylvania State University, University Park, Pennsylvania

Simon Bogdansky Department of Drug Delivery, Nova Pharmaceutical Corporation, Baltimore, Maryland

Henry Brem Departments of Neurosurgery, Ophthalmology, and Oncology, The Johns Hopkins University School of Medicine, Baltimore, Maryland

Mark Chasin Department of Technology Development, Nova Pharmaceutical Corporation, Baltimore, Maryland

Daan J. A. Crommelin Department of Pharmaceutics, Faculty of Pharmacy, University of Utrecht, Utrecht, The Netherlands

Abraham Domb Department of Drug Delivery, Nova Pharmaceutical Corporation, Baltimore, Maryland

Stuart Grossman Department of Oncology, The Johns Hopkins Oncology Center, Baltimore, Maryland

Jorge Heller Controlled Release and Biomedical Polymers Department, SRI International, Menlo Park, California

Joachim Kohn Department of Chemistry, Rutgers, The State University of New Jersey, New Brunswick, New Jersey

Robert Langer Department of Chemical Engineering, Massachusetts Institute of Technology, Cambridge, Massachusetts

Cato Laurencin Division of Health Sciences and Technology, Massachusetts Institute of Technology, Cambridge, Massachusetts; Department of Orthopedic Surgery, Massachusetts General Hospital, Harvard Medical School, Boston, Massachusetts

Kam Leong Department of Biomedical Engineering, The Johns Hopkins University, Baltimore, Maryland

Danny H. Lewis Stolle Research and Development Corporation, Decatur, Alabama

Edith Mathiowitz Department of Chemical Engineering, Massachusetts Institute of Technology, Cambridge, Massachusetts

Ulla K. Nässander Department of Pharmaceutics, Faculty of Pharmacy, University of Utrecht, Utrecht, The Netherlands

Pierre A. M. Peeters* Department of Pharmaceutics, Faculty of Pharmacy, University of Utrecht, Utrecht, The Netherlands

Colin G. Pitt Pharmaceutics and Drug Delivery, Amgen Inc., Thousand Oaks, California

Eyal Ron[†] Massachusetts Institute of Technology, Cambridge, Massachusetts

Randall V. Sparer Inter$_x$ Research Corporation, Merck, Sharp & Dohme Research Laboratories, Lawrence, Kansas

Gert Storm[‡] Department of Pharmaceutics, Faculty of Pharmacy, University of Utrecht, Utrecht, The Netherlands

Gaylen M. Zentner Inter$_x$ Research Corporation, Merck, Sharp & Dohme Research Laboratories, Lawrence, Kansas

Current Affiliation:
*Institute for Pharmaceutical and Biomedical Consultancy, Pharma Bio-Research International B. V., Assen, The Netherlands
†Genetics Institute, Cambridge, Massachusetts
‡Liposome Technology, Inc., Menlo Park, California

Biodegradable Polymers as Drug Delivery Systems

1

Controlled Release of Bioactive Agents from Lactide/Glycolide Polymers

DANNY H. LEWIS   Stolle Research and Development Corporation,
Decatur, Alabama

I.   INTRODUCTION

For more than two decades, the delivery of bioactive agents from
polymeric materials has attracted the considerable attention of investigators throughout the scientific community.   Polymer chemists, chemical engineers, pharmaceutical scientists, and entomologists are among
those seeking to design predictable, controlled delivery systems for
bioactive agents ranging from insulin to rodenticides.   This challenge
has been a formidable one to date as evidenced by the small number
of fully developed products based on this concept.   One must realize,
however, that only in the past 10 years with the onsurge of biotechnology has research in controlled drug delivery benefited from the
intense dedication of resources.   With the availability of new molecules,
often with short biological half-lives and relatively high molecular
weights, the need for reliable controlled release systems has been
apparent.
      The trend in drug delivery technology has been toward biode-
gradable polymer excipients requiring no follow-up surgical removal
once the drug supply is depleted.   The advantages of biodegradable
polymers have been described (1-3).   Unfortunately, investigators
seeking advanced drug delivery systems are severely limited in can-
didate polymeric materials as evidenced by the relatively small number
of systems described in this text.   Historically, designers of drug
delivery systems have "borrowed" polymeric materials originally developed for other applications.   Only one or two synthetic polymers
have been developed specifically for use in controlled release formulations.

The most widely investigated and advanced polymers in regard
to available toxicological and clinical data are the aliphatic polyesters
based on lactic and glycolic acids. The family of homo- and copolymers derived from these monomers has received considerable attention
since about 1973 as excipients for drug delivery. Features such as
biocompatibility, predictability of biodegradation kinetics, ease of
fabrication, and regulatory approval in commercial suture applications
have attracted investigators to lactic/glycolic polymers. During the
1970s, a wealth of literature on polyglycolic acid sutures became
available. Much of this pioneering work resulted from studies at
U.S. suture companies including American Cyanamid and Johnson
and Johnson, and also at the U.S. Army Institute of Dental Research
(4,5). Those early studies clearly demonstrated the nontoxic nature
of the polymers and provided biodegradation data for various types
of implants. Although the bulk of this work was aimed at suture applications, the potential for drug delivery from these polymers became quite obvious.

The biodegradable polyesters have attracted attention in a variety
of biomaterial applications including tracheal replacement (6), ventral
herniorrhaphy (7), ligament reconstruction (8,9), dental repairs (101),
fracture repair (11-15), and surgical dressings (16). Among the first
reports of polylactic acid used for controlled release were those of
Boswell (17), Yolles (18), Wise (19), Sinclair (20), and Beck (21).
These research teams were seeking delivery systems for such agents
as narcotic antagonists, contraceptive hormones, and other conventional drug compounds. Early efforts were directed to the homopolymer of lactic acid rather than the copolymers. This was primarily
due to the limited availability of the glycolide comonomer. Recently,
the full range of monomers and polymers has become rather easily
accessible through major chemical companies. DuPont now provides
the lactide/glycolide polymers under the trade name Medisorb. This
availability of materials has greatly broadened the scope of possibilities for designing drug delivery systems.

II. SYNTHESIS

The homo- and copolymers of lactic and glycolic acids are synthesized
by the ring-opening melt condensation of the cyclic dimers, lactide
and glycolide (22,23). Due to the asymmetrical β carbon of lactic
acid, D and L stereoisomers exist, and the resulting polymer can be
either D, L, or racemic DL. The polymerizations are usually conducted over a period of 2-6 hr at about 175°C in the melt. Organo-
tin catalysts are normally utilized with stannous chloride and stannous
octoate being the most common. Other catalysts such as p-toluene
sulfonic acid and antimony trifluoride have been successfully employed
on a limited basis. Lauryl alcohol is often added to control molecular
weight during synthesis.

   As with most polymerizations, monomer purity is highly critical
in the synthesis of polylactides. Differential scanning calorimetry
(DSC) purity of 99.9% or greater is usually required with the starting lactide and glycolide materials. Low monomer acidity is also a
critical parameter. Free acid of 0.05% or less is normally required
for achieving a high molecular weight polymer. Of equal importance,
however, are the environmental conditions, particularly humidity levels, in the processing areas. Most failed glycolide polymerizations
can be traced to high levels of humidity or high monomer acidity.

III. POLYMER CHARACTERISTICS

A well-proven advantage of the lactide/glycolide polymers is no doubt
the available versatility in polymer properties and performance characteristics. For wide applications in controlled drug delivery, it is
imperative that a range of rates and durations of drug release be
achievable. A broad spectrum of performance characteristics with
the polylactides can be obtained by careful manipulation of four key
variables: monomer stereochemistry, comonomer ratio, polymer chain
linearity, and polymer molecular weight. Because the mechanism of
biodegradation is simple hydrolysis of the ester linkages, it is apparent how each of these factors plays an important role in the in
vivo performance of the lactide/glycolide materials. Crystallinity and
water uptake are key factors in determining the rates of in vivo degradation.
   The racemic poly(DL-lactide) DL-PLA is less crystalline and lower
melting than the two stereoregular polymers, D-PLA and L-PLA. Further, the copolymers of lactide and glycolide are less crystalline than
the two homopolymers of the two monomers. In addition, the lactic
acid polymer, because of the methyl group, is more hydrophobic than
the glycolide polymer.
   Table 1 provides a summary of the glass transition temperatures
of several lactide/glycolide polymers. Tg values range from about
40 to 65°C. Poly(L-lactide) has the highest Tg at about 65°C.
   Rate of hydration of the polymeric materials has been shown to
be an important consideration in regard to drug release. Gilding
and Reed (24) demonstrated that water uptake increases as the glycolide ratio in the copolymer increases. The extent of block or random structure in the copolymer can also affect the rate of hydration
and the rate of degradation (25). Careful control of the polymerization conditions is required in order to afford reproducible drug release behavior in a finished product. Kissel (26) showed drastic
differences in water uptake between various homopolymers and copolymers of caprolactone, lactide, and glycolide.
   Siemann (27) recently determined the solubility parameters and
densities of a group of biodegradable polyesters. Solubility parameters

TABLE 1   Glass Transition Temperatures of Aliphatic Polyesters

Polymer	Glass transition temp., Tg(°C)	Melting temp., Tm(°C)
Poly(L-lactide)	60−67	172−174
Poly(DL-lactide)	57−59	None
Poly(glycolide)	36	230
85:15 DL-lactide/glycolide copolymer	45	None
25:75 DL-lactide/L-lactide copolymer	60	None
70:30 L-lactide/glycolide copolymer	58	None
Polycaprolactone	−65	63

were in the 16.2−16.8 range while the densities of the polymers tested were 1.15−1.29 g/cm^3. The solubility parameters were comparable to those of polystyrene (17.6−19) and polyisoprene (16−17).

Solubility of the polymers in common organic solvents is an important factor in regard to fabrication of drug delivery systems. The homopolymers from DL-, D-, and L-lactide are quite soluble in halogenated hydrocarbons, ethyl acetate, tetrahydrofuran (THF), dioxane, and a few other solvents. At glycolide contents of less than 50%, lactide/glycolide copolymers display characteristics similar to those of the lactide homopolymers. Polyglycolic acid and the glycolide-rich copolymers are quite insoluble materials. Exotic solvents such as hexafluoroisopropanol are often used in analytical procedures as solvents for those polymers.

Occasionally in the synthesis of the copolymers, insoluble material is produced. This results from polymer containing blocks of polyglycolide rather than the desired random structure. Obviously, such compositions would have considerable effect on the performance of controlled release formulations utilizing those polymers. This problem is particularly evident when one is seeking to utilize the 50:50 glycolide/lactide copolymer as a biodegradable excipient. However, with carefully controlled polymerization conditions, useful 50:50 polymer is readily produced.

A novel lactide composition involving interlocking segments of poly(L-lactide) with segments of poly(D-lactide) has been reported (28). The segmental interlocking can afford a novel crystalline phase which has a crystalline melting point higher than that of either component. The advantages of such compositions are expected to be in improved mechanical properties of fabricated devices. Although not yet extensively tested, these polymers may offer some advantages in drug delivery as well because of crystallinity and possible water uptake differences.

IV. BIODEGRADATION

Biodegradation of the aliphatic polyesters occurs by bulk erosion. The lactide/glycolide polymer chains are cleaved by hydrolysis to the monomeric acids and are eliminated from the body through the Krebs cycle, primarily as carbon dioxide and in urine. Because the rate of hydrolysis of the polymer chain is dependent only on significant changes in temperature and pH or presence of catalyst, very little difference is observed in rate of degradation at different body sites. This is obviously an advantage in regard to drug delivery formulations.

The role of enzymatic involvement in the biodegradation of the lactide/glycolide polymer has been somewhat controversial. Most early literature concluded that bioerosion of these materials occurred strictly through hydrolysis with no enzymatic involvement. Other investigators, including Williams (29,30), Herrmann (31), and Reed (32), suggest that enzymes do play a significant role in the breakdown of the lactide/glycolide materials. Much of this speculation is based on differences observed between in vivo and in vitro degradation rates. Holland et al. (33) concluded that little enzyme involvement is expected in the early stages with polymers in the glassy state, whereas enzymes can play a significant role for polymers in the rubbery state.

Occasionally, an unusual combination of polymeric excipient and bioactive agent can lead to unexpected biodegradation profiles. Maulding (34) found that the tertiary amino compound thioridazine (free base) accelerated the degradation rate of poly(DL-lactide) in vitro and in vivo. Drug release profiles and gel permeation chromatography (GPC) data showed that the amine influenced the rate of hydrolysis of the polymer. A unique basic functional group is apparently required for this behavior as several other amine-containing compounds have been incorporated into lactide/glycolide polymers without this enhancement in the rate of biodegradation.

Many investigators have studied the in vivo degradation kinetics of lactide/glycolide materials (5,35-39). There has been some confusion in the interpretation of results primarily because of lack of consistency in nomenclature and careful attention in describing the specific stereoisomers evaluated. Nevertheless, the overall degradation kinetics are fairly well established for the entire family of homopolymers and copolymers. At the present, this common knowledge of the in vivo lifetimes of various lactide/glycolide polymers is a primary reason for their popularity.

Ethoxylation of the carboxylic acid end groups of aliphatic polyesters significantly changes the biodegradation rate as well as the
crystallinity of these materials (41).

The most pertinent degradation studies in regard to providing
useful data for designers of drug delivery systems are those of Miller
(35), Pitt (36), and Tice (38). Other investigators (39) reported
studies with various suture materials and surgical implants; however,
these data, although related, are not as directly applicable to drug
release.

The earlier work of Miller (35), Cutright (37), and Brady (5)
on nonmedicated implants provided an excellent basis for further
studies on specific controlled release formulations such as the determination of the biodegradation rates of lactide/glycolide drug-loaded
microspheres (38). Those studies were done with ^{14}C-labeled polymers produced from DL-lactic acid and glycolide. The final formulations tested in rats were microspheres loaded with ^{3}H-labeled steroid
and ^{14}C polymer as the matrix. The microspheres were administered
intramuscularly and animals were serially sacrificed over a period of
about a year.

Figure 1 shows the results of those experiments. It is important
to recognize that in this experiment the polymer was present in a
form affording high surface area when compared to other studies
where rods, pellets, plates, fibers, and so forth were used. It is
often quite difficult to compare exact degradation times from various
independent studies due to differences in implant surface area and

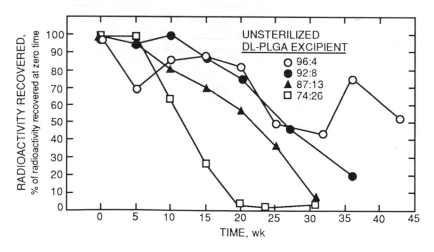

FIGURE 1 Biodegradation of unsterilized NET microcapsules (63–125
μm) as determined by radioactive recoveries from injection sites of
rats. (From Ref. 38.)

porosity. One should exercise caution in the interpretation of various degradation data.

Pitt (36) determined that the first stage in the biodegradation process was only a decrease in molecular weight caused by random hydrolytic cleavage of the ester linkage. The second stage was onset of weight loss and a change in the rate of chain scission.

It can be summarized (Table 2) that the 50:50 lactide/glycolide copolymer has the fastest degradation rate of the DL-lactide/glycolide materials, with that polymer degrading in about 50–60 days. The 65:35, 75:25, and 85:15 DL-lactide/glycolides have progressively longer in vivo lifetimes, with the 85:15 lasting about 150 days in vivo, whereas poly(DL-lactide) requires about 12–16 months to biodegrade completely. Poly(L-lactide), being more crystalline and less hydrophilic, can be found in vivo at 1-1/2 to 2 years. A review of the biodegradation literature of lactide/glycolides was published (33).

Recently, Brich and coworkers (40) reported the synthesis of lactide/glycolide polymers branched with different polyols. Polyvinylalcohol and dextran acetate were used to afford polymers exhibiting degradation profiles significantly different from that of linear polylactides. The biphasic release profile often observed with the linear polyesters was smoothened somewhat to a monophasic profile. Further, the overall degradation rate is accelerated. It was speculated that these polymers can potentially afford more uniform drug release kinetics. This potential has not yet been fully demonstrated.

Developers of controlled release formulations have employed polymers produced from both L-lactide and DL-lactide. In terms of clinical studies, however, it appears that perhaps the DL-lactide formulations have been somewhat more successful. It is unclear if this is due to the DL-lactide materials being less crystalline and more permeable to most drugs or perhaps more sophisticated techniques and

TABLE 2 Biodegradation of Lactide/Glycolide Polymers

Polymer	Approximate time for biodegradation (months)[a]
Poly(L-lactide)	18–24
Poly(DL-lactide)	12–16
Poly(glycolide)	2–4
50:50 (DL-lactide-co-glycolide)	2
85:15 (DL-lactide-co-glycolide)	5
90:10 (DL-lactide-co-caprolactone)	2

[a]Biodegradation times vary depending on implant surface area, porosity, and molecular weight.

approaches followed with those polymers, or simply to the greater availability of the DL-lactide polymers.

The 50:50 DL-lactide/glycolide copolymer is the excipient of choice for many drug delivery systems designed for a 30-day duration of action. Vischer et al. (42) demonstrated that microspheres of 50:50 poly(DL-lactide-co-glycolide) containing lypressin could no longer be detected histologically or by ^{14}C tracer at day 63 in rats. For consecutive injections the 50:50 polymer offers the advantage that the polymer is essentially depleted by 50–60 days from the first treatment and no buildup of residual polymer occurs.

Several investigators have utilized the 85:15 DL-lactide/glycolide copolymer for 90-day delivery of bioactive agents. This polymer is essentially bioeroded by about 150 days, thus making the 85:15 a useful matrix for 90-day systems.

The interaction of biodegradable microspheres of lactide/glycolide polymers with blood constituents has been studied in vitro (43,44). Polyglycolide and a 90:10 glycolide/L-lactide suture polymer (Polyglactin) were tested in the form of microspheres about 1.6 μm in diameter. The microspheres were found to interact with cultured monocytes-macrophages. Biodegradation or digestion was evident within 78 hr. These results support the feasibility of targeting drug to macrophages with biodegradable polymers.

V. FABRICATION TECHNIQUES

One of the reasons for the popularity of the lactide/glycolide materials in drug delivery is their relative ease of fabrication. Three general types of drug delivery systems based on these polymers have been investigated: microparticles, implants, and fibers. The lactide/glycolide polymers are generally low-melting thermoplastics with good solubility in common solvents, polyglycolic acid and glycolide-rich copolymers being the exceptions. These favorable characteristics have allowed investigators considerable flexibility in the fabrication of drug delivery formulations.

A. Microparticles

Microspheres and microcapsules of lactide/glycolide polymers have received the most attention in recent years. Generally, three microencapsulation methods have been employed to afford controlled release formulations suitable for parenteral injection: (1) solvent evaporation, (2) phase separation, and (3) fluidized bed coating. Each of these processes requires lactide/glycolide polymer soluble in an organic solvent.

A solvent evaporation procedure particularly useful for entrapment of water-insoluble agents in the biodegradable polymers has been developed (45). Basically, the process involves dissolving the polymer in a volatile organic solvent, adding the drug to that organic phase, emulsifying the organic phase in water, and finally removing the solvent under vacuum to form discrete, hardened monolithic microspheres. The reproducibility and scale-up of this process was demonstrated recently (46) with a 90-day contraceptive formulation. In vitro release profiles on phase II and phase III clinical supplies prepared more than 2 years apart are shown in Fig. 2. Several thousand doses were prepared for the phase III trial initiated in 1988. Figure 3 shows the reproducibility of six individual batches of microspheres produced by the solvent evaporation method. Other studies have been reported with similar processes (47).

Phase separation microencapsulation procedures are suitable for entrapping water-soluble agents in lactide/glycolide excipients. Generally, the phase separation process involves coacervation of the polymer from an organic solvent by addition of a nonsolvent such as silicone oil. This process has proven useful for microencapsulation of water-soluble peptides and macromolecules (48).

Processes based on fluidized bed coating have been developed (49). In this process, the bioactive agent is dissolved in an organic solvent along with the polymer. This solution is then processed through a Wurster air suspension coater apparatus to form the final microcapsule product. A solvent partition technique based on continuous injection of a polymer-drug solution into flowing mineral oil has been reported (50).

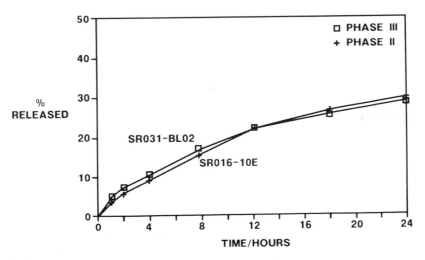

FIGURE 2 Comparison of in vitro release profiles of phase II and phase III clinical batches of 90-day norethisterone microspheres.

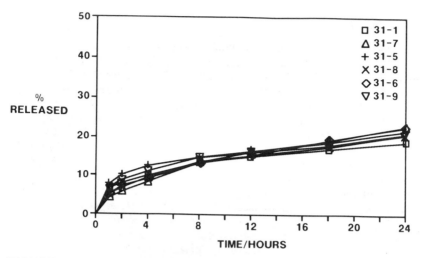

FIGURE 3 Batch-to-batch reproducibility of 90-day norethisterone
microspheres as determined by in vitro release profiles.

There are literally dozens of patents and literature references
addressing the microencapsulation of bioactive agents in polymers.
It appears that none of these process techniques is dominant, at
least with the lactide/glycolide materials. Researchers have considerable choices available in regard to fabrication of microspheres from
these polymers. The most commonly used procedures employ relatively
mild conditions of pH and temperature and are usually quite compatible
with the bioactive agents to be entrapped, including proteins and
other macromolecules. Only in the case of live virus and living cell
encapsulation have serious deactivation problems been encountered
and those problems were due to solvents used in the process.

B. Implants

A variety of macroimplants based on lactide/glycolide polymers have
been fabricated and studied in recent years. Compression molding,
injection molding, and screw extrusion have been utilized to load a
wide range of bioactive agents in the biodegradable polymers. Implants a few millimeters to several centimeters in size have been
tested for drug delivery.

Although implants are rather easily fabricated, problems have
occurred in instances when care is not taken in maintaining an ultra-
dry processing environment. It is extremely important to thoroughly
dry the bulk polymer and the bioactive agent, usually at ambient
temperature under vacuum, prior to processing. Dry nitrogen

blankets over critical process equipment such as extrusion feedhoppers are essential. The limiting factor, in regard to melt processing of implants for drug delivery, is of course the heat stability of the active agent. Most of the lactide/glycolide copolymers are injection-molded at temperatures between 140 and 175°C. L-lactide requires 190–220°C for most molding operations. Many of the macromolecules of current interest are unstable at these conditions.

Residual monomer in the bulk polymer is a detrimental factor as is high moisture content. Monomer levels greater than 2–3% by weight often cause substantial degradation of lactide/glycolide copolymers in injection molding operations. Decreases from about 150,000 to less than 40,000 weight-average molecular weight have been observed in molding 50:50 and 85:15 lactide/glycolide polymers with monomer contents of 5–10%. This loss in molecular weight in the presence of acid monomer, elevated temperature, and pressure is to be expected in view of the theory of the autocatalytic hydrolysis of the polymer chains from acid end groups normally present.

C. Fibers

Controlled release fiber systems based on aliphatic polyesters were investigated by Dunn and Lewis (51–54). The feasibility of hollow fibers spun from poly(L-lactide) and containing the contraceptive steroid levonorgestrel has been demonstrated (55).

Fabrication of drug-containing fibers is a natural progression when one considers the extensive history of lactide/glycolides in suture applications. The lactide/glycolide polymers are easily melt-spun into mono- or multifilament products at relatively low temperatures.

Two basic types of drug-loaded fibers have been reported. Fibers in which the drug is dissolved or dispersed throughout the polymer matrix yield a monolithic system and the expected first-order release kinetics (51,53). Monolithic formulations such as these can readily be produced with bench scale equipment such as melt rheometers commonly found in polymer laboratories. The active agent and polymer are simply blended and extruded under pressure at the lowest possible temperature. Up to 50–60% drug by weight can often be incorporated into the biodegradable fibers. Fibers fabricated in this manner exhibit very poor mechanical strength and usually require postspinning drawing to increase tensile properties. Fortunately, in most drug delivery applications, mechanical strength is of secondary importance.

Reservoir or coaxial fibers can be produced from the glycolide/lactide polymers (53,55). Two different methods have been investigated. Dunn (53) utilized a melt-spinning technique in which the drug was introduced during the spinning process as a suspension or solution in a suitable lumen fluid. Eenink (55) developed a dry-wet phase inversion process for poly(L-lactide) fibers. In this process, the drug must be added to the hollow fiber after the fibers are produced. This step appears rather tedious but offers the possibility of utilizing many drugs normally unstable under spinning conditions. Also, the solvent-based spinning procedure is less harmful to the heat- and moisture-sensitive lactide/glycolide polymers. A serious disadvantage is the environmental and handling problems with solvents and nonsolvents. Residual solvent in the finished fiber is also a potential concern.

Major advantages of drug-loaded fibers include ease of fabrication, high surface area for drug release, wide range of physical structures possible, including monolithic and reservoir devices, and localized delivery of the bioactive agent to the target.

Disadvantages would certainly include the limitation in drug loading (although hollow fibers can be used to achieve higher loadings), drug instability under fiber-spinning conditions (especially proteins), and the fact that fibers normally will require surgical procedures for administration.

The concept of fibrous polymer formulations was extended to the delivery of aquatic herbicides (56). Several herbicides including Diquat, Fluridone, and Endothal were spun into biodegradable polycaprolactone. Monolithic fibers and a modified monolithic system were produced with levels of herbicide from 5 to 60% by weight. Laboratory and field trials showed efficacious delivery of the active agent. Fibers provided both targeted localized delivery and controlled release of the herbicide to the aquatic weed.

D. Sterilization Techniques

Sterilization of the finished drug delivery formulation is an important consideration often overlooked in the early design of lactide/glycolide delivery systems. Aseptic processing and terminal sterilization are the two major routes of affording an acceptably sterile product. Both of these methods are suitable for products based on lactide/glycolide polymers if proper care is exercised in processing or selection of the treatment procedures.

Because of the good solubility of lactide/glycolide polymers in organic solvents, it is feasible to filter-sterilize drug/polymer solutions in a clean room environment under Good Manufacturing Practice (GMP) conditions. Viscosities of the solutions can normally be adjusted to allow practical filtration pressures and processing times. Usually solutions containing 20% or less of polymer by weight can be processed through 0.2 μm filters. The degree of the difficulty of this procedure depends on the molecular weight of the polymer.

Aseptic processing is particularly useful with microencapsulated products, which almost always involve solutions of the polymer in organic solvents. Occasionally, bioactive molecules sensitive to

terminal sterilization methods are encountered, leaving aseptic processing as the only alternative. Ethylene oxide was found to soften
or plasticize some lactide compositions. Residual ethylene oxide is
also a potential problem.

γ Irradiation has proven useful for many lactide/glycolide formulations. This again is an extension of technology from the biodegradable suture field. The appropriate dose of γ irradiation must
be determined for each new drug delivery product (57). The D
value, the amount of irradiation in Mrads needed to achieve a 90%
reduction in the microorganism population, can be determined from
Eq. (1):

$$D = \frac{U}{\log a - \log b} \tag{1}$$

where U = radiation dose in Mrads
     a = initial number of microorganisms
     b = number of surviving microorganisms after receiving U
        amount of radiation.
While the minimum irradiation dose (MID) to achieve a 10^{-6} probability of sterility assurance may be calculated by Eq. (2):

$$MID = (\log A - \log 10^{-6}) \times D \tag{2}$$

where

    log A = total bioburden level in the product
    D = D value of the most radioresistant isolate

Various studies have been made on the effects of radiation on
lactide/glycolide polymers (24,38,58). Gilding and Reed (24) reported
the effect of γ rays on Dexon sutures. Those results confirmed that
deterioration of the sutures occurs but that random chain scission
is not the primary mechanism. Number average-molecular weight Mn
showed a dramatic decrease at doses above 1.0 Mrad. Thus, unzip-
ping of the polymer chain appeared to be the more dominant process,
at least in the case of polyglycolide.

Experiments in rats (38) have shown that γ irradiation decreases
the inherent viscosity of lactide/glycolide copolymers and increases
the biodegradation rate. The in vivo lifetime of a 92:8 DL-lactide/
glycolide copolymer was decreased from about 40 weeks to about 30
weeks after treatment with 2 Mrad of γ irradiation.

Recent studies on 85:15 and 50:50 copolymers show decreases of
about 15–20% in inherent viscosity after treatment with 2.0–2.5 Mrad
of γ rays. For example, an 85:15 DL-lactide/glycolide copolymer
showed a drop in inherent viscosity from 0.71 to 0.56 dl/g (30°C in
chloroform) upon treatment with 2.41 Mrad of γ radiation. In most

cases, doses of radiation in the 1.5- to 3.0-Mrad range can be utilized without serious degradation of the polymers. However, it is
well established that drug release kinetics are often changed (increased) after irradiation with γ. Therefore, it is imperative that
sterilization procedures be considered quite early in the development
of drug delivery systems based on lactide/glycolide polymers.

Spenlehauer (59) reported that in the case of cisplatin-loaded
microspheres, irradiation only changes the processing considerations
and does not influence drug release. This observation is in conflict
with other literature reports showing increases in drug release rates
(60,61).

VI. DRUG DELIVERY SYSTEMS

The lactide/glycolide polymers are presently the most widely investigated biodegradable excipients for controlled drug delivery. Several
major pharmaceutical companies have extensive development programs
based on these polymers. The majority of these research programs
are aimed at injectable microsphere formulations, although implantable
rods and pellets are also being investigated. Applications cover both
human and veterinary medicine. Because of the wide range of biodegradation profiles available with lactide/glycolide polymers, durations of action of drug delivery products range from a few days to
about 2 years. Figure 4 provides an example of the range of systems
possible with these polymers.

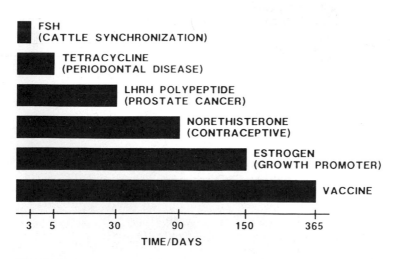

FIGURE 4 Controlled release systems based on lactide/glycolide
polymers with duration of action of a few days to 1 year.

Until the early 1980s, most bioactive agents of interest in drug delivery were classified simply as water-soluble or water-insoluble and low or high molecular weight. The fabrication process to be used and the difficulty in achieving the desired drug release kinetics are directly related to these drug characteristics. The water solubility of drugs has remained a dominant factor in the design of controlled released systems.

Within the past few years, however, another subclassification has become necessary. New biologically active agents derived from recombinant DNA and hybridoma techniques have presented a major challenge to designers of drug delivery systems. Many of these agents are water-soluble proteins of relatively high molecular weight, often in the 40,000–100,000 range. In addition, three-dimensional structures, critical to the bioactivity, stability, and safety of the molecules, are common. Thus, not only are solubility and molecular weight critical factors but also retention of these complex structures. These molecules, usually proteins, can be inactivated by both chemical and physical mechanisms. Fragmentation, oxidation, dimerization, unfolding, aggregation, and adsorption are just a few of the routes to loss in activity (62).

Because of the extreme differences between conventional pharmaceuticals and the protein molecules in terms of formulation techniques and drug release kinetics, the two categories will be discussed separately here.

A. Classical Pharmaceutical Agents

Classical or conventional pharmaceutical agents in combination with lactide/glycolide polymers have been widely studied since about 1973. In general, these compounds are bioactive agents usually produced by synthetic chemistry, with molecular weights of less than a few hundred and relatively stable structures. Examples include steroid hormones, antibiotics, narcotic antagonists, anticancer agents, and anesthetics.

1. Steroid Systems

The earliest reports of controlled release steroids were those of Jackanicz (63), Yolles (64), Anderson (65), and Wise (66). Most of those early studies were based on poly[(L+)-lactic acid]. Implants and granular particles were fabricated with progesterone, norgestrel, and norethisterone. In vivo urinary excretion studies were conducted on [^{14}C]progesterone beads (64). The reported results were somewhat questionable as only 20% of the original implanted drug could be accounted for.

None of the early systems have enjoyed significant clinical success (67). This is most probably due to two factors. Most of the formulations were based on crystalline poly(L-lactide) rather than the glycolide/DL-lactide copolymers presently used. Further, the fabrication techniques, primarily grinding and ball-milling of films to produce crude particles, were difficult to reproduce and did not afford precise controlled release drug kinetics in vivo.

The most successful lactide/glycolide formulations in regard to clinical results have been those of Beck and coworkers (21,68-73). Norethisterone, levonorgestrel, testosterone, testosterone propionate, progesterone, norgestimate, ethinyl estradiol, and estradiol benzoate have been formulated in lactide/glycolide copolymers and homopolymers for human and animal trials. These formulations are based on spherical, free-flowing particles comprising about 50% by weight of the active steroid. The microspheres are suspended at the time of treatment in an aqueous injection vehicle and injected intramuscularly. Conventional syringes and needles (21-gauge) are employed for the injection. Normally, $1-3$ cm^3 of vehicle is adequate for delivery of the suspension.

The steroid-loaded formulations are prepared by a patented solvent evaporation process (45,46). Basically, the wall-forming polymer and the steroid are added to a volatile, water-immiscible solvent. The dispersion or solution is added to an aqueous solution to form an oil-in-water emulsion. The volatile solvent is then removed to afford solid microparticles. The microparticles are usually subdivided with sieves to isolate fractions of the desired diameters. It is imperative that a reliable and reproducible microencapsulation procedure be used to fabricate long-acting formulations.

The in vitro release profiles of many microsphere formulations including steroids can be determined by an ethanol/water model (74). By adjusting the ethanol/water ratio in the receiving fluid, the rate and duration of release can be optimized to afford a rapid evaluation tool for developmental and quality control purposes. The model is not intended to have one-to-one correlation with in vivo results. . The effects of microsphere size distribution, drug/polymer ratio, and microsphere quality can be easily demonstrated in this laboratory model. Furthermore, as animal data and human clinical trial results are available the model becomes quite useful as a quality control method (46).

Although poly(DL-lactide) was used successfully in a 180-day norethisterone product (21), the glycolide/DL-lactide copolymers have been the excipients of choice for most of the steroid systems. The 90-day systems for norethisterone, progesterone, testosterone, norgestimate, levonorgestrel, and ethinyl estradiol all employ the 85:15 DL-lactide/glycolide copolymer. This polymer was shown to biodegrade in about 150 days in rat muscle (38). The 30-day systems, including a norethisterone formulation, are based on a 50:50 DL-lactide/glycolide copolymer which degrades in about 60 days.

The mechanism of steroid release from the glycolide/lactide copolymers is a combination of initial leaching/diffusion and later bioerosion of the matrix. It is unlikely that true Fickian diffusion is a
major factor in the mechanism of release. However, microsphere
formulations exhibit relatively high surface areas and ultrathin membranes as compared to other implants and macrodevices. Numerous
clinical and animal studies (68-73) have shown a biphasic release pattern from the injectable formulations. The elevated serum drug levels
early on are probably due to leaching and diffusion of drug from
near the surface of the particles. The levels show a decline and as
a substantial bioerosion of the matrix occurs, a second elevated drug
level is observed. In practice, these two regions have been made
to merge by careful adjustment of the polymer and microsphere characteristics.

The rate and duration of steroid release is affected by (1) polymer composition, (2) drug/polymer ratio; (3) microsphere size distribution, and (4) microsphere quality (75). The ratio of glycolide to
lactide in the copolymer has been found to be more dominant than the
polymer molecular weight in the design of controlled release formulations. Microspheres of smaller size provide in vivo drug profiles of
higher levels and shorter durations because of greater surface area.

The 90-day female contraceptive (Fig. 5) based on norethisterone
as the active hormone and 85:15 DL-lactide/glycolide as the excipient
has undergone successful phase I and II clinical trials in various
geographic areas. Those trials in women of child-bearing age demonstrated safety and efficacy of a prototype formulation in about 300
subjects (76). In the phase II trial, there were no pregnancies in
the higher dose (100 mg) group over 540 women-months and one in
the lower dose (65 mg) group over 532 women-months (77).

A testosterone microsphere system for treatment of hypogonadol
males has been developed and clinically evaluated (61,78). This
formulation is based on a glycolide/DL-lactide copolymer and natural
testosterone (Fig. 6). Because testosterone is not a very potent
compound, about 600 mg of the drug is needed in adult males over
a 90-day period. The performance of the testosterone system in baboons is shown in Fig. 7. Similar formulations have also been used
in the control of the wild horse population of the western United
States. Stallions were injected with a testosterone microsphere formulation designed to inhibit sperm production over a 6-month period
(79).

The steroid microsphere systems are probably the most successful drug delivery formulations thus far based on lactide/glycolide
polymers. Several of these products appear to be on track for
human and animal applications in the 1990s. The success of these
formulations is due to the known safety of the polymer, the reproducibility of the microencapsulation process, reliability in the treatment procedure, and in vivo drug release performance (80).

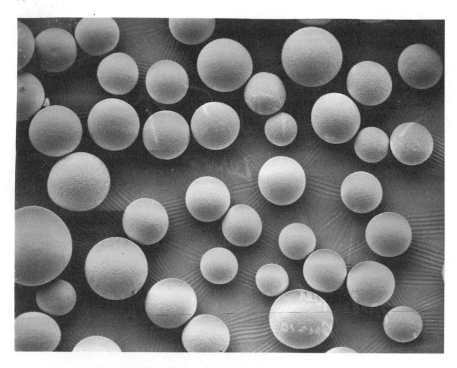

FIGURE 5 Scanning electron micrograph of 90-day norethisterone microspheres 45–90 μm in diameter and containing 47% by weight drug.

2. Narcotic Antagonist Systems

Several narcotic antagonists, including naloxone, naltrexone, L-methadone, and cyclazocine, have been incorporated in lactide homopolymers and lactide/glycolide copolymers. Cyclozocine was incorporated in poly(L-lactide) in the form of films (81,82). Lamination of drug-polymer films with a drug-free film created a reservoir device and eliminated the burst observed with the monolithic films originally tested.

Naltrexone in combination with lactide/glycolide copolymer has been investigated (83-87). Chiang (85) reported the clinical evaluations of a bead preparation containing 70% naltrexone and 30% of a 90:10 lactide/glycolide copolymer. Each subject received a 10-mg i.v. dose of naltrexone and a 63-mg dose by subcutaneous implantation of the beads. Average plasma naltrexone levels were maintained at 0.3–0.4 ng/ml for approximately 1 month. Two out of three subjects experienced a local inflammatory reaction at the site of implantation. This unexplained problem prevented further clinical testing of

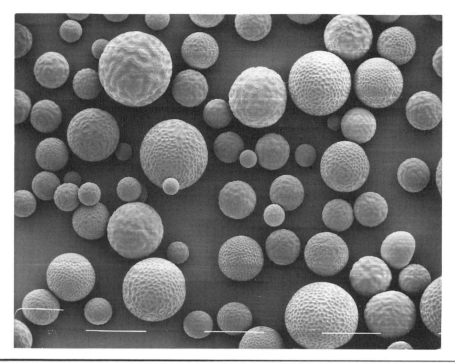

FIGURE 6   Scanning electron micrograph of 90-day testosterone microspheres based on 85:15 lactide/glycolide copolymer.

that particular formulation.   This incident is likely related to some
unique aspect of that product, as no similar incidences have been
reported with lactide/glycolide polymers.
   A microfluidized bed coating process was used to encapsulate
naltrexone free base in a 65:35 L-lactide/glycolide copolymer (86).
Uncoated microspheres released 80% of the drug in 2 days in phosphate buffer.   Coating the microspheres with additional copolymer
slowed the release kinetics in the in vitro release model.   In vivo
studies in rats demonstrated that morphine action was blocked for
60 days following a single injection of the controlled release naltrexone
formulation.   L-methadone microspheres and microcapsules were formulated from poly(L-lactide acid), poly(glycolic acid-co-L-lactic acid),
and poly(ε-caprolactone-co-L-lactic acid) (88).   Microspheres were
prepared by solvent evaporation techniques while the microcapsules
were formulated by an air suspension coating procedure.   In vitro
release of L-methadone in buffered water was very rapid (2 days)
whereas a pH 7.4 buffer afforded zero-order release kinetics over
several days.   This was attributed to the low water solubility of L-

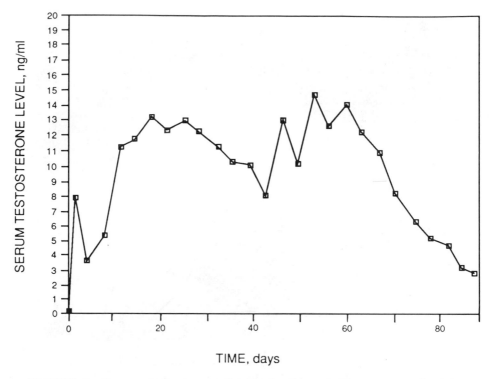

FIGURE 7   Serum testosterone levels in three baboons treated with
a single injection of testosterone-loaded microspheres.

methadone causing a boundary layer at the polymer-water interface.
The buffer resulted in protonation of the drug in the aqueous phase,
maximizing the drug concentration gradient.

The investigators studied various blends of the three polymers
in order to control the rate of chain scission and thus influence the
induction period and onset of drug release.   None of the blends provided the desired 1-week zero-order kinetics.   However, blends of
different microsphere types did show promise in vitro (88).

3.   Antimalarial Systems

Various antimalarial drugs have been studied in biodegradable delivery
systems.   Wise (89) reported the use of a lactide/glycolide copolymer
and also poly(L-lactic acid) for release of drugs such as quinazoline
and sulfadiazine.   Although in vitro data and experiments in mice
were somewhat encouraging, these early formulations failed to reach
significant clinical status.

Recently, Tsakala et al. (90) formulated pyrimethamine systems based on several lactide/glycolide polymers. These studies were conducted with both microspheres (solvent evaporation process) and implants (melt extrusion process). In vitro studies indicated that pyrimethamine-loaded implants exhibited apparent zero-order release kinetics in aqueous buffer whereas the microspheres showed an initial high burst and considerably more rapid drug release. In vivo studies in P. berghi infected mice confirmed that the microspheres did not have adequate duration of release for practical application. However, the implants offer promise for future clinical work as more than 3 months protection was observed in animals.

4. Anticancer Systems

Although a few anticancer agents have been combined with lactide/glycolide polymers, it is somewhat surprising that this has not been one of the most investigated applications. Controlled release formulations offer the potential of reducing the drug toxicity, which is almost always a serious problem in cancer chemotherapy. Cisplatin (59,91-93), mitomycin (94), and adriamycin (95) have been studied in biodegradable delivery systems.

Poly(DL-lactide) was used as the excipient in microspheres of CCNU, a nitrosourea, prepared by a solvent evaporation procedure (96,97). PLA-CCNU microspheres 3.0 μm in diameter were injected i.v. and leukemia cell survival was determined by spleen colony assay. A 100-fold decrease in leukemia cell survival was observed with the microspheres in both spleen and liver compared to untreated controls. Promising results were also obtained with Lewis lung carcinoma in mice. These studies showed that 2- to 4-μm microspheres were preferentially targeted to the lungs.

Extensive studies have been reported with cisplatin in the field of chemoembolization (59,98). Microspheres prepared by a solvent evaporation procedure were characterized in vitro and critical processing parameters in regard to drug release kinetics were identified. It was concluded that the presence of glycolic units in the polymer backbone and radiosterilization appear to be secondary factors for determining release profiles. Factors which modify drug crystal distribution in the microspheres were believed to be more important. These conclusions are not entirely consistent with findings from studies with other drug systems by various investigators. Nevertheless, the approach is highly encouraging in the preliminary stages.

Lactic acid oligomer microspheres containing aclarubicin have been studied for selective lymphatic delivery. Low (less than 10,000 molecular weight oligomers were used to produce microspheres designed to release drug over a 30-day period (99). Additives have been used to alter the release rate of aclarubicin-loaded poly(lactide) microspheres (100). Mitomycin C was incorporated into poly(lactic acid) to produce microcapsules about 95 μm in diameter (94). The microcapsules exhibited a dose-dependent drug release pattern with higher drug loaded formulations showing a faster release rate than lower loaded samples. The formulation showed antiproliferative activity against the growth of K562 human erythroleukemia cells.

Strobel et al. (101) reported a unique approach to delivery of anticancer agents from lactide/glycolide polymers. The concept is based on the combination of misonidazole or adriamycin-releasing devices with radiation therapy or hyperthermia. Prototype devices consisted of orthodontic wire or sutures dip-coated with drug and polymeric excipient. The device was designed to be inserted through a catheter directly into a brain tumor. In vitro release studies showed the expected first-order release kinetics on the monolithic devices.

5. Antibiotic Systems

The formulation of antibiotics in lactide/glycolide polymers is quite appealing at first thought. In practice, however, this class of drugs has proven rather challenging. Three problems have been encountered by investigators in this field: large daily doses of antibiotic are often required, many of the antibiotics are relatively unstable, and most are water-soluble. Fortunately, delivery of the antibiotic by controlled release formulations, often directly to the target organ or organism, has in some cases reduced the total daily dose needed. The instability of the drugs has been more of an experimental hurdle in conducting in vitro release studies than in actual application problems.

The feasibility of antibiotic delivery systems based on lactide/glycolide polymers was demonstrated a decade ago (102). Poly(DL-lactide-co-glycolide) comprising molar ratios of lactide/glycolide in the range of 50:50 to 90:10 were investigated with ampicillin, gentamicin, polymyxin B, and chloramphenicol. Phase separation techniques and solvent evaporation procedures were used to formulate microspheres of each drug. In vivo studies in rats were conducted with induced infections with Streptococcus pyogenes and Staphylococcus aureus. Wounds were treated topically with ampicillin-loaded microspheres (68:32 DL-lactide/glycolide). At 7 days postinfection, the test groups were free of infection and microbial assays showed no cultivatable organisms. Further studies confirmed these initial observations (103). Although parenteral antibiotic formulations have not received much attention, specialty applications have been studied. For example, compositions designed to release antibiotics in the periodontal cavity are being investigated (104,105). Agents such as ampicillin and tetracycline are encapsulated in lactide/glycolide copolymers designed to release drug over a 1- to 2-week period. The microspheres are suspended in a biocompatible gel and extruded into the periodontal pocket with a syringe fitted with a thin tube. Figure 8 shows a typical tetracycline microsphere formulation based on a 50:50 glycolide/DL-lactide copolymer. These microsphere formulations exhibit rapid first-order release kinetics. Fibers of biodegradable polymers have also been utilized to delivery tetracycline to the periodontal pocket (51).

Antibiotic-loaded composites of poly(DL-lactic acid) and hydroxyapatite have been studied as a filler in the repair of bone defects (106). The drug dideoxykanamycin was formulated at 20% by weight in a 50:50 PLA/hydroxyapatite composite. Cyclindrical composites were implanted into the penetrated tibeas of rats. Antibiotic was present at the implantation site at least 8 weeks postimplantation. The rate of drug release was affected by the molecular weight of PLA used in the composite.

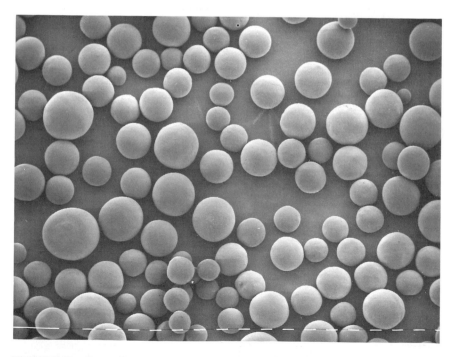

FIGURE 8 Scanning electron micrographs of tetracycline-loaded microspheres for treatment of periodontal disease.

6. Antiinflammatory Systems

Several antiinflammatory compounds have been formulated in lactide/
glycolide polymers (107-111). Methylprednisolone microspheres based
on an 85:15 DL-lactide/glycolide copolymer were developed for intra-
articulate administration (111). The microspheres, prepared by a
solvent evaporation procedure, are 5–20 μm in diameter and are designed to release low levels of the steroid over a extended period in
the joint. Controlled experiments in rabbits with induced arthritis
showed that the microspheres afforded an antiinflammatory response
for up to 5 months following a single injection.

Hydrocortisone microspheres (108,109) and films (110) based on
poly(lactic acid) have been investigated. A cage implant technique
was used to study the performance of monolithic poly(DL-lactide)
films loaded with hydrocortisone acetate (110). Films 1.5 × 0.6 cm
were inserted into titanium wire-mesh cages 3.5 × 1.0 cm. The
cages were implanted in the backs of rats and the inflammatory ex-
udate was sampled periodically. The white cell concentration in the
samples was lower than that of controls at all times during the 21-
day test.

7. Anesthetics Systems

Lidocaine (112), xylocaine, and dibucaine (113) have been formulated
in homo- and copolymers of lactide and glycolide. The goal of these
studies has been relatively short-term (24-hr) controlled release of
the anesthetic. Injectable microcapsules of lidocaine hydrochloride
were produced by an air suspension coating technique and administered i.m. to rabbits (112). Serum levels of lidocaine indicated an
initial rise over the first 2 hr and then a gradual decline with clearance after about 8–10 hr.

8. Other Applications

Lactide/glycolide polymers have been investigated for delivery of
agents in applications outside the pharmaceutical field. For example,
the microbiocidal properties of chlorine dioxide disinfectants have
been improved by formulating a long-acting chlorine dioxide system
based on lactide/glycolide copolymers. Blends of microspheres based
on 50:50 and 87:13 copolymers were developed to afford the release
of chlorine dioxide over several months (114).

The use of polylactides for delivery of insect hormone analogs
and other veterinary compounds (115,116) has been studied. Microspheres, pellets, and reservoir devices based on polyglycolide, poly-
(DL-lactide), poly(L-lactide), and various copolymers have been
used to deliver methoprene and a number of juvenile hormone analogs.

A melatonin microsphere system has been developed for increasing
the production rate of wool and fur. Figure 9 shows the results of

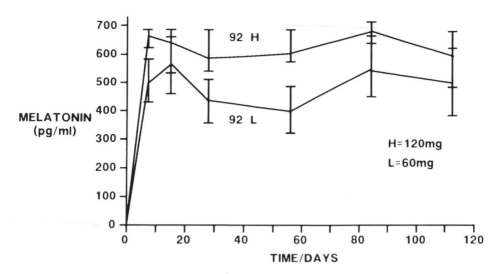

FIGURE 9   Serum melatonin levels in goats treated with a single injection of biodegradable microspheres containing melatonin.

a single treatment of melatonin-loaded microspheres in goats.   Good
dose response was observed in that model.

B.   Bioactive Macromolecules

Polypeptides, viral and bacterial antigens, and many other macromolecules such as those derived from recombinant DNA technology
represent a challenge of greater magnitude for designers of drug
delivery systems for the reasons previously stated.   Because of the
tremendous clinical and commercial potential, combinations of these
agents with lactide/glycolide polymers have received considerable
attention (33,117).   These investigations have afforded mixed results.
Serious problems in achieving long-term (greater than 2 weeks) release have been encountered in several cases where the macromolecule
lost bioactivity in vivo after a few days.   On the other hand, promising lactide/glycolide systems for long-term delivery of luteinizing
hormone releasing hormone (LHRH) polypeptides, viral and bacterial
antigens, somatostatin, and RNA have been developed.

1.   Polypeptides

Calcitonin, LHRH, lypressin, and somatostatin have been formulated
in lactide/glycolide copolymers as injectable, controlled release formulations (117).   Various agonistic and antagonistic analogs have been
studied.   Generally these compounds are hydrophilic polypeptides,

1200–3000 in molecular weight, with relatively short biological half-lives. The molecular weight is sufficiently high to prevent signifi-cant molecular diffusion in the lactide/glycolide matrix. Thus, most polypeptide delivery systems are based on erosion of the polymer to afford long-term release. Most of the low molecular weight (less than 5000) polypeptides are considered quite stable in the presence of lactide/glycolide excipients and their acidic bioerosion byproducts.

   Kent et al. first reported on the feasibility of controlled release LHRH analogs from a 50:50 lactide/glycolide copolymer (118,119) (Fig. 10). A phase separation microencapsulation technique was used to formulate nafarelin acetate in 50:50, 69:31, and 45:55 DL-lactide/glycolide copolymers (120).

   The influence of polymer composition and molecular weight on the estrus-suppressing activity of the microspheres in female rats was determined. A triphasic release pattern was observed. An in-itial rapid release over the first few days was attributed to diffusion from superficial areas of the microspheres. A second phase of lower release levels then occurs and continues until the onset of the third, major phase of release. This phase is facilitated by the onset of erosion of the polymer matrix.

   The triphasic pattern was more pronounced with copolymers of higher lactide content (longer bioerosion time). The 45:50 copolymer afforded efficacy in the rat over 5 weeks. The studies confirmed the homogeneous (bulk) rather than heterogeneous (surface) erosion of the polymer. Subsequent studies (121) with 3-mm implants

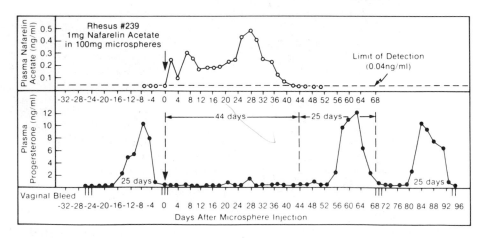

FIGURE 10 Plasma profile of nafarelin acetate and biological response parameter in female Rhesus monkeys after injection with 100 mg of 55:45 lactide/glycolide microspheres. (From Ref. 118.)

prepared by co-melt extrusion further substantiated this conclusion.
A series of DL-lactide/glycolide copolymers and the lactide homopolymer were screened in vitro and in vivo with nafarelin. More than 8
months release was achieved with certain implants.

United States Patent 4,767,628 assigned to Imperial Chemical In-
dustries describes a similar lactide/glycolide delivery system for LHRH
polypeptide (122,123). A multiphase release pattern is again postulated. The first phase occurs by diffusion of drug through aqueous
polypeptide domains linked to the exterior surface of the matrix.
The second phase occurs when the polymer undergoes significant
biodegradation. Broad molecular weight compositions were proposed
as one method of reducing the triphasic nature of these release patterns.

Binding of the basic peptide to terminal polymer carboxylate
groups may play a significant role in the release mechanism (124).
The polymer may behave as a weak acid ion exchange resin. Branched
copolymers have also been proposed as an approach to achieving
"smoother" release patterns for polypeptide systems (40).

Several other investigators (125-128) studied LHRH delivery systems based on lactide/glycolide polymers. Although there have been
some slight differences, these systems have generally been formulated
in a similar way to those reported earlier (118,119). Very low molecular weight polylactide has been studied in LHRH implants (129).
Veterinary applications such as ovulation control in heifers and mares
have been studied with GnRH entrapped in lactide/glycolide microcapsules (130). Extensive human clinical studies are in progress
with LHRH systems and one product is registered in the United States.

2. Vaccines

Antigen- or antibody-containing compositions for active or passive
immunization have been designed with lactide/glycolide polymers.
Vaccines offer an excellent opportunity for controlled release technology because many immunizations require a primary treatment followed by repeated booster regimens. In terms of patient numbers
worldwide, these products represent hundreds of millions of treatments annually. In addition, there is a large number of veterinary
vaccines which appear suitable for long-acting delivery.

A method for delivering antibody or antigen to the female reproductive tract has been described (131,132). Lactide/glycolide polymers were used to microencapsulate bacteria (Pneumococcus), herpes
simplex viral antigens, and bovine HcG. Rabbits were vaginally
administered a 5-mg quantity of each microcapsule formulation. Vaginal washes obtained 2 weeks posttreatment demonstrated that intra-
vaginal installation of the biodegradable microcapsules caused immunization against the various antigens administered. Estrogen and progestins have also been formulated in similar reproductive tract systems.

Viral and bacterial antigens have been delivered over 6–12 months from controlled release microspheres based on lactide/glycolide copolymers. Such formulations have been utilized in dairy cattle to hyperimmunize animals for production of various biologicals in dairy milk (133). In this approach, killed viruses or bacteria are entrapped in the biodegradable copolymers by a phase separation microencapsulation procedure. Animals are injected intramuscularly with antigen-loaded microspheres suspended in an aqueous injection vehicle. Antibody titers in bovine milk were shown to reach levels at least as high as those achieved by conventional immunization techniques which require frequent booster injections with antigen. Figure 11 shows a typical immune response following a single treatment with a lactide/glycolide antigen system. Similar technology is presently being investigated for long-term release of many other antigens of veterinary interest. Several of these products provide 30–40 days duration of action.

A human contraceptive vaccine based on lactide polymers is currently being developed. The antigen is a 37-amino-acid peptide of B-HCG conjugated to diphtheria toxoid. The antigen is administered wtih microencapsulated muramyl dipeptide as an adjuvant. Studies in rabbits have shown 9–12 months of elevated antibody liter following

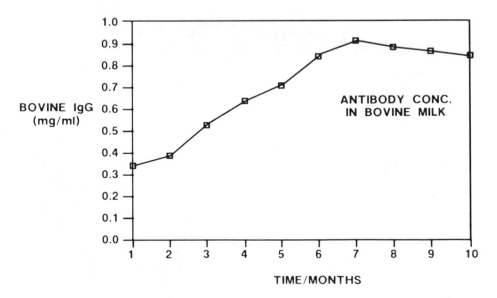

FIGURE 11   IgG antibody levels in bovine milk following a single immunization with long-acting bacterial antigen microspheres.

a single injection of microspheres (Fig. 12) of the antigen in lactide/
glycolide polymers.

   Muramyl dipeptide derivatives have also been microencapsulated
in lactide/glycolide copolymers for use alone as an immuno potentiator.
L-lactide/glycolide copolymers were used to deliver MDP-B30, a lipophilic compound, from very small microspheres (less than 5 μm in
diameter). The amount of MDP-B30 required for tumor growth inhibitory activity of mouse peritoneal macrophages was 2000 times less
for the controlled release MDP-B30 microspheres than for the unen-
capsulated drug (134).

3. Insulin

Insulin (molecular weight 7000) has been formulated in controlled release microbeads and pellets (135,136). A solvent evaporation microencapsulation procedure was used to produce microspheres with up
to 20% by weight insulin. Solvent-casting techniques were used to
prepare pellets. The investigations demonstrated that the PLA

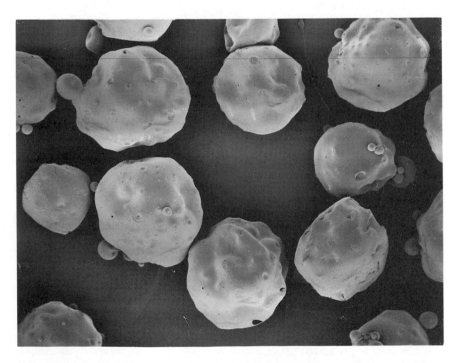

FIGURE 12 Scanning electron micrograph of contraceptive vaccine
microspheres based on lactide/glycolide polymer.

formulations were effective in lowering blood glucose levels in diabetic rats for about 18 days. No deactivation of the micromolecule
or inflammation at the implant site or other adverse effects was observed (135).

4. Other Macromolecules

Lactide/glycolide polymers have been investigated for delivery of
several other macromolecules. Synthetic double-stranded RNA, polyisosinic acid/polycytidylic acid, a potent inducer of interferon, was
formulated in a 53:47 copolymer of DL-lactide-co-glycolide. The
microspheres were evaluated in mice challenged with Right Valley
fever virus. More than 16 days protection was afforded versus only
3 days for controls (137).

A unique method of formulating delivery systems based on starch/
PLA systems was studied (138). In that approach, the goal was to
provide a better matrix for delivery of high molecular weight hydrophilic molecules. A hydrophilic material, starch, was combined
through graft polymerization to PLA. The carbolactic polymers were
then used to entrap bovine serum albumin in microspheres.

Considerable effort has been devoted to the development of controlled release formulations for growth hormone, especially porcine
and bovine growth hormone. Systems capable of delivering 5−25 mg
of hormone per day for 30−60 days are sought. Lactide/glycolide
systems have been extensively explored in proprietary programs for
these applications. The results of these development projects have
been rather disappointing thus far. Systems based on lactide/glycolide
polymers have been able to provide at very best only about 2 weeks
duration of biological activity as measured by growth promotion in
rats or increased milk production in dairy cattle. However, radioimmunoassay (RIA) and enzyme-linked immunosorbent assay (ELISA)
methods have shown in some cases significant serum hormone levels
at 30−40 days posttreatment. A complex and as yet unexplained interaction apparently occurs in vivo with the acidic polymer and the
hormone. In vitro release studies have shown that bovine growth
hormone can become insoluble when incorporated in poly(lactide) films.
Delivery systems based on more hydrophobic and less acidic natural
and synthetic waxes have proven more feasible for 2- to 4-week delivery of growth hormone to animals (139).

Interferon has also been a troublesome protein for lactide/glycolide
systems. Again, in major proprietary studies, problems have been
encountered in maintaining biological activity for longer than 5−10
days in vivo. As observed with growth hormone, ELISA determina-
tions show the presence of interferon several days beyond the last
biological activity as determined by CPE (cytopathic effect) technique.
However, for veterinary applications, such as shipping fever in cattle,
5- to 10-day delivery systems for animal interferon show considerable

promise. Injectable microsphere formulations based as DL-lactide/glycolide are under study for those applications.

VII. SUMMARY

The feasibility of lactide/glycolide polymers as excipients for the controlled release of bioactive agents is well proven. These materials have been subjected to extensive animal and human trials without evidence of any harmful side effects. When properly prepared under GMP conditions from purified monomer, the polymers exhibit no evidence of inflammatory response or other adverse effects upon implantation. The availability of commercial polymer from large, reputable companies has accelerated the investigation of these materials for drug delivery. The future for lactide/glycolides in delivery of conventional pharmaceuticals, vaccines, and some lower molecular weight polypeptides looks very promising. Several products are already moving through the regulatory registration process in the United States, Europe, and Japan. It is clear, however, that there are limitations on use of the aliphatic polyesters for protein delivery. The incompatibility of some macromolecules with lactide/glycolides has been observed. As with other polymeric excipients, lactide/glycolides will serve some but not all applications.

REFERENCES

1. Heller, J., Bioerodible systems, in Medical Applications of Controlled Release, Vol. 1 (R. S. Langer and D. L. Wise, eds.), CRC Press, Boca Raton, 1984, pp. 70-98.
2. Hsieh, D. S. T., Controlled Release Systems: Fabrication Technology, Vol. 2, CRC Press, Boca Raton, 1988.
3. Baker, R., Controlled Release of Biologically Active Materials, John Wiley and Sons, New York, 1987.
4. Frazza, E. J., and Schmitt, E. E., A new absorbable suture, J. Biomed. Mater. Res. Symp., 1, 43 1971.
5. Brady, J. M., Cutright, D. E., Miller, R. A. and Battistone, G. C., Resorption rate, route of elimination, and ultrastructure of the implant site of polylactic acid in the abdominal wall of the rat, J. Biomed. Mater. Res., 7, 155, 1973.
6. Mendak, S. H., Jr., Jensik, R. J., Haklin, M. F., and Roseman, D. L., The evaluation of various bioabsorbable materials on the titanium fiber metal tracheal prosthesis, Ann. Thorac. Surg., 38, 488, 1984.
7. Greenstein, S. M., Murphy, T. F., Rush, B. F., Jr., and Alexander, H., The experimental evaluation of a carbon-polylactic acid mesh for a ventral herniorrhaphy, Curr. Surg., 41, 358, 1984.

8. Bercovy, M., Goutallier, D., Voisin, M. C., Geiger, D.,
    Blanquaert, D., Gaudichet, A., and Patte, D., Carbon-PGLA
    prostheses for ligament reconstruction. Experimental basis and
    short-term results in man, Clin. Orthop., 196, 159, 1985.
9. King, J. B., and Bulstrode, C., Polylactate-coated carbon
    fiber in extra-articular reconstruction of the unstable knee,
    Clin. Orthop., 196, 139, 1985.
10. Brekke, J. H., Bresner, M., and Reitman, M. J., Effect of
    surgical trauma and polylactate cubes and granules on the in-
    cidence of alveolar osteitis in mandibular third molar extraction
    wounds, Can. Dent. Assoc. J., 52, 315, 1986.
11. Rokkanen, P., Bostman, O., Vainionpaa, S., Vihtonen, K.,
    Tormala, P., Laiho, J., Kilpikari, J., and Tamminmaki, M.,
    Biodegradable implants in fracture fixation: Early results of
    treatment of fractures of the ankle, Lancet, 1, 1422, 1985.
12. Christel, P., Chabot, F., and Vert, M., In vivo fate of bio-
    resorbable bone plates of long-lasting poly(L-lactic acid), Proc.
    2nd World Congress on Biomaterial, 279, 1984.
13. Rudolf, R. M., Boering, G., Roseman, F., and Leenslay, J.,
    Resorbable poly(L-lactide) plates and screws for the fixation
    of zygomatic fracture, J. Oral Maxillofac. Surg., 45, 751, 1987.
14. Wehrenberg, R. J., Lactic acid polymers: Strong degradable
    thermoplastics, Mater. Eng., 3, 163, 1981.
15. Vert, M., Cristel, P., and Chabot, F., Bioresorbable plastic
    materials for bone surgery, in Macromolecular Biomaterials
    (C. W. Hastings and P. Ducheyne, eds.), CRC Press, Boco
    Raton, 1984, pp. 120-142.
16. Brekke, J. H., Bresner, M., and Reitman, M. J., Polylactic
    acid surgical dressing material. Postoperative therapy for
    dental extraction wounds, Can. Dent. Assoc. J., 52, 599,
    1986.
17. Boswell, G. A., and Scribner, R. M., U.S. Patent 3,773,919,
    1973.
18. Yolles, S., Eldridge, J. E., Leafe, T. D., Woodland, J. H.,
    Blake, D. R., and Meyer, F. J., Long-acting delivery systems
    for narcotic antagonists, Adv. Exp. Med. Biol., 47, 177, 1973.
19. Wise, D. L., McCormick, G. F., Willet, G. P., and Anderson,
    L. C., Sustained release of an antimalarial drug using a co-
    polymer of glycolic/lactic acid, Life Sci., 19, 867, 1976.
20. Sinclair, R. G., Slow-release pesticide system polymers of
    lactic and glycolic acids as ecologically beneficial, cost-effective
    encapsulating materials, Environ. Sci. Technol., 7, 955, 1973.
21. Beck, L. R., Cowsar, D. R., Lewis, D. H., Gibson, J. W.,
    and Flowers, C. E., Jr., New long-acting injectable micro-
    capsule contraceptive system, Am. J. Obstet. Gynecol., 135,
    419, 1979.

22. Kulkarni, R. K., Moore, E. G., Hegyelli, A. F., and Leonard, F., Biodegradable polylactic acid polymers, J. Biomed. Mater. Res., 5, 169, 1971.

23. Dittrich, V. W., and Schulz, R. C., Kinetics and mechanism of the ring-opening polymerization of L-lactide, Angew. Makromol. Chem., 15, 109, 1971.

24. Gilding, D. K., and Reed, A. M., Biodegradable polymers for use in surgery: poly(glycolic)/poly(lactic acid) homo- and copolymers: 1, Polymer, 20, 1459, 1979.

25. Dunn, R. L., English, J. P., Strobel, J. D., Cowsar, D. R., and Tice, T. R., Preparation and evalution of lactide/glycolide copolymers for drug delivery, in Polymers in Medicine III (C. Migliaresi, ed.), Elsevier, Amsterdam, 1988, pp. 149-159.

26. Kissel, T., Demirdere, A., Siemann, U., and Sucker, H., Permeability and release properties of biodegradable polymers, Proc. Int. Symp. Control. Rel. Bioact. Mater., 12, 179, 1985.

27. Siemann, U., Densitometric determination of the solubility parameters of biodegradable polyesters, Proc. Int. Symp. Control. Rel. Bioact. Mater., 12, 53, 1985.

28. Murdock, J. R., and Loomis, G. L., U.S. Patent 4,719,246, 1988.

29. Williams, D. F., and Most, E., Enzyme-accelerated hydrolysis of poly(glycolic acid), J. Bio. Eng., 1, 231, 1977.

30. Williams, D. F., Enzyme hydrolysis of polylactic acid, Eng. Med., 10, 5, 1981.

31. Herrman, J. B., Kelly, R. J., and Higgins, G. A., Polyglycolic acid sutures, laboratory and clinical evaluation of a new absorbable suture material, Arch. Surg., 100, 1970.

32. Reed, A. M., In vivo and in vitro studies of biodegradable polymers for use in medicine and surgery, PH.D. Thesis, University of Liverpool, 1978.

33. Holland, S. J., Tighe, B. J., and Gould, P. L., Polymers for biodegradable medical devices, I. The potential of polyesters as controlled macromolecular release systems, J. Control. Rel., 4, 155, 1986.

34. Maulding, H. V., Tice, T. R., Cowsar, D. R., Fong, J. W., Pearson, J. E., and Nazareno, J. R., Biodegradable microcapsules: Acceleration of polymeric excipient hydrolytic rate by incorporation of a basic medicament, J. Control. Rel., 3, 103, 1986.

35. Miller, R. A., Brady, J. M., and Cutright, D. E., Degradation rates of oral resorbable implants (polylactates and polyglycolates): Rate modification with changes in PLA/PGA copolymer ratios, J. Biomed. Mater. Res., 11, 711, 1977.

36. Pitt, C. G., Gratzel, M. M., Kimmel, G. L., Surles, J., and Schindler, A., Aliphatic polyesters. 2. The degradation of

poly(D,L-lactide), poly(e-caprolactone) and their copolymers
in vivo. Biomaterials, 2, 215, 1981 .

37. Cutright, D. E., Bienvenido, P., Beasly, J. D., and Larsen,
     W. J., Degradation rates of polymers and copolymers of poly-
     lactic and polyglycolic acids, Oral Surg. Oral Med. Oral Pathol.,
     37, 142, 1974.

38. Tice, T. R., Lewis, D. H., Dunn, R. L., Meyers, W. E.,
     Casper, R. A., and Cowsar, D. R., Biodegradation of micro-
     capsules and biomedical devices prepared with resorbable poly-
     esters, Proc. Int. Symp. Control. Rel. Bioact. Mater., 9, 21,
     1982.

39. Chu, C. C., Degradation phenomena of two linear aliphatic
     polyester fibers used in medicine and surgery, Polymer, 26,
     591, 1985.

40. Brich, Z., Nimmerfall, F., Kissel, T., and Bantle, S., Branched
     ter-polyesters: Synthesis, characterization, in vitro and in
     vivo degradation behaviors, Proc. Int. Symp. Control. Rel.
     Bioact. Mater., 15, 95, 1988.

41. Pitt, C. G., and Gu, Z. W., Modification of the rates of chain
     cleavage of poly(E-caprolactone) and related polyesters in the
     solid state, J. Control. Rel., 4, 283, 1987.

42. Visscher, G. E., Robinson, R. L., Maulding, H. V. Fong,
     J. W., Pearson, J. E., and Argentieri, G. J., Biodegradation
     of and tissue reaction to 50:50 poly(DL-lactide-co-glycolide)
     microcapsules, J. Biomed. Mater. Res., 19, 349, 1985.

43. Kanke, M., Morlier, E., Geissler, R., Powell, D., Kaplan, A.,
     and DeLuca, P. P., Interaction of microsheres with blood con-
     stituents. II. Uptake of biodegradable particles by macrophages,
     J. Parent. Sci. Technol., 40, 114, 1986.

44. Kanke, M., Geissler, R. G., Powell, D., Kaplan, A., and De-
     Luca, P. P., Interaction of microspheres with blood constituents.
     III. Macrophage phagocytosis of various types of polymeric drug
     carriers, J. Parent. Sci. Technol., 42, 157, 1988.

45. Tice, T. R., and Lewis, D. H., U.S. Patent 4,389,330, 1983.

46. Lewis, D. H., Beck, L. R., Forman, T. D., Manek, T. A.,
     and Pope, V. Z., Reproducibility studies on microencapsulation
     of steroids in biodegradable polymers, Proc. Int. Symp. Control.
     Rel. Bioact. Mater., 15, 266, 1988.

47. Fong, J. W., U.S. Patent 4,384,975, 1983.

48. Fong, J. W., U.S. Patent 4,166,800, 1979.

49. Nuwayser, E. S., U.S. Patent 4,623,585, 1986.

50. Leelarusamee, N., Howard, S. A., Malango, C. J., Ma, J. K.,
     A method of the preparation of polylactic acid microcapsules of
     controlled particle size and drug loading, J. Microencapsul.,
     5, 147, 1988.

51. Dunn, R. L., Lewis, D. H., and Goodson, J. M., Monolithic fibers for controlled delivery of tetracycline, Proc. Int. Symp. Control. Rel. Bioact. Mater., 9, 157, 1982.
52. Dunn, R. L., English, J. P., Stoner, W. C., Jr., Potter, A. G., and Perkins, B. H., Biodegradable fibers for the controlled release of tetracycline in treatment of peridontal disease, Proc. Int. Symp. Control. Rel. Bioact. Mater., 14, 289, 1987.
53. Dunn, R. L., Lewis, D. H., and Beck, L. R., Fibrous polymer for the delivery of contraceptive steroids to the female reproductive tract, in Controlled Release of Pesticides and Pharmaceuticals (D. H. Lewis, ed.), Plenum Press, New York, 1981, pp. 125-146.
54. Dunn, R. L., Lewis, D. H., and Cowsar, D. R., A fibrous delivery system for quinacine: A fallopian tube occluding agent, in Proc. 8th Int. Symp. Control. Rel. Bioact. Mater., 8, 14, 1981.
55. Eenink, M. J. D., Feijen, J., Oligslanger, J., Albers, J. H. M., Rieke, J. C., and Greidonus, P. J., Biodegradable hollow fibers for the controlled release of hormones, J. Control. Rel., 6, 225, 1987.
56. Lewis, D. H., and Dunn, R. L., Aust. Patent 19378/83, 1983.
57. Bussey, D. M., Kane, M. P., and Tsuji, K., Sterilization of corticosteroids by [60]CO irradiation, J. Parenteral Sci. Technol., 37, 51, 1983.
58. Gupta, M. C., and Deshmukh, V. G., Radiation effects on poly(lactic acid), Polymer, 827, 1983.
59. Spenlehauer, G., Vert, M., Benoit, J. P., Chabot, F., and Veillard, M., Biodegradable cisplatin microspheres prepared by the solvent evaporation method: Morphology and release characteristics, J. Control. Rel., 7, 217, 1988.
60. Beck, L. R., and Pope, V. Z., Controlled-release delivery systems for hormones, Drugs, 270, 528, 1984.
61. Lewis, D. H., Tice, T. R., and Beck, L. R., Overview of controlled release systems for male contraception, in Male Contraception: Advances and Future Prospects (G. I. Zatuchni, ed.), Harper and Row, Philadelphia, 1986, pp. 336-345.
62. Jones, R. E., Challenges in formulation design and delivery of protein drugs, Proc. Int. Symp. Control. Rel. Bioact. Mater., 14, 96, 1987.
63. Jackanicz, J. M., Nash, H. A., Wise, D. L., and Gregory, J. B., Polylactic acid as a biodegradable carrier for contraceptive steroids, Contraception, 8, 227, 1973.
64. Yolles, S., Leafe, T., Sartori, M., Torkelson, M., Ward, L., and Boettner, F., Controlled release of biologically active agents in Controlled Release Polymeric Formulations (D. R. Paul and F. W. Harris, eds.), American Chemical Society, Washington, D.C., 1976, Chap. 8.

65. Anderson, L. C., Wise, D. L., and Howes, J. F., An injectable sustained release fertility control system, Contraception, 13, 375, 1976.

66. Wise, D. L., Gregory, J. B., Newberne, D. M., Bartholow, L. C., and Stanbury, J. B., Results on biodegradable cylindrical subdermal implants for fertility control, in Polymeric Delivery Systems (R. J. Kostelnik, ed.), Gordon and Breach, New York, 1978, Chap. 8.

67. Zatuchni, G. I., Labbok, M. H., and Scrarra, J. J. (eds.), Research Frontiers in Fertility Regulation, Harper and Row, Hagerstown, 1981.

68. Beck, L. R., and Tice, T. R., in Long-Acting Steroid Contraception (R. Mishell, Jr., ed.), Raven Press, New York, 1983, pp. 175-199.

69. Beck, L. R., Ramos, R. A., Flowers, C. E., Lopez, G. Z., Lewis, D. H., and Cowsar, D. R., Clinical evaluation of injectable biodegradable contraceptive system, Am. J. Obstet. Gynecol., 140, 799, 1981.

70. Beck, L. R., Pope, V. Z., Flowers, C. E., Jr., Cowsar, D. R., Tice, T. R., Lewis, D. H., Dunn, R. L., Moore, A. R., and Gilley, R. M., Poly(DL-lactide-co-glycolide)/norethisterone microcapsules: An injectable biodegradable contraceptive, Biol. Reprod., 28, 186, 1983.

71. Beck, L. R., Flowers, C. E., Jr., Pope, V. Z., Wilborn, W. H., and Tice, T. R., Clinical evaluation of an improved injectable microcapsule contraceptive system, Am. J. Obstet. Gynecol., 147, 815, 1983.

72. Beck, L. R., Pope, V. Z., Cowsar, D. R., Lewis, D. H., and Tice, T. R., Evaluation of a new three-month injectable contraceptive microsphere system in primates (baboon), Contracept. Deliv. Syst., 1, 79, 1980.

73. Beck, L. R., Pope, V. Z., Tice, T. R., and Gilley, R. M., Long-acting injectable microsphere formulation for the parenteral administration of levonorgestrel. Adv. Contracept., 1, 119, 1985.

74. Lewis, D. H., Meyers, W. E., Dunn, R. L., and Tice, T. R., The use of in vitro release methods to guide the development of controlled release formulations, Proc. Int. Symp. Control. Rel. Bioact. Mater., 9, 61, 1982.

75. Lewis, D. H. and Tice, T. R., Polymeric considerations in the design of microencapsulated contraceptive steroids, in Long-Acting Contraceptive Delivery Systems (G. I. Zatuchni, ed.), Harper and Row, Philadelphia, 1983, pp. 77-95.

76. Population Reports, Population Information Program, The Johns Hopkins University, Baltimore, No. 3, March 1987.

77. Grubb, G., Welch, J., Cole, L., Goldsmith, A., and Riveria, R. A., Comparative evaluation of the safety and contraceptive effectiveness of 65 mg and 100 mg of 90-day norethindrone microspheres for controlled release steroid contraception: Phase II study, Fertil. Steril., 51, 803, 1989.

78. Asch, R. H., Heitman, T. O., Gilley, R. M., and Tice, T. R., Preliminary results on the effects of testosterone microcapsules, in Male Contraception: Advances and Future Prospects (G. I. Zatuchni, ed.), Harper and Row, Hagerstown, 1986, pp. 347-360.

79. Tice, T. R., Personal Communication.

80. Zatuchni, G. I., Goldsmith, A., Shelton, J. D., and Sciarra, J. J. (eds.), Long-Acting Contraceptive Delivery Systems, Harper and Row, Philadelphia, 1984.

81. Woodland, J. H. R., Yolles, S., Blake, D. A., Helrich, M., and Meyer, F. J., Long-acting delivery systems for narcotic antagonists. I. J. Med. Chem., 16, 16, 1973.

82. Yolles, S., and Sartori, M. F., Degradable polymers for sustained drug release, in Drug Delivery Systems (R. L. Juliano, ed.), Oxford Univ. Press, New York, 1980, Chap. 3.

83. Wise, D. L., Schwope, A. D., Harrigan, S. E., McCarty, D. A., and Hower, J. F., Sustained delivery of narcotic antagonists from lactic.glycolic acid copolymer implants, in Polymeric Delivery Systems (R. J. Kostelnik, ed.), Gordon and Breach, New York, 1978, Chap. 4.

84. Schwope, A. D., Wise, D. L., and Hower, J. F., Lactic/glycolic acid polymers as narcotic antagonist delivery systems, Life Sci., 17, 1877, 1975.

85. Chiang, C. N., Hollester, L. E., Kishimoto, A., and Barnett, G., Kinetics of a naltrexone sustained-release preparation, Clin. Pharmacol. Ther., 36, 51, 1984.

86. Nuwayser, E. S., Gay, M. H., DeRoo, D. J., and Blaskovish, P.D., Sustained release injectable naltrexone microspheres, Proc. Int. Symp. Control. Rel. Bioact. Mater., 15, 1988.

87. DuPont Magazine, 83(1), 6, 1989.

88. Cha, Y., and Pitt, C. C., A one-week subdermal delivery system for L-methadone based in biodegradable microspheres, J. Control. Rel., 7, 69, 1988.

89. Wise, D. L., McCormick, G. J., Willet, G. P., and Anderson, L. C., Sustained release of an antimalarial drug using a copolymer of glycolic/lactic acid, Life Sci., 19, 867, 1976.

90. Tsakala, M., Gillard, J., Roland, M., Chabot, F., and Vert, M., Pyrimethamine sustained release systems based on bioresorbable polyesters for chemoprophylaxis of rodent malaria, J. Control. Rel., 5, 233, 1988.

91. Hecquet B., Chabot, F., Delatone, G., Fournier, C., Hilati,
     S., Cambies, L., Depadt, G., and Vert, M., In vivo sustained
     release of cisplatin from bioresorbable implants in mice, Anti-
     cancer Res., 6, 1251, 1986.
92. Hecquet, B., Chabot, F., Delatorre Gonzalez, J. C., Fournier,
     C., Hilali, S., Cambier, L., Depadt, G., and Vert M., In vivo
     sustained release of cisplatin from bioresorbable implants in
     mice, Anticancer Res., 6, 1251, 1986.
93. Puisieux, F., Spenlehauer, G., Benoit, J. P., Veillard, M.,
     and Fizamer, C., Preparation, in vitro and in vivo evaluation
     of cisplatin microspheres, Proc. Int. Symp. Control. Rel. Bio-
     act. Mater., 13, 167, 1986.
94. Tsai, D. C., Howard, S. A., Hogan, T. F., Malangu, C. J.,
     Kandzari, S. J., and Ma, J. K., Preparation and in vitro eval-
     uation of polylactic acid-mitomycin c microcapsules, J. Micro-
     encapsul., 3, 181, 1986.
95. Juni, K., Ogata, J., Nakano, M., Ichihara, T., Mori, K.,
     and Akagi, M., Preparation and evaluation in vitro and in vivo
     of polylactic acid microspheres containing doxorubicin, Chem.
     Pharm. Bull., 33, 313, 1985.
96. Bissery, M. C., Valeriote, F., and Thies, C., Fate and effect
     of CCNU-loaded microspheres made of poly(D,L) lactide (PLA)
     or poly B-hydroxybutyrate (PHB) in mice, Proc. Int. Symp.
     Control. Rel. Bioact. Mater., 12, 181, 1985.
97. Bissery, M. C., Valeriote, F., and Thies, F., In vitro lomustine
     release from small poly(B-hydroxybutyrate) and poly(D,L-lactide)
     microspheres, Proc. Int. Symp. Control. Rel. Bioact. Mater.,
     11, 25, 1984.
98. Spenlehouer, G., Benoit, J. P., and Veitlard, M., Formation
     and characterization of poly(D,L-lactide) microspheres for
     chemoembolization, J. Pharm. Sci., 25, 750, 1986.
99. Wada, R., Hyon, S. H., Ikada, Y., Nakao, Y., Yoshikawa,
     H., and Muranishi, S., Lactic acid oligomer microspheres con-
     taining an anticancer agent for selective lymphatic delivery.
     I. In vitro studies, J. Bioact. Compat. Polym., 3, 126, 1988.
100. Juni, K., Ogata, J., Matsui, N., Kubota, M., and Nakano,
     M., Modification of the release rate of aclarubicin from poly-
     lactic acid microspheres by using additives, Chem. Pharm.
     Bull., 33, 1734, 1985.
101. Strobel, J. D., Laughin, T. J., Ostroy, F., Lilly, M. D.,
     Perkins, B. H., and Dunn, R. L., Controlled-release systems
     for anticancer agents, Proc. Int. Symp. Control. Rel. Bioact.
     Mater., 14, 261, 1987.
102. Lewis, D. H., Dappert, T. O., Meyers, W. E., Pritchett, G.,
     and Suling, W. J., Sustained release of antibiotics from bio-
     degradable microcapsules, Proc. Int. Symp. Control. Rel.
     Bioact. Mater., 7, 129, 1980.

103. Tice, T. R., Rowe, C. E., and Setterstrom, J. A., Development of microencapsulated antibiotics for topical administration to wounds. Proc. Int. Symp. Control. Rel. Bioact. Mater., 11, 6, 1984.

104. Tice, T. R., Rowe, C. E., Gilley, R. M., Setterstrom, J. A., and Mirth, D. D., Development of microencapsulated antibiotics for topical administration, Proc. Int. Symp. Control. Rel. Bioact. Mater., 13, 169, 1986.

105. Baker, R. W., Krisko, E. A., Kochinke, F., Grassi, M., Armitage, G., and Robertson, P., A controlled release drug delivery system for the periodontal pocket, Proc. Int. Symp. Control. Rel. Bioact. Mater., 15, 238, 1988.

106. Ikada, Y. Hyon, S. H., Jamshidi, K., Higashi, S., Yamamuro, T., Katutani, Y., and Kitsugi, T., Release of antibiotics from composites of hydroxyapatite and poly(lactic acid), in Advances in Drug Delivery Systems (J. M. Anderson and S. W. Kim, eds.), Elsevier, New York, 1986, pp. 179-186.

107. Ratcliffe, J. H., Hunneyball, I. M., Smith, A., Wilson, C. G., and Davis, S. S., Preparation and evaluation of biodegradable polymeric systems for the intra-articular delivery of drugs, J. Pharm. Pharmacol., 36, 431, 1984.

108. Leelarassama, N., Howard, S. A., Malanga, C. J., Cuzzi, L. A., Hogan, T. F., Kandzari, S. J., and Ma, J. K., Kinetics of drug release from polylactic acid-hydrocortisone microcapsules, J. Microencapsul., 3, 171, 1986.

109. Cavalier, M., Benoit, J. P., and Thies, C., The formation and characterization of hydrocortisone-loaded poly((+/-)-lactide) microspheres, J. Pharm. Pharmacol., 38, 249, 1986.

110. Spilizewski, K. L., Marchant, R. E., Hamlin, C. R., Anderson, J. M., Tice, T. R., Dappert, T. O., and Meyers, W. E., The effect of hydrocortisone acetate loaded poly(DL-lactide) films on the inflammatory response, in Advances in Drug Delivery Systems (J. M. Anderson and S. W. Kim, eds.), Elsevier, New York, 1986, pp. 197-203.

111. Tice, T. R., Lewis, D. H., Cowsar, D. R., and Beck, L. R., U.S. Patent 4,542,025, 1985.

112. Williams, D. L., Nuwayser, E. S., Creeden, D. E., and Gay, M. H., Microencapsulated local anesthetics, Proc. Int. Symp. Control. Rel. Bioact. Mater., 11, 69, 1984.

113. Wakiyama, N., Juni, K., and Nakano, M., Preparation and evaluation in vitro and in vivo of polylactic acid microspheres containing dibucaine, Chem. Pharm. Bull., 30, 3719, 1982.

114. Gilley, R. M., Pledger, K. L., and Tice, T. R., Development of a long-acting biocidal formulation, Proc. Int. Symp. Control. Rel. Bioact. Mater., 14, 232, 1987.

115. Jaffe, H., Hayes, D. K., Lutra, R. P., Shuka, P. G., Amarnath, N., and Chaney, N. A., Controlled release microspheres of insect hormone analogues, Proc. Int. Symp. Control. Rel. Bioact. Mater., 12, 282, 1985.

116. Jaffe, H., Miller, J. A., Giang, P. A., Hayes, D. K., and Stroud, B. H., Implantable System for delivery of insect growth regulation to livestock, in Controlled Release of Pesticides and Pharmaceuticals (D. H. Lewis, ed.), Plenum Press, New York, 1980, pp. 303-310.

117. Maulding, H. V., Prolonged delivery of polypeptides by microcapsules, J. Control. Rel., 6, 167, 1987.

118. Kent, J. S., Sanders, L. M., Tice, T. R., and Lewis, D. H., Microencapsulation of the peptide nafarelin acetate for controlled release, in Long-Acting Contraceptive Delivery Systems (G. I. Zatuchni, A. Goldsmith, J. D. Shelton, and J. J. Sciarra, eds.), Harper and Row, Philadelphia, 1984, pp. 169-179.

119. Kent, J. S., Lewis, D. H., Sanders, L. M., and Tice, T. R., U.S. Patent 4,675,189, 1987.

120. Sanders, L. M., Kent, J. S., McRae, G. I., Vickery, B. H., Tice, T. R., and Lewis, D. H., Controlled release of an LHRH analogue from poly-d,l-lactide co-glycolide microspheres, J. Pharm. Sci., 73, 1294, 1984.

121. Sanders, L. M., Kell, B. A., McRae, G. I., and Whitehead, G. W., Poly(lactic-co-glycolic) acid: Properties and performance in controlled release delivery systems of LHRH analogues, Proc. Int. Symp. Control. Rel. Bioact. Mater., 12, 177, 1985.

122. Hutchinson, F. G., U.S. Patent 4,767,628, 1988.

123. Hutchinson, F. G., Furr, B. J. A., and Churchill, J. R., Biodegradable polymers for the delivery of polypeptides and proteins, Ziekenhuisformacie, 4, 54, 1988.

124. Lawter, J. R., Brizzolara, N. S., Lanzilotti, M. G., and Morton, G. O., Drug release from poly(glycolide-co-lactide) microcapsules, Proc. Int. Symp. Control. Rel. Bioact. Mater., 14, 99, 1987.

125. Asch, R. H., Rojas, F. J., Bartke, A., Schalley, A. V., Tice, T. R., Klemeke, H. G., Siler-Khodr, T. M., Bray, R. W., and Hogan, M. P., Prolonged suppression of plasma LH levels in male rats after a single injection of an LHRH agonist in DL-lactide/glycolide microcapsules, J. Androl., 6, 83, 1985.

126. Tice, T. R., Meyers, W. E., Schalley, A. V., and Redding, T. W., Inhibition of rat prostate tumors by controlled release of [D-Trp[6]]-LHRH from injectable microcapsules, Proc. Int. Symp. Control. Rel. Bioact. Mater., 11, 88, 1984.

127. Ogawa, Y., Yamamoto, M., Takada, S., Okada, H., and Shimamoto, T., Controlled-release of leuprolide acetate from

polylactic acid or copoly(lactic/glycolic) acid microcapsules: Influence of molecular weight and copolymer ratio of polymer, Chem. Pharm. Bull., 36, 1502, 1988.

128. Redding, T. W., Schalley, A. V., Tice, T. R., and Meyers, W. E., Long-acting delivery systems for polypeptides. Inhibition of rat prostate tumors by controlled release of [D-Trp[6]]-LHRH from injectable microcapsules, Proc. Natl. Acad. Sci. USA, 81, 5845, 1984.

129. Asano, M., Yoshida, M., Kaetsu, I., Imai, K., Yuassa, H., Yaminaka, E., Jap. Patent 61,172,813, 1985.

130. Kesler, D. J., Cruz, L. C., McKenzie, J. A., and Henderson, E. A., The effect of GNRH burst release on the efficacy of GnRH microcapsules, Proc. Int. Control. Rel. Bioact. Mater., 13, 54, 1986.

131. Beck, L. R., Flowers, C. F., Jr., Cowsar, D. R., and Tanquary, A. C., U.S. Patent 4,756,907, 1988.

132. Beck, L. R., Flowers, C. F., Jr., Cowsar, D. R., and Tanquary, A. C., U.S. Patent 4,732,763, 1988.

133. Stolle Milk Biologics International, Cincinnati, Ohio, Product Bulletin, 1988.

134. Tabata, Y. and Ikada, Y., Activation of macrophage in vitro to acquire antitumor activity by a muramyl dipeptide derivative encapsulated in microspheres composed of lactide copolymer, J. Control. Rel., 6, 189, 1987.

135. Kwong, A. K., Chou, S., Sun, A. M., Sefton, M. V., and Goosen, M. E. A., In vitro and in vivo release of insulin from poly(lactic acid) microbeads and pellets, J. Control. Rel., 4, 47, 1986.

136. Lin, S. Y., Ho, L. T., Chiou, H. L., Microencapsulation and controlled release of insulin from polylactic acid microcapsules, Biomater. Med. Devices Artif. Organs, 86, 187, 1985.

137. Tice, T. R., Pledger, K. L., Gilley, R. M., Hollingshead, M. G., Westbrook, L. S., Shannon, W. M., and Kende, M., Development of injectable, controlled-release poly(I.C) microcapsules for the inhibition of viral replication, Proc. Int. Symp. Control. Rel. Bioact. Mater., 14, 275, 1987.

138. Schroder, U., Lager, C., and Norrlow, O., Carbo-lactic microspheres; Graft polymerization of PLA to starch microspheres, Proc. Int. Control. Rel. Bioact. Mater., 14, 238, 1987.

139. Steber, W., Fishbein, R., and Cady, S. M., Eur. Patent Appl. 87111217, 1987.

2

Polyanhydrides as Drug Delivery Systems

MARK CHASIN and ABRAHAM DOMB   Nova Pharmaceutical Corporation, Baltimore, Maryland

EYAL RON,* EDITH MATHIOWITZ, and ROBERT LANGER   Massachusetts Institute of Technology, Cambridge, Massachusetts

KAM LEONG   The Johns Hopkins University, Baltimore, Maryland

CATO LAURENCIN   Massachusetts Institute of Technology, Cambridge, and Massachusetts General Hospital, Harvard Medical School, Boston, Massachusetts

HENRY BREM   The Johns Hopkins University School of Medicine, Baltimore, Maryland

STUART GROSSMAN   The Johns Hopkins Oncology Center, Baltimore, Maryland

I.   INTRODUCTION

As is described elsewhere in this book, a large number of biodegradable polymers have been investigated for potential use as drug delivery systems. To maximize control over the release process, it has generally been considered desirable to have a polymeric system which degrades only from the surface. Achieving such a heterogeneous degradation requires that the rate of hydrolytic degradation at the surface of the polymeric system be much faster than the rate of water penetration into the bulk of the matrix. Such a feature may also aid in the delivery of water-labile drugs by making it more difficult for water to interact with these substances until they are released. In designing a biodegradable system that would erode in a controlled heterogeneous manner without requiring any additives,

Current Affiliation:
*Genetics Institute, Cambridge, Massachusetts

we have suggested that due to the high liability of the anhydride
linkage, polyanhydrides may be promising candidates. This chapter
reviews the chemistry, formulation procedures, safety considerations,
and release kinetics of polyanhydride drug delivery systems.

II. CHEMISTRY: Synthesis

We examined several approaches for synthesizing polyanhydrides, including melt polycondensation, dehydrochlorination, and dehydrative
coupling. Extensive details of these new polymer synthesis techniques and numerous polymerization conditions for a wide variety of
polyanhydrides were previously described (1).
    One major drawback of all previous work on polyanhydrides was
their low molecular weight, which made them impractical for many applications. We therefore conducted a systematic study to determine
the mechanism of polymerization and factors affecting the polymer
molecular weight (2). The highest molecular weight polymers were
obtained using melt polymerization techniques, by operating under
conditions which optimized the polymerization process while at the
same time minimizing the depolymerization process. By carefully
understanding the factors affecting both mechanisms, quite high
molecular weights of polyanhydrides were obtained, as determined
by gel permeation chromatography (GPC) and viscosity measurements.
For example, by reacting pure individually prepared prepolymers to
produce P(CPP-SA), a copolymer of biscarboxyphenoxy propane and
sebacic acid in a 20:80 molar ratio, a molecular weight (M_W) of 116,800
([n] = 0.92) was achieved. This is in sharp contrast to the molecular weight of 12,030 obtained when an unisolated and unpurified
prepolymer mixture was used. Factors other than the purity of the
starting materials which were found to be critical to achieving high
molecular weight polyanhydrides were the reaction temperature and
times, and the rapid removal of the acetic acid byproduct by maintaining an appropriate vacuum during the polymerization reaction.
    We also studied the use of catalysts in polyanhydride synthesis
by the melt polycondensation method (2). Since the polymerization
reaction is an anhydride interchange which involves a nucleophilic
attack on a carbonyl carbon, a catalyst which will increase the electron deficiency of the carbonyl carbon will facilitate the polycondensation. Over 20 coordination catalysts were examined in the synthesis of 20:80 (CPP-SA) copolymer. Significantly higher molecular
weights were achieved using shorter reaction times by utilizing cadmium acetate, earth metal oxides, and $ZnEt_2$-H_2O. The molecular
weights ranged from 140,935 to 245,010 with catalysts, in comparison

to the material described above, which had a molecular weight of 116,800 when produced without catalysts. When acidic p-toluene-sulfonic acid (p-TSA) or basic 4-dimethylaminopyridine (4-DMAP) catalysts were tested, the acid catalyst p-TSA did not show any effect while the basic catalyst (4-DMAP) caused a decrease in molecular weight. Since the polymerization and depolymerization reactions involve anhydride interchange, which leads to a high molecular weight polymer with the removal of acetic anhydride as the condensation product (polymerization) and internal ring formation (depolymerization), catalysts affect both reactions. Optimizing the reaction time in the presence of catalysts is therefore critical to achieving high molecular weight polymers (2).

We also developed two approaches for one-step solution polymerization of polyanhydrides at ambient temperature. In the first approach, highly pure polymers (>99.7%) were obtained by using sebacoyl chloride, phosgene, or diphosgene as coupling agent, and poly-(4-vinylpyridine) or K_2CO_3 as insoluble acid acceptor. The second approach for one-step synthesis of pure polyanhydride was the use of an appropriate solvent system during chemical polymerization, where the polymer is extensively soluble but the corresponding polymerization byproduct (e.g., $Et_3N \cdot HCl$) is insoluble (3).

III. FORMULATIONS

A. Molding Procedures

Drug-incorporated matrices were formulated either by compression or by injection molding. The polymers were ground in a Micro mill grinder and sieved into a particle size range of 90–150 μm. For compression molding, model drugs sieved to the same size range were mixed with the polymer manually and the mixture pressed into circular discs in a carver test cylinder outfit at 30 kpsi and 5°C above the polymers' T_g for 10 min. Polymers that had glass transition temperatures below 30°C were molded at room temperature. Alternatively, injection molding was performed in a SCI Mini Max injection molder. A molding temperature of 10°C above the T_m was used. Before the final molding, the polymer-drug mixture was extruded once in the molder for better mixing (4). The choice of molding procedure to be used must be tailored to each drug to be incorporated. Factors to be considered must include the thermal stability of the drug, the potential reactivity of the drug with the polymer chosen, the T_m and T_g values of the polymer of interest, the importance of porosity on the release kinetics of the drug from the molded product, and the cost limitations and manufacturing scale of the final product.

B. Microsphere Fabrication Techniques

For many drug delivery applications, the preferred method of delivery
of the dosage form is by injection. For controlled release applications,
the most frequently used approach to allow this method of administration is to prepare microspheres of the polymer containing the drug
to be delivered. Several different techniques have been developed
for the preparation of microspheres from polyanhydrides.

1. "Hot-melt" Microencapsulation

In the hot-melt microencapsulation process, the drug, encapsulated
as solid particles, was mixed with melted polymer. The mixture was
suspended in a nonmiscible solvent that was heated to 5°C above the
melting point of the polymer and stirred continuously. Stirring was
done using an overhead stirrer and a four-blade impeller. Once the
emulsion was stabilized, it was cooled until the core material solidified. The solvent used in this process was silicon or olive oil. In
some cases the drug can be used without sieving but, in general, a
particle size of less than 50 μm was found to be optimum and substantially improved the drug distribution within the microspheres.
After cooling, the microspheres were washed by decantation with
petroleum ether to give a free-flowing powder. They were then
sieved, dried, and stored in a freezer. Size distribution can be
controlled by the stirring rate; the yield is 70–90%. The process
was quite reproducible with respect to yield, size, and loading distribution, if the same molecular weight of polymer was used. Less
than 5% error was observed (5).

2. Microsphere Preparation by Solvent Removal

Several variations of the solvent removal technique were developed
(6,7). For the PCPP-SA, 20:80, M_w = 16,000, microspheres were
prepared as follows: 1 g polymer was dissolved in 1 ml methylene
chloride, drug or dye was suspended in the solution, mixed, dropped
into silicon oil containing 1–5% of Span 85, and stirred at a known
stirring rate. Stirring was done using an overhead stirrer and a
three-blade impeller. After 1 hr, petroleum ether was introduced
and stirring was continued for another hour. The microspheres
were isolated by filtration, washed with petroleum ether, dried overnight in a lyophilizer, sieved, and stored in a freezer.

When trying to apply the same method to PCPP-SA polymers of
higher molecular weight or with higher percentages of CPP, the
above process results in rod formation rather than microspheres.
In this case a different method was used: 2 g polymer was dissolved
in 10 ml of methylene chloride, drug was added, and the mixture was
suspended in silicon oil containing Span 85 and a known amount of
methylene chloride. The amount of methylene chloride depended on
the type and molecular weight of the polymer used. For example,

for PCPP-SA, 20:80, with a molecular weight of 30,000–40,000, the
ratio between the silicon oil and the methylene chloride was 4:1, and
for PCPP-SA, 50:50 with a molecular weight of 40,000, this ratio was
1:1. After dropping the polymer solution into the silicon oil, petroleum ether was added and the stirring continued for 2 hr. The microspheres were isolated by filtration, washed with petroleum ether,
dried overnight in a lyophilizer, and stored in a freezer. All drugs
used were sieved to sizes lower than 50 μm. Almost no precipitation
on the stirrer was observed. The process is reproducible to within
5% with respect to yield and size distribution, if polymers of the
same molecular weight are used (6).

IV. KINETICS OF DRUG RELEASE

A. Degradation Characteristics

Within a series of closely related polyanhydride copolymers, the relative ratios of the two monomers have a marked effect on the rate of
degradation of the resulting polymer. An example is shown in Fig.
1 (4). In disc form, when prepared by compression molding, the
more hydrophobic polymers, PCPP and PCPP-SA, 85:15, displayed
constant erosion kinetics over 8 months. By extrapolation, 1-mm-
thick discs of PCPP will completely degrade in over 3 years. The
degradation rates were increased by copolymerization with sebacic
acid. An increase of 800 times was observed when the sebacic acid
concentration reached 80%. By altering the CPP-SA ratio, nearly
any degradation rate between 1 day and 3 years can be achieved
(4).

The polyanhydrides in general degrade more rapidly in basic
media than in acidic media (4). This effect is shown in Fig. 2.
Pure PCPP was used for this experiment to magnify the effect. At
pH 7.4, pure PCPP degrades in about 3 years, as discussed above.
However, this rate increases markedly as the pH rises, and at pH
10.0, this material degrades in just over 100 days. At very acidic
pH values, many of the polyanhydrides virtually do not erode at all.

The effect of different backbones on erosion rates was demonstrated in a study of the homologous poly[(p-carboxyphenoxy)alkane]
series. As the number of methylene groups in the backbone increased
from 1 to 6, thus decreasing the reactivity of the anhydride linkage
and rendering the polymer more hydrophobic, the erosion rates underwent a decrease of three orders of magnitude (4).

B. Newer Polyanhydrides

Several new series of polyanhydrides with advantageous properties
for a variety of applications were also synthesized (8). The first
are aliphatic-aromatic homopolyanhydrides of the structure

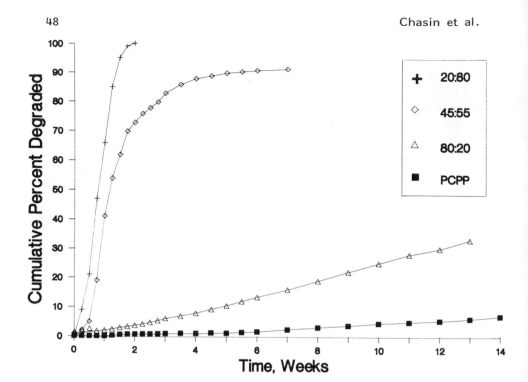

FIGURE 1   Rate of polyanhydride degradation versus time.   PCPP
and SA copolymers were formulated into 1.4-cm-diameter disks 1 mm
thick by compression molding, and placed into a 0.1 M pH 7.4 phosphate buffer solution at 37°C.   The cumulative percentage of the
polymer which degraded was measured by absorbance at 250 nm.

$-(OOC-C_6H_4-O(CH_2)_x-CO-)_n$, where x varies from 1 to 10.   These
polyanhydrides display a zero-order hydrolytic degradation profile
and drug release profile, as can be seen in Fig. 3.   Analogs of this
structure have been used for the preparation of a series of polyanhydrides which degrade for periods from 2 to 10 weeks.   The rate
of degradation is a function of the length of the aliphatic chain.   As
can be seen in Fig. 4, the zero-order degradation profile is maintained independently of the time over which the particular polyanhydride degrades.

     The second type of polymer, unsaturated polyanhydrides of the
structure $[-(OOC-CH=CH-CO)_x-(OOC-R-CO)_y-]_n$, have the advantage of being able to undergo a secondary polymerization of the
double bonds to create a crosslinked matrix.   This is important for
polymers requiring great strength, for instance.   These polymers
were prepared from the corresponding diacids polymerized either by

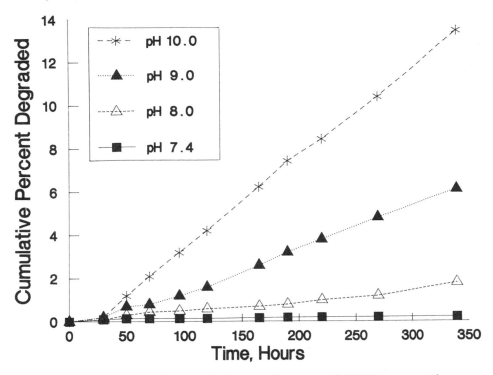

FIGURE 2 pH dependance of the erosion rate of PCPP versus time. Discs of PCPP were formulated into 1.4-cm-diameter discs 1 mm thick by compression molding, and placed into 0.1 M phosphate buffers at various pH values at 37°C. The cumulative percentage of the polymer which degraded was measured by absorbance at 250 nm.

melt polycondensation or by polymerization in solution. Molecular weights of up to 44,000 were achieved for polyanhydrides of p-carboxyphenoxyalkanoic acid and fumaric acid (9).

C. Release Characteristics

The release behavior depends on both the choice of polymer and on the formulation procedure. The best results were obtained with injection-molded samples. For prototypical drugs like p-nitroaniline, the drug release pattern followed closely that of the polymer degradation over a period of 9 months for PCPP. The correlation between release and degradation was still maintained in the more hydrophilic PCPP-SA, 20:80, where both processes were completed in 2 weeks. Compression molding can also be used for these polymers, but the correlation between drug release and polymer degradation is not as

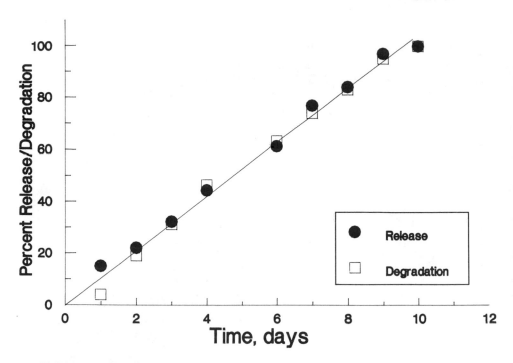

FIGURE 3   Release of p-nitroaniline and the degradation of PCPA
versus time.   Disks (1.4 cm in diameter and 1 mm thick) of poly-
(carboxyphenoxyacetic acid) (PCPA) were prepared containing 5%
(w/w) p-nitroaniline by compression molding, and degraded in 0.1 M
pH 7.4 phosphate buffer at 37°C.   The cumulative release of p-
nitroaniline and degradation of PCPA were measured by absorbance
at 380 and 235 nm, respectively.

good as injection-molded materials for the less hydrophobic polymers
(4,6).

The incorporation and release kinetics from polyanhydride matrices of a number of drugs have been studied.   Representative examples of several of these are described below.

1.   BCNU (Carmustine)

BCNU has been incorporated into PCPP-SA 20:80 in two different
ways (10).   These involve either (1) triturating the dry powdered
BCNU with similarly treated polymer and pressing weighed aliquots
of the mixture in a Carver press, or (2) codissolving polymer and
BCNU in methylene chloride, evaporating the solvent, and pressing
the resulting material as in method 1.   The former method produces
a solid mixture of BCNU and polymer, while the latter method

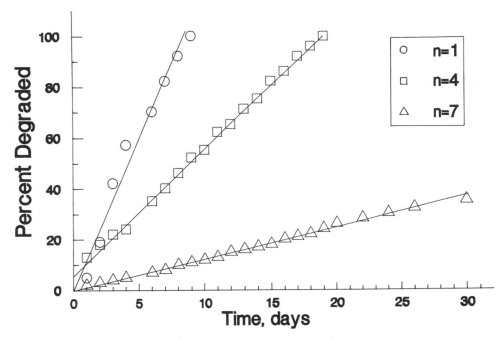

FIGURE 4   Degradation of poly(carboxyphenoxyacetic acid) (n = 1),
poly(carboxkyphenoxyvaleric acid) (n = 4), and poly(carboxyphenoxyoctanoic acid) (n = 7) versus time.   Discs (1.4 cm in diameter and
1 mm thick) of the appropriate poly(carboxyphenoxyalkanoic acid)
were prepared by compression molding, and degraded in 0.1 M pH
7.4 phosphate buffer at 37°C.   The cumulative degradation of the
polymers was measured by absorbance at 235 nm.

produces a solid solution of BCNU in the polymer.   These are referred to here as the trituration or solution methods.
    These two methods produce different release profiles in vitro.
Figure 5 demonstrates the release kinetics of BCNU from wafers
loaded with 2.5% BCNU pressed from materials produced using these
two methods.   The wafers containing tritiated BCNU were placed into
beakers containing 200-ml aliquots of 0.1 M phosphate buffer, pH
7.4, which were placed in a shaking water bath maintained at 37°C.
The shaking rate was 20 cycles/min to avoid mechanical disruption
of the wafers.   The supernatant fluid was sampled periodically, and
the BCNU released was determined by liquid scintillation spectrom-
etry.   The BCNU was completely released from the wafers prepared
by the trituration method within the first 72 hr, whereas it took
just about twice as long for the BCNU to be released from wafers

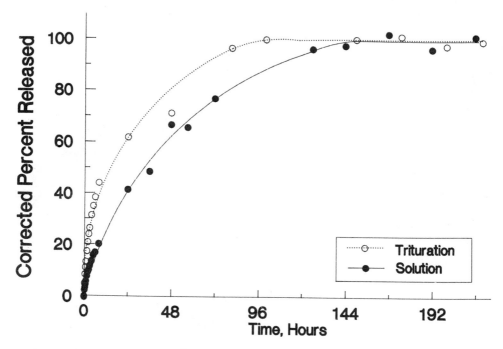

FIGURE 5 Release of [³H]BCNU from compression-molded discs of PCPP-SA, 20:80. Disks 3 mm in diameter and 1 mm thick containing 2.5% [³H]BCNU were prepared by compression molding using either the trituration or solution methods described in the text. In vitro release of [³H]BCNU was measured as described in the text.

produced by the solution method. Essentially the same pattern was observed for wafers loaded with 10% BCNU by weight.

The difference in rates of release of BCNU from wafers produced by the trituration or solution methods is also seen in vivo (11,14), as is shown in Fig. 6. Wafers of PCPP-SA 20:80 were prepared by either the solution or trituration methods, as described above, and were implanted into the brains of rabbits. The animals were sacrificed at various times after implantation and the brains were removed, fixed, and processed for quantitative autoradiography. To quantitate the percentage of the brain exposed to BCNU released from these wafers, the following calculation was performed. The percentage of the brain in which the radioactivity from the tritiated BCNU released from the wafers exceeded the background counts by at least two standard deviation units was plotted as a function of time following implantation in Fig. 6. A control set of rabbits had a solution of BCNU injected directly into the same location in the

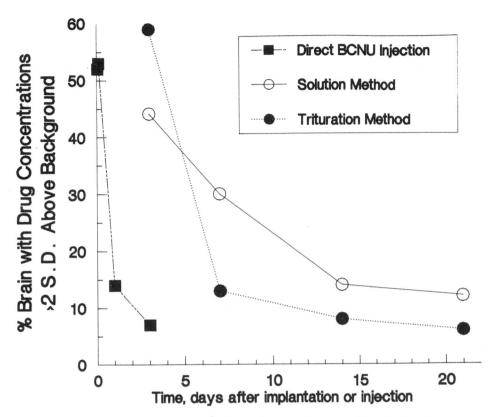

FIGURE 6   Distribution of [³H]BCNU-associated radioactivity time in
rabbit brain.   The distribution of radioactivity was measured using
quantitative autoradiography techniques following implantation of
discs described for Fig. 5 into the brains of rabbits.   Details of the
experimental procedure are described in the text.

brain as the implants.   The rate of release of BCNU from wafers
prepared by the solution method was slower than that from wafers
prepared by the trituration method, in agreement with the in vitro
results.   At the longest time measured, 21 days after implantation,
these animals had significant levels of BCNU in an area of brain almost twice as large as those receiving the wafers produced by the
trituration method.

Figure 6 also shows that the use of this polymeric delivery system, also known as the BIODEL® polymeric drug delivery system,
for BCNU greatly increases the time over which the brains of these
animals are exposed to significant BCNU concentrations.   The brains

of animals which received a single injection of the same amount of
BCNU contained in the wafers were almost free of BCNU by 3 days
after injection.   Since it has been shown conclusively both in vitro
and in vivo that the key determinant of the ability of a drug to kill
cancer cells depends on the product of the drug's concentration and
the time over which the drug and the cancer cells are in contact
(12,13), it seems reasonable to conclude that the greatly increased
time over which BCNU is delivered to the brain using PCPP-SA 20:80
should increase the efficacy of BCNU.

Since these studies utilized autoradiographic techniques, it was
important to determine the chemical nature of the material measured
using radiochemical procedures.   Using punch biopsies of the brain
slices which were measured autoradiographically, it was shown using
thin-layer chromatography that about 30% of the radioactivity was
associated with unchanged BCNU (14).   It was therefore concluded
that the measurements described above accurately reflect brain concentrations of BCNU.

From these same autoradiography studies, it is also possible to
determine the local brain concentrations of BCNU which can be
achieved using the BIODEL® delivery system.   The brain adjacent to
the surface of these wafers (signified by zero on the distance axis
in Fig. 7) is exposed to concentrations of BCNU of approximately
6.5 mM at 3 days following implantation.   Even as far as 10 mm from
this surface, the local concentrations of BCNU are approximately 200
μM.   This concentration range is certainly much higher than the
brain concentrations achieved through a single intravenous administration of BCNU, which is the standard treatment for this disease.

2.   Cortisone Acetate

Cortisone acetate has been incorporated into several polyanhydrides
(15).   The rates of release of cortisone acetate from microcapsules
of poly(terephthalic acid), poly(terephthalic acid-sebacic acid) 50:50,
and poly(carboxyphenoxypropane-sebacic acid) 50:50 are shown in
Fig. 8.   These microcapsules were produced by an interfacial condensation of a diacyl chloride in methylene chloride with the appropriate dicarboxylic acid in water, with or without the crosslinking
agent trimesoyl chloride.   This process produces irregular microcapsules with a rough surface.   The release rates of cortisone acetate from these microcapsules varied correspondingly with the rate
of degradation of the respective polyanhydrides.   It can be expected
that the duration of release of cortisone acetate from solid microspheres, such as those produced by the hot-melt process, would be
considerably longer.

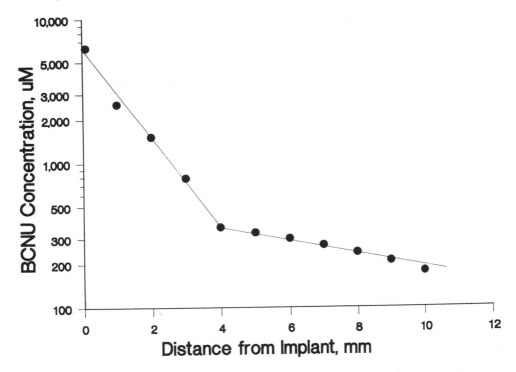

FIGURE 7 Movement of [³H]BCNU-associated radioactivity through rabbit brain. Radioactivity resulting from [³H]BCNU was measured at various distances from the implantation site following implantation of discs described for Fig. 5 into the brains of rabbits. Details of the experimental procedure are described in the text.

3. Angiogenesis Inhibitors

The controlled release from PTA-SA 50:50 of several drugs known to inhibit the formation of new blood vessels in vivo, cortisone and heparin, is shown in Fig. 9 (15). The inhibitors of angiogenesis delivered in vivo using this polyanhydride were shown to prevent new blood vessel growth for over 3 weeks, following the implantation of the VX2 carcinoma into rabbit cornea (15).

4. Bethanechol

The memory deficits characteristic of Alzheimer's disease have not yet been successfully treated. There is reason to believe that potent acetylcholinesterase-resistant cholinomimetics might be effective in treating these deficits, but systemic administration of agents of this type, such as bethanechol, does not adequately deliver such

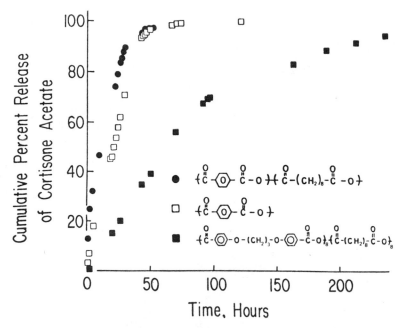

FIGURE 8   Release of cortisone acetate from 10% loaded microspheres of various polyanhydrides.   The microspheres were prepared by an interfacial condensation.   Details of the experimental procedure are described in the text.

drugs to the brain.   Howard et al. (16) studied the delivery of bethanechol directly to the brains of rats which had been lesioned to produce memory deficits and measured the performance scores of these rats in a radial maze test.   As can be seen in Fig. 10, both the sham-operated group and the group receiving unloaded polymer showed persistent poor spatial memory whereas the rats receiving polymer containing bethanechol displayed significant improvement within 10 days of implantation, demonstrating that intracerebral implantation of such a controlled release form of a potent cholinomimetic in a polyanhydride matrix has potential for treating neurodegenerative diseases.

5.   Incorporation and Release of Proteins

A number of complex molecules such as proteins have been incorporated into the polyanhydrides, including insulin, enzymes, chondrogenic stimulating proteins, and a protein synthesized by genetic engineering techniques.

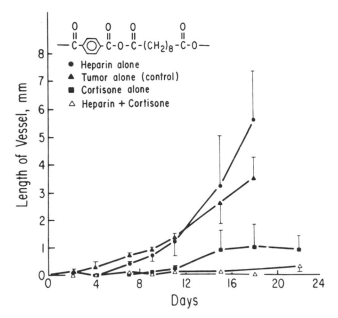

FIGURE 9 Influence of angiogenesis inhibitors on blood vessel growth. Several inhibitors of angiogenesis were released from PTA-SA 50:50 in vivo. The effect of these agents on the growth of blood vessels around the VX2 carcinoma implanted into rabbit corneas was then determined as described in the text.

6. Insulin

Insulin has been incorporated into microspheres of PCPP-SA 50:50 (5,17). The loading of insulin was 15% in these microspheres, which were between 850 and 1000 µm in diameter. The pattern of release of insulin in vitro from these microspheres is shown in Fig. 11. Much of the insulin is released over the first 1–2 days, but significant amounts continue to be released for 4–5 days.

The temporal correlation between in vitro and in vivo release of insulin was quite good. To induce diabetes, two groups of rats were administered 65 mg/kg of streptozotocin in 0.1 M citrate buffer, pH 4.5. Within several days, the animals had become diabetic, as evidenced by blood glucose levels of approximately 400 mg/dl, and substantial output of glucose in their urine. One group of these animals was then injected subcutaneously with 40–50 mg of 15% insulin-loaded PCPP-SA 50:50 microspheres, 850–1000 µm in diameter. A third group of animals receiving no treatment served as a control.

As can be seen in Fig. 12, the injection of insulin-containing microspheres decreased the blood glucose values in the diabetic rats

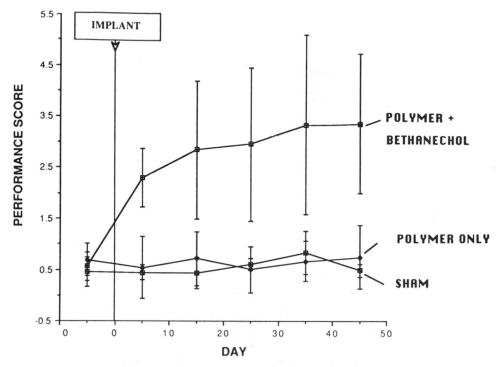

FIGURE 10 Effect of bethanechol on memory in lesioned rats. Bethanechol was released from PCPP-SA 50:50 implanted intracerebrally
in rats. The effect of the bethanechol released on the performance
of lesioned rats in a radial maze test was performed as described in
the text.

to the levels seen with the normal, untreated rats for up to 5 days.
Not shown here is that the same effect was seen on urinary glucose
levels; they returned to normal for about the same length of time
(5,17). In addition, in vitro release studies using these microspheres
showed that by HPLC analysis, the insulin released was indistinguish-
able from unincorporated insulin.

Once the blood glucose values in the treated animals had returned
to the high, diabetic levels, a second injection of insulin-containing
microspheres again reduced these levels to normal for about 5 more
days. It is therefore possible to incorporate labile biological products into the polyanhydrides and to release them in a biologically
active form. At the same time, this release can be sustained over
a period of time in a controlled fashion.

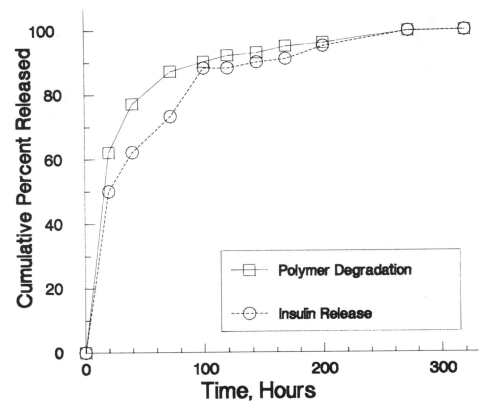

FIGURE 11   Release of insulin from microspheres of PCPP-SA 50:50 versus time in vitro.   Experimental details were as described in the text.

7.   Alkaline Phosphatase

Alkaline phosphatase, an enzyme with a molecular weight of approximately 86,000, has been incorporated into a polyanhydride matrix using compression molded PCPP-SA 9:91.   Five percent loaded wafers, 50 mg each, were perpared, and measured 1.4 cm in diameter, with a thickness of 0.5 mm.   Release experiments were then conducted using techniques similar to those described for carmustine above.   As can be seen in Fig. 13, the alkaline phosphatase was released in a well-controlled manner over a prolonged period of time, just over a month, from this polyanhydride.

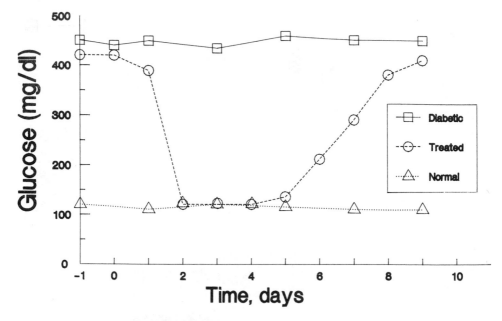

FIGURE 12 Effect of insulin released from microspheres of PCPP-SA 50:50 injected subcutaneously in streptozotocin-diabetic rats. Details were as described in the text.

8. β-Galactosidase

As in the alkaline phosphatase example above, β-galactosidase, an enzyme with a molecular weight of approximately 360,000, has also been incorporated into a polyanhydride and released in a well-controlled fashion. As is shown in Fig. 14, the release of β-galactosidase was quite linear over most of the time examined, and was complete, reaching 100% release in about 800 hr. This experiment utilized 5% loaded, compression-molded wafers of PCPP-SA 9:91, 1.4 cm in diameter and 0.5 mm thick, weighing 50 mg.

9. Bovine Growth Hormone

Bovine growth hormone, a difficult protein for which to develop controlled release systems due to its propensity toward self-aggregation and inactivation, has successfully been incorporated into polyanhydride matrices (18). The growth hormone was colyophilized with sucrose, dry-mixed with finely powdered polyanhydride, and then compression molded into 1.4-cm-diameter wafers, 1 mm thick. As is shown in Fig. 15, release of bovine growth hormone was well controlled over a prolonged period of time. The assay for bovine

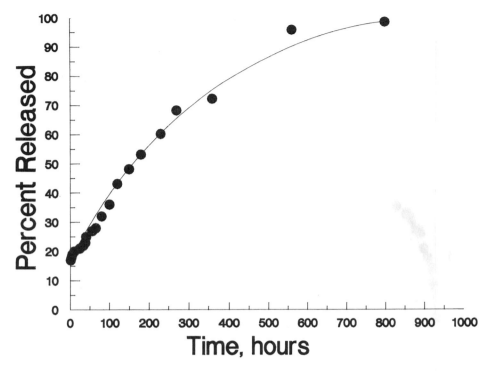

FIGURE 13   Release of alkaline phosphatase from compression-molded discs of PCPP-SA 9:91.   Details were as described in the text.

growth hormone used in these experiments was an HPLC assay specific for the monomeric form of the hormone and has been verified by radioimmunoassay.

10.   Chondrogenic Stimulating Proteins

Although water-soluble proteins isolated from bone matrix demonstrated chondrogenic stimulating activity in stage 24 chick limb bud cultures, they were not able to induce cartilage or bone growth in vivo in mice intramuscularly (19).   Similarly, the polyanhydrides had no effect when tested alone in vivo.   When the water-soluble proteins from bone matrix were incorporated into the polyanhydrides and then implanted between muscle beds in the thighs of mice, induction of cartilage and bone was demonstrated.   This is a unique example of a potential drug whose biological activity in vivo requires sustained, controlled, local delivery rather than a delivery system merely serving to enhance an otherwise moderate biological effect.

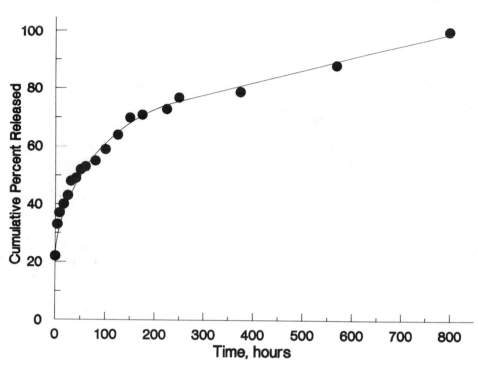

FIGURE 14   Release of β-galactosidase from compression-molded discs of PCPP-SA 9:91.   Details were as described in the text.

These examples of incorporation of a variety of drugs and proteins is meant to be representative, not inclusive, and suggests that the polyanhydrides are capable of delivering a wide range of drugs and proteins for prolonged periods of time from a variety of different dosage forms.

D.   Stability

The stability of polyanhydrides composed of the diacids sebacic acid (SA), bis(p-carboxyphenoxy)methane (CPM), 1,3-bis(p-carboxyphenoxy)propane (CPP), 1,6-bis(p-carboxyphenoxy)hexane (CPH), and phenylenedipropionic acid (PDP), in solid state and in organic solutions, was studied over a 1-year period.   Aromatic polyanhydrides such as poly(CPM) and poly(CPH) maintained their original molecular weight for at least a year in both solid state and solution (20).

In contrast, aliphatic polyanhydrides such as poly(SA) and poly-(PDP) decreased in molecular weight over time.   The decrease in molecular weight shows first-order kinetics, with activation energies

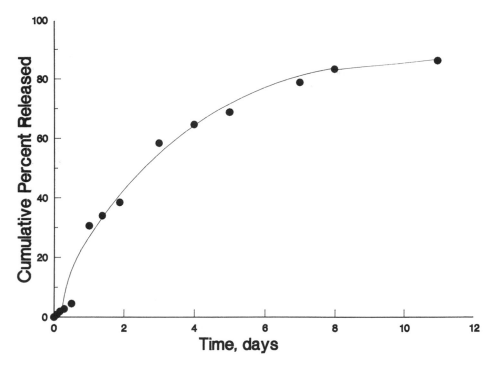

FIGURE 15   Release of bovine growth hormone from compression-
molded discs of PCPH.   Details were as described in the text.

of 7.5 kcal/mol–°K, as shown in Table 1.   The decrease in molecular
weight was explained by an internal anhydride interchange mechanism,
as revealed from elemental and spectral analysis (IR and $^{13}C, ^{1}H$-NMR).
The depolymerization in solution can be catalyzed by metals.   Among
several metals tested, copper and zinc were the most effective.   Interestingly, the stability of the polymers in the solid state or in organic solutions does not, in many cases, correlate with their hydrolytic stability (20). (See Fig. 16).
     Stability studies under anhydrous conditions were also conducted
using the aliphatic-aromatic homopolyanhydrides.   These polymers
were completely stable in the solid state in vacuo for over 6 months
at 25°C (8).

E.   Polymer-Drug Interactions

Studies were also done to explore possible polymer-drug interactions.
In particular, we were concerned that anhydrides could react with
drugs containing active groups such as amines.   Infrared spectroscopy

Chasin et al.

TABLE 1 Stability of Various Polyanhydrides Versus Time in Solution

Polymer	Depolymerization[a] rate constant (t^{-1})[b]						Activation energy (kcal/mol deg)
	−20°C	0°C	21°C	37°C	45°C		
Poly(SA)	0.0071	0.0324	0.0763	0.1325	0.2247		8.08
Poly(CPP-SA) 20:80	0.0077	0.0251	0.0777	0.1535	0.1535		8.27
Poly(CPP-SA) 50:50	0.0006	0.0041	0.0395	0.0743	0.1014		7.27[c]
Poly(CPH)	0.0000	0.0000	0.0000	0.0000	0.0000		—
Poly(CPM)	0.0000	0.0000	0.0000	0.0000	0.0000		—

[a]Depolymerization was in chloroform solution with a polymer concentration of 10 mg/ml. The depolymerization rate constant was determined by viscosity measurements over 24 hours. Details as described in the text.
[b]Calculated on the basis of first order kinetics.
[c]Determined from the rate constants at 21, 37, and 45°C.

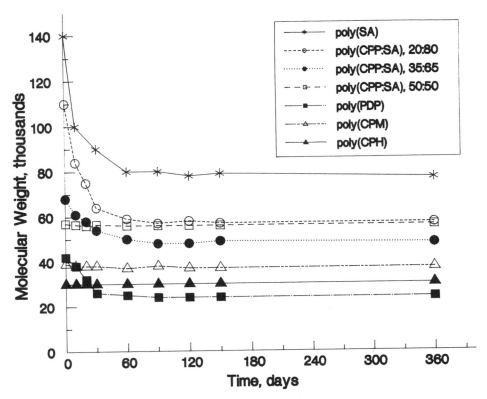

FIGURE 16 Stability of various polyanhydrides versus time. The molecular weight of various polyanhydrides stored in vacuo in glass ampules at room temperature was measured by GPC at various times. Details as described in the text.

showed that chemical interaction between the drug and the polyanhydride was possible when high temperatures ($>150°C$) were used in
formulating devices. Whether interaction occurred depended on the
reactivity of the drug. With p-nitroaniline as a marker, there was
no observable amide bonding even using these temperatures. When
p-bromoaniline was tested, which has a basicity constant 700 times
higher, clustered bands with moderate intensity began to appear in
the amide region. Using an amine which is 10^5 times more basic
than p-nitroaniline, p-phenylenediamine, we observed definite amide
formation.

Using the same spectroscopic analysis, there was no evidence of
interaction when the polymer and drug were compression-molded at
37°C. In analyzing the degradation products of the compression-
molded sample, only the free drug was retrieved. In release studies,
no new product formation and essentially complete recovery of the
intact free drug was detected by chromatographic analysis (HPLC).
The above results indicate the interaction would occur only in the
fabrication step, and only at elevated temperatures. This could be
solved by synthesizing a polymer of low T_m that could be easily
injection-molded (8).

F. Drug Safety Evaluation of the Polyanhydrides

A series of biocompatibility studies was performed on several polyanhydrides. As evaluated by mutation assays (21), the degradation
products of the polymer were nonmutagenic and noncytotoxic. In
vitro tests measuring teratogenic potential were also negative. Growth
of two types of mammalian cells in tissue culture was also not affected
by these polymers (21); both the cellular doubling time and cellular morphology were unchanged when either bovine aorta endothelial cells or
smooth muscle cells were grown directly on the polymeric substrate.

Implants in the rabbit corneas exhibited no observable inflammatory characteristics over a period of 6 weeks. Compared to other
previously tested polymers, the inertness of these polyanhydrides
rivals that of the biocompatible poly(hydroxyethyl methacrylate) and
ethylene-vinyl acetate copolymer. Histological examination of the removed corneas also revealed the absence of inflammatory cells (21).

Additional evidence of biocompatibility was provided from subcutaneous implantation tests in rats (21-23). After a period of 6
months, only very slight tissue encapsulation was seen around im-
plantations of PCPP. After 56 days, no tissue encapsulation was
seen around implants of PCPP-SA 20:80. No inflammation was apparent in the tissues adjacent to the implant from histological evaluations of PCPP, and only very slight inflammation was seen following
implantation of PCPP-SA 20:80. No changes were seen in measures
of blood chemistry and hematology of the rats during the degradation
of these polymers in vivo, nor did gross and microscopic postmortem
analysis reveal any abnormalities (21-23).

The biocompatibility of the PCPP-SA 20:80 with
studied to evaluate the safety of the polymer and it
products in the rat, and to compare it to two stan
which have been extensively studied and proven to be no
mildly inflammatory to neural tissue (24,25). These materials are
Gelfoam (absorbable gelatin sponge) and Surgicel (another commonly
used material in brain surgery). The animals were sacrificed using
CO_2 asphyxiation, one group per day on days 3, 6, 10, 15, 21, 28,
and 36 following surgery. Histological evaluation of the tissue demonstrated a small rim of necrosis around the implant, and a mild to
marked cellular inflammatory reaction limited to the area immediately
adjacent to the polymer implant site, slightly more marked than Surgicel at the earlier time points, but noticeably less marked than Surgicel at the later times. The reaction to Gelfoam was essentially
equivalent to sham-operated control animals. Using PCPP-SA 50:50
in rabbit brain, even less of an inflammatory reaction was observed;
the polymer was essentially equivalent to Gelfoam (26).

The biocompatibility of the PCPP-SA 20:80 with the monkey brain
was studied to measure the direct effects of the polymer, polymer
degradation products, and polymer containing BCNU on the monkey
brain (27,28). This was considered to be the best experimental
model of the human brain. Blood samples were periodically obtained
from each monkey for blood chemistry and hematology analyses. On
the ninth day following surgery, the animals had both non-contrast-
and contrast-enhanced computer tomography (CT) studies performed.
On the twelfth day following surgery, magnetic resonance imaging
(MRI) studies were performed. At the conclusion of the experiment,
samples of the brain and some 35 other tissues were prepared for
histological examination.

No abnormalities were noted in any of the computer-assisted
tomography scans or magnetic resonance images, nor in the blood
chemistry or hematology evaluations. No systemic effects of the
implants were noted on histological examination of any of the tissues
examined. No unexpected or untoward reactions to the treatments
were observed.

G. Clinical Studies

PCPP-SA 20:80 with the cancer chemotherapeutic agent BCNU incorporated has been studied in man for the treatment of glioblastoma
multiforme, a universally fatal form of brain cancer. In these studies, patients undergoing reoperation for the removal of the bulk of
the tumor have had the surgical cavity lined with the polymer containing BCNU. Following surgery, the BCNU is released directly
onto adjoining cancer cells that may not have been removed during
surgery. The safety of this material implanted into these patients
has been demonstrated. No systemic side effects of doses of BCNU

which would produce marked effects on the hemopoietic system when injected intravenously have been observed. Further studies designed to measure the efficacy of this approach to the treatment of brain cancer are currently under way.

ACKNOWLEDGMENTS

The support of NIH grant NS1058-01, the Andrew W. Mellow Foundation, and the Johns Hopkins University Faculty Development Award are gratefully acknowledged.

REFERENCES

1. Leong, K. W., Simonte, V., and Langer, R., Synthesis of polyanhydrides: Melt-polycondensation, dehydrochlorination, and dehydrative coupling, Macromolecules, 20, 705-712, 1987.
2. Domb, A., and Langer R., Polyanhydrides. I. Preparation of high molecular weight polyanhydrides, J. Polym. Sci., 25, 3373-3386, 1987.
3. Domb, A., Ron, E., and Langer, R., Polyanhydrides. II. One step polymerization using phosgene or diphosgene as coupling agents, Macromolecules, 21, 1925-1929, 1988.
4. Leong, K. W., Brott, B. C., and Langer, R., Bioerodible polyanhydrides as drug-carrier matrices. I. Characterization, degradation and release characteristics, J. Biomed. Mater. Res., 19, 941-955, 1985.
5. Mathiowitz, E., and Langer R., Polyanhydride microspheres as drug carriers. I. Hot-melt microencapsulation, J. Control. Rel., 5, 13-22, 1987.
6. Mathiowitz, E., Saltzman, M., Domb, A., Dor, Ph., and Langer, R., Polyanhydride microspheres as drug carriers. II. Microencapsulation by solvent removal, J. Appl. Polym. Sci., 35, 755-774, 1988.
7. Bindschaedler, C., Leong, K. W., Mathiowitz, E., and Langer, R., Polyanhydride microsphere formulation by solvent extraction, J. Pharm. Sci., 77, 696-698, 1988.
8. Domb, A., Gallardo, C., and Langer, R., Poly(anhydrides). 3. Poly(anhydrides) based on aliphatic-aromatic diacids, Macromolecules. In Press.
9. Domb, A., and Langer, R., Polyanhydrides: Stability and novel composition, Makromol. Chem. Macromol. Symp., 19, 189-200, 1988.
10. Chasin, M., Lewis, D., and Langer, R., Polyanhydrides for controlled drug delivery, BioPharm Manufact., 1, 33-46, 1988.

11. Grossman, S. A., Reinhard, C. S., Brem, H., Brundrette, R., Chasin, M., Tamargo, R., and Colvin, O. M., The intracerebral delivery of BCNU with surgically implanted biodegradable polymers: A quantitative autoradiographic study, Proc. Am. Soc. Clin. Oncol., 7, 84, 1988.

12. Alberts, D. S., and van Daalen Wetters, T., The effect of phenobarbital on cyclophosphamide antitumor activity, Cancer Res., 36, 2785-2789, 1976.

13. Alberts, D. S., Chen, H.-S. G., and Salmon, S. E., In vitro drug assay: Pharmacologic considerations, Prog. Clin. Biol. Res., 48, 197-207, 1980.

14. Grossman, S. A., Reinhard, C., Colvin, O. M., Chasin, M., Brundrett, R., Tamargo, R., and Brem, H., The intracerebral distribution of BCNU delivered by surgically implanted bioerodable polymers. Submitted.

15. Leong, K. W., Kost, J., Mathiowitz, E., and Langer, R., Polyanhydrides for controlled release of bioactive agents, Biomaterials, 7, 364-371, 1986.

16. Howard, M. A., Gross, A., Grady, M. S., Langer, R., Mathiowitz, E., Winn, H. R., and Mayberg, M. R., Intracerebral drug delivery in rats reverses lesion-induced memory deficits, J. Neurosurg. In Press.

17. Mathiowitz, E., Leong, K., and Langer, R., Macromolecular drug release from biodegradable polyanhydride microspheres, 12th Int. Symp. Control. Rel. Bioact. Mater., 183-184, 1985.

18. Ron, E., Turek, T., Mathiowitz, E., Chasin, M., and Langer, R., Release of Polypeptides from Poly(anhydrides) Implants, Proc. Contr. Rel. Soc., 338, 1989.

19. Lucas, P. A., Laurencin, C., Syftestad, G. T., Domb, A., Goldberg, V. M., Caplan, A. I., and Langer, R., Ectopic induction of cartilage and bone by water-soluble proteins from bovine bone using a polyanhydride delivery vehicle, J. Control. Rel. In Press.

20. Domb, A., and Langer, R., Solid state and solution stability of poly(anhydrides) and poly(esters), Macromolecules, 21, 1925-1929, 1988.

21. Leong, K. W., D'Amore, P., Marletta, M., and Langer, R., Bioerodible polyanhydrides as drug-carrier matrices: II. Biocompatibility and chemical reactivity, J. Biomed. Mat. Res., 20, 51-64, 1986.

22. Laurencin, C., Domb, A., Morris, C., Brown, V., Chasin, M., McConnell, R., Lange, N., and Langer, R., Poly(anhydride) administration in high doses in vivo: Studies of biocompatibility and toxicology, J. Biomed. Mat. Res., In Press.

23. Laurencin, C. T., Domb, A. J., Morris, C. D., Brown, V. I.,
     Chasin, M., McConnell, R. F., and Langer, R., High dosage
     administration of polyanhydrides in vivo: Studies of biocom-
     pabibility and toxicology, Proc. Int. Symp. Control. Rel. Bio-
     act. Mater., 14, 140-141, 1987.
24. Tamargo, R. J., Epstein, J. I., Reinhard, C., Chasin, M.,
     and Brem, H., Brain biocompatibility of a biodegradable poly-
     mer capable of sustained release of micromolecules, Abstracts
     of the 1988 Annual Meeting of the American Association of Neu-
     rological Surgeons, p. 399, 1988.
25. Tamargo, R. J., Epstein, J I., Reinhard, C. S., Chasin, M.,
     and Brem, H., Brain biocompatibility of a biodegradable con-
     trolled-release polymer in rats, J. Biomed. Mater. Res., 23,
     253-266, 1989.
26. Brem, H., Kader, A., Epstein, J. I., Tamargo, R., Domb, A.,
     Langer, R., and Leong, K., Biocompatibility of bioerodible
     controlled release polymers in the rabbit brain, Selective Cancer
     Therapeutics, 5, 55-65, 1989.
27. Brem, H., Ahn, H., Tamargo, R. J., Pinn, M., and Chasin,
     M., A biodegradable polymer for intracranial drug delivery:
     A radiological study, Abstracts of the 1988 Annual Meeting of
     the American Association of Neurological Surgeons, 1988, p.
     349.
28. Brem, H., Tamargo, R. J., Pinn, M., and Chasin, M., Bio-
     compatibility of a BCNU-loaded biodegradable polymer: A tox-
     icity study in primates, Abstracts of the 1988 Annual Meeting
     of the American Association of Neurological Surgeons, 1988,
     p. 381.

3
Poly- ε -Caprolactone and Its Copolymers

COLIN G. PITT   Amgen Inc., Thousand Oaks, California

I. INTRODUCTION

The propensity of ε-caprolactone to undergo ring-opening polymerization was first established by Carothers (1) in his classic studies of polyesters during the early 1930s. The identity of poly-ε-caprolactone (PCL) as a "biodegradable polymer" subsequently emerged as the result of an extensive effort by scientists at Union Carbide (2) to identify synthetic polymers degraded by microorganisms in the environment. It was suggested that PCL be used in biodegradable packaging designed to reduce environmental pollution; as one example, PCL was evaluated as a component of biodegradable containers for aerial planting of conifer seedlings (3). Microbial degradation of PCL remains an active research area (4). The successful use of polymers of lactic acid and glycolic acid as biodegradable drug delivery systems and as biodegradable sutures led naturally to an evaluation of other aliphatic polyesters, and to the discovery of the degradability of PCL in vivo (5). The homopolymer itself is degraded very slowly when compared with polyglycolic acid and poly-glycolic acid-co-lactic acid, and is most suitable for long-term delivery systems such as Capronor, a 1-year contraceptive (6). However, the biodegradability can be enhanced greatly by copolymerization. This fact, coupled with a high permeability to many therapeutic drugs and a lack of toxicity, has made PCL and its derivatives well suited for controlled drug delivery. Another property of PCL, which has stimulated much research, is its exceptional ability to form compatible

71

blends with a variety of other polymers (7,8). This too adds to the
diversity of PCL-derived materials that may be used in drug delivery
applications.

This review of PCL and its copolymers is largely drawn from the
nonpatent literature and focuses primarily on aspects relevant to drug
delivery. Methods of polymerization are considered at some length
because of the impact on polymer structure and morphology, which
in turn determine the permeability and biodegradability of the product.

II. CHEMISTRY

A. Monomer Synthesis

ε-Caprolactone is manufactured by oxidation of cyclohexanone with
peracetic acid in an efficient continuous process (Fig. 1) (9). It is
available from a number of supply houses and may be dried by dis-
tillation from diphenylmethane-4,4'-diisocyanate (10), calcium hydride
(11,12; see however, Ref. 13), or sodium metal (13). Derivatives
of ε-caprolactone are most conveniently prepared by Bayer-Villiger
peroxidation of the corresponding ketone. Caro's acid and percar-
boxylic acids, preferably m-chloroperbenzoic acid, have been used
most frequently. The method has been applied to alkyl- and aryl-
substituted ε-caprolactones (14,15), to a t-butyldimethylsiloxy substituted ε-caprolactone which is a source of a hydroxylated PCL (16),
and to bislactones used for crosslinking (17). Ring closure of the
cesium salt of ω-halocarboxylic acids is reported to be a high-yield
synthesis of macrocyclic lactones, including the dimer of ε-caprolactone (18).

B. Polymer Synthesis

The general subject of lactone polymerization has been reviewed (7,
19). Polymerization of ε-caprolactone can be effected by at least
four different mechanisms categorized as anionic, cationic, coordination, and radical. Each method has unique attributes, providing

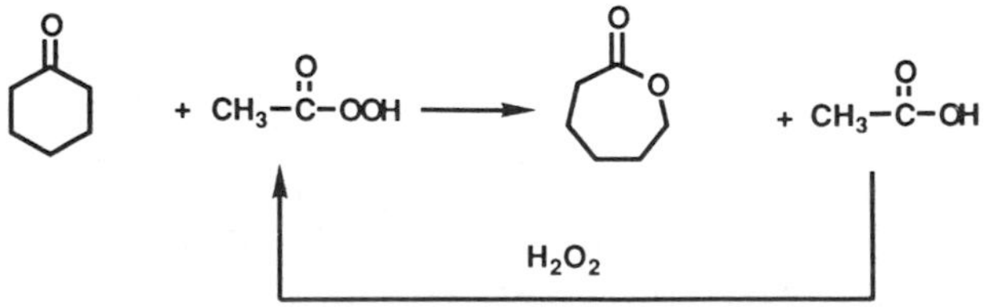

FIGURE 1 Continuous process for the manufacture of ε-caprolactone
by oxidation of cyclohexanone with peracetic acid.

different degrees of control of molecular weight and molecular weight distribution, end-group composition, and the chemical structure and sequence (block versus random) distribution of copolymers. Each of these characteristics, in turn, is important in defining the permeability and degradability of the polymer.

1. Anionic Polymerization

The anionic process is exemplified by the use of alcohols, tertiary amines, and carboxylates. Both the initiation and propagation steps are believed to involve uncatalyzed nucleophilic cleavage of the ester alkyl-oxygen bond (Fig. 2) and are relatively slow. As an example, ethylene glycol-initiated polymerization of ε-caprolactone requires 35 hr at 190°C for quantitative conversion (7). Rates are substantially increased by using the more nucleophilic alkali metal alcoholates; the sodium salt of diethylene glycol reduces the conversion time to 17 hr at 120°C (20). The effectiveness of the sodium salts of other diols is reported to diminish with increasing hydrocarbon chain length. Polymerization in the presence of KOH, KCl, NaCl, KCNS, NaOAc, and t-BuOK is accelerated by the addition of the alkali metal chelators, dibenzo-18-crown-6 and dicyclohexyl-18-crown-6, consistent with a nucleophilic mechanism (21,22). Lithium t-butoxide and (s- or n-)butyl lithium are very effective initiators in benzene (11) but are less effective in tetrahydrofuran (THF) (23); polymerization to high molecular weight (>50,000) PCL is quantitative in less than 6 min in benzene.

Polymerization of ε-caprolactone with potassium t-butoxide in THF is a living chain cycle equilibrium system (23,24). The predominant product at equilibrium is the cyclic dimer, formed by backtiting degradation of the initially formed linear polymers. This and similar intermolecular and intramolecular ester interchange processes (Fig. 3) effect the dispersity of the polymer. While the initial molecular weight distribution is narrow ($M_w/M_n \sim 1.2$) and consistent with a ring-opening polymerization process, ester interchange results in an increase in dispersity with time. An exception to this behavior is observed with the graphitide KC_{24}. Polymerization in the presence of this initiator is heterogeneous, and can be accomplished at 0°C in xylene or THF without significant ester interchange (25).

The anionic method of polymerization is most useful for the synthesis of low molecular weight hydroxy-terminated oligomers and polymers that are to be further processed. For example, the treatment of hydroxy-terminated oligomers with isocyanates has been used to obtain polyester-urethanes (9,20), while triblock copolymers (PCL-PEG-PCL) are prepared by initiating the polymerization of ε-caprolactone with the disodium alcoholate from polyethylene glycol (26). A related strategy has been employed to prepare a triblock copolymer of styrene and ε-caprolactone by initiating the polymerization

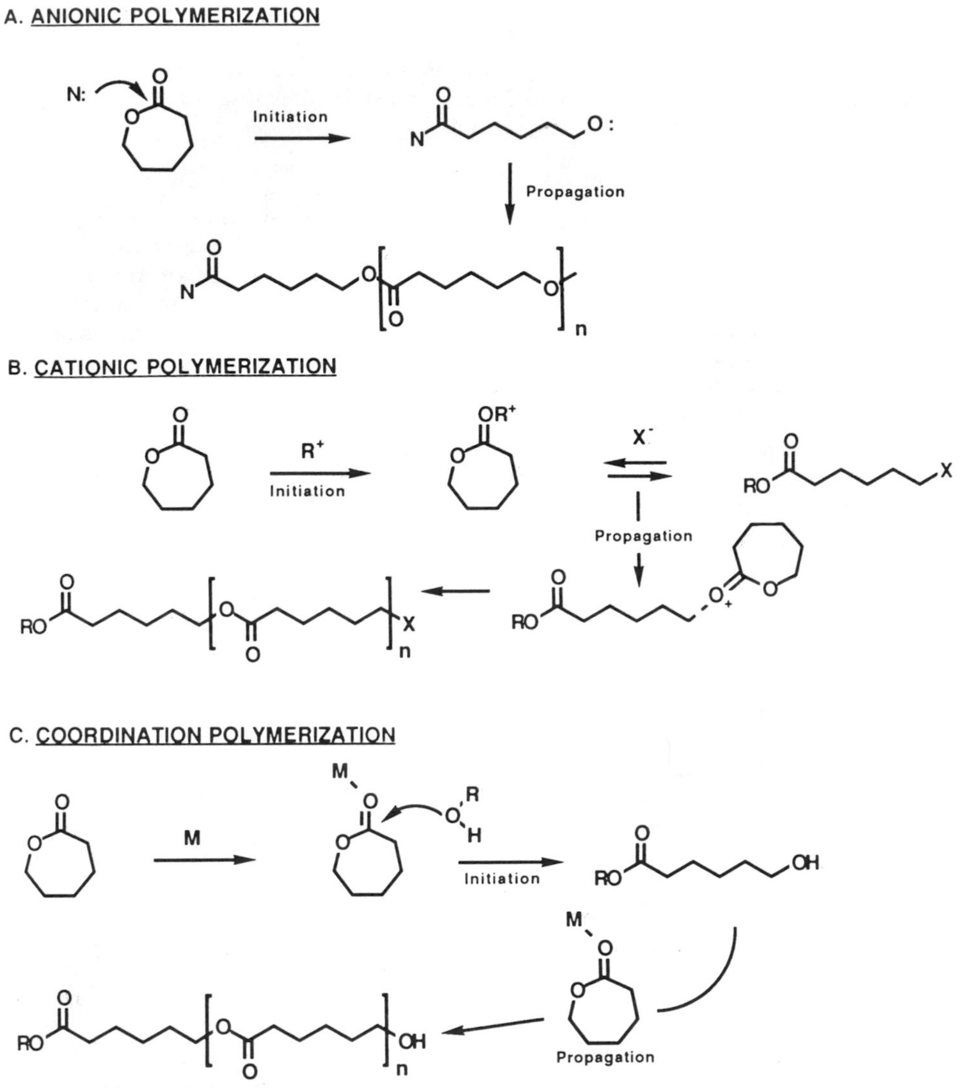

A. ANIONIC POLYMERIZATION

B. CATIONIC POLYMERIZATION

C. COORDINATION POLYMERIZATION

FIGURE 2 Anionic, cationic, and coordination mechanisms of poly-merization of ε-caprolactone and related lactones.

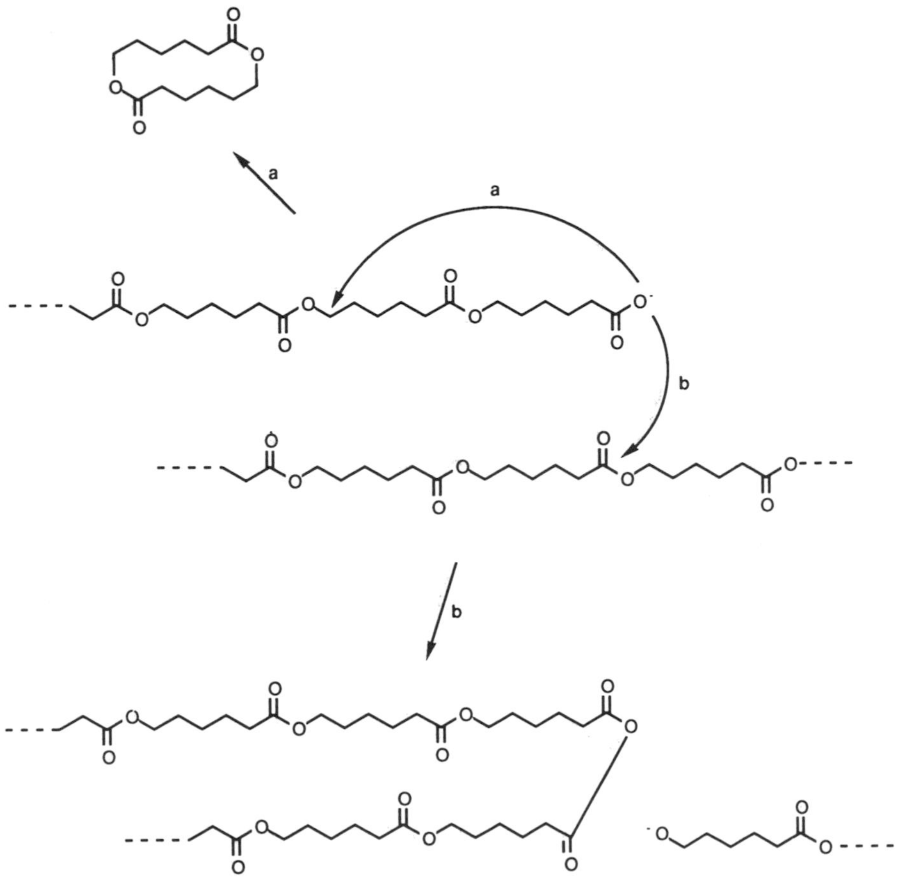

FIGURE 3 The transesterification and chain cycle equilibration of PCL and related polyesters.

of ε-caprolactone with the living polystyrene dianion (27). Treatment of hydroxy-terminated PCL with an excess of maleic anhydride has been used to introduce unsaturated end groups. This unsaturation provided a means of peroxide-initiated crosslinking of PCL (28). A number of these structures are illustrated in Fig. 4.

2. Cationic Polymerization

Various classes of cationic initiators have been used to polymerize lactones: protic acids, Lewis acids, acylating agents, and alkylating agents. These initiators are often difficult to handle experimentally

MONOFUNCTIONAL INITIATORS

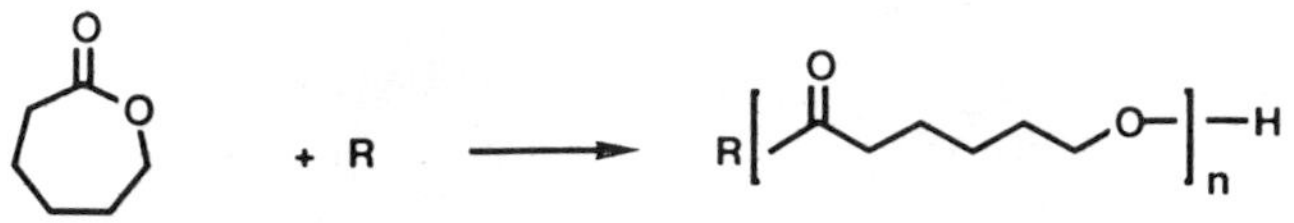

R = H_2O, ROH, RONa, RNH_2, RCOONa

DIFUNCTIONAL INITIATORS

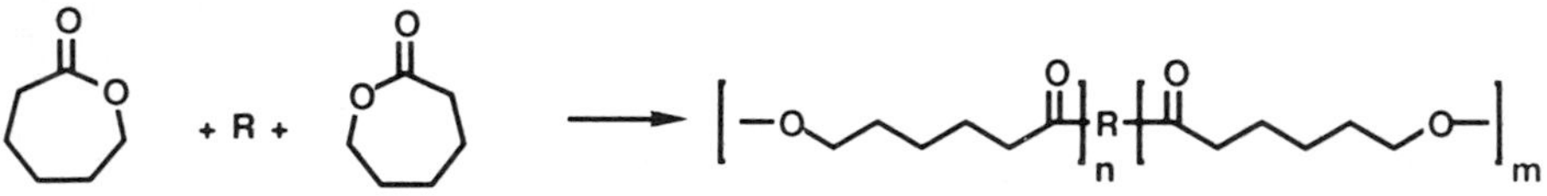

R = $NaO(CH_2CH_2O)_nNa$

$Na(CHPhCH_2)_nR(CH_2CHPh)_mNa$ polystyrene dianion,
                                        polystyrene-butadiene anion
$HO(CH_2)_3(SiMe_2O)_n(CH_2)_3OH$ polydimethylsiloxane

POLYFUNCTIONAL INITIATORS

R= CH_2OH-CHOH-CH_2OH glycerol

CH_2OH-$(CHOH)_2$-CH_2OH erythritol

CH_2OH-$(CHOH)_3$-CH_2OH sorbitol

CH_2OH-$(CHOH)_4$-CH_2OH xylitol

FIGURE 4   Different skeletal structures of PCL and its copolymers derived from the polymerization of ε-caprolactone using mono- and polyfunctional initiators.

and can promote rapid degradation of the polymer. The primary advantage of cationic polymerization is its ability to obtain otherwise inaccessible copolymers.

Cationic polymerization of ε-caprolactone was first demonstrated with the allyl carbocation derived from AgPF$_6$ and allyl chloride (29). Significant (75%) conversion was achieved after 3 days, and the same initiator was shown to be applicable to the synthesis of graft copolymers with polyvinylchloride (PVC) and polychloroprene. While molecular weights were not reported in this case, PCL with a M_W of 10,000 was obtained using P(CN)$_3$ in acetonitrile. The versatility of cationic polymerization is illustrated by the use of the latter catalyst to obtain a copolymer of ε-caprolactone with THF (30). Diphenyliodonium salts in the presence of Cu(II) effect polymerization of ε-caprolactone in 1 hr at 70°C (90% conversion, [η] 0.45 dl/g) (31). Graphite intercalation compounds of SbCl$_5$, FeCl$_3$, and AlBr$_3$ are reported to polymerize ε-caprolactone to oligomers and polymers with molecular weights less than 15,000 (32).

Diglycolide and ε-caprolactone have been copolymerized in the presence of FeCl$_3$, BF$_3$·Et$_2$O, and FSO$_3$H at 70–150°C and the yields, degree of transesterification, and block lengths compared with the results obtained with anionic and coordination initiators (33). In the cited study it was shown that, in contrast to anionic initiation, ε-caprolactone is polymerized more rapidly than diglycolide. The first stage is ε-caprolactone polymerization, followed by diglycolide polymerization, and finally randomization of the block structure by transesterification.

Alkyl sulfonates are very effective cationic initiators of ε-caprolactone, although only the more reactive methyl triflate and methyl fluorosulfate result in a high conversion. The mechanism of polymerization in the presence of these initiators is believed to involve methylation of the exocyclic carbonyl oxygen, followed by partial ring opening of the activated lactone by the counteranion (Fig. 2) (34). Because very rapid depolymerization occurred at higher temperatures, it was necessary to control the temperature within the narrow range of 50 ± 10°C. Even so, the M_W of the polymer was no greater than 15,000 because of rapid degradation by the living cationic end group.

Trimethylsilyl triflate has also been used as an initiator. High yields of PCL were obtained in 1,2-dichloroethane at 50°C and M_W values up to 50.000 were attained after 192 hr (135).

3. Coordination Polymerization

Coordination polymerization is the most versatile method of preparing PCL and its copolymers, affording high molecular weights and conversions, and either block or random copolymers depending on the conditions. As with the preceding classes of initiators, the product is a living polymer. Di-n-butyl zinc, stannous chloride and octoate, and alkoxides and halides of Mg, Al, Zn, Ti, Zi, and Sn are among the more frequently used. Mechanistic studies of alkyl zinc, alkyl magnesium, and alkyl aluminum, and main group II, III, and IV alkoxides have been reported (36,37). In some cases, it is clear that trace amounts of water or another nucleophile is the initiator and the coordination compound serves as a catalyst.

Transfer and termination reactions are the least significant with aluminum catalysts. For example, the half-life of polymerization of ε-caprolactone at 10°C in toluene in the presence of $Zn(OAl(OPr^i)_2)_2$ is 23 min ($[CL] = 1.0$ M, $[Zn] = 1.83 \times 10^{-2}$) (38,39). The slow rate of transesterification relative to chain growth results in a narrow molecular weight distribution, $M_w/M_n = 1.1$, and the value of M_n is determined by the ratio of [monomer]/[initiator]. Transesterification-induced broadening of the molecular weight distribution to $M_w/M_n = 1.95$ only occurred when the conversion was greater than 60% (40) or when the resulting polymer was heated to 120°C (under argon) for longer time periods (41). Heating at 120°C in air caused auto-oxidative degradation of the polymer in the presence of this catalyst.

The living nature of PCL obtained in the presence of $Zn(OAl(OPr^i)_2)_2$ has been used to prepare both di- and triblock copolymers of ε-caprolactone and lactic acid (42,43). Treatment of the initial living PCL with dilactide afforded a PCL-PLA diblock with $M_w/M_n = 1.12$, with each block length determined by the proportions of the reactants, i.e., the ratio of [monomer]/[Zn]. While the living diblock copolymer continued to initiate dilactide polymerization, it failed to initiate ε-caprolactone polymerization. To obtain a PCL-PLA-PCL triblock, it was necessary to treat the living PCL-PLA-OAlR$_2$ intermediate with ethylene oxide, then activate the hydroxy-terminated PCL-PLA-$(OCH_2CH_2)_n$OH with a modified Teyssie catalyst (Fig. 5).

A porphinatoaluminum alkoxide is reported to be a superior initiator of ε-caprolactone polymerization (44,45). A living polymer with a narrow molecular weight distribution ($M_w/M_n = 1.08$) is obtained under conditions of high conversion, in part because steric hindrance at the catalyst site reduces intra- and intermolecular transesterification. Treatment with alcohols does not quench the catalytic activity although methanol serves as a coinitiator in the presence of the aluminum species. The immortal nature of the system has been demonstrated by preparation of an AB block copolymer with ethylene oxide. The order of reactivity is ε-lactone $>$ β-lactone.

With stannous octoate-promoted polymerization, the metal species is believed to function as the catalyst and water (added or endogenous), or alcohol, serves as the initiator (Fig. 2). This mechanism is supported by recent kinetic studies of PCL polymerization in the presence of triphenyltin acetate (46). After an induction period, polymerization is zero order with respect to monomer and near first

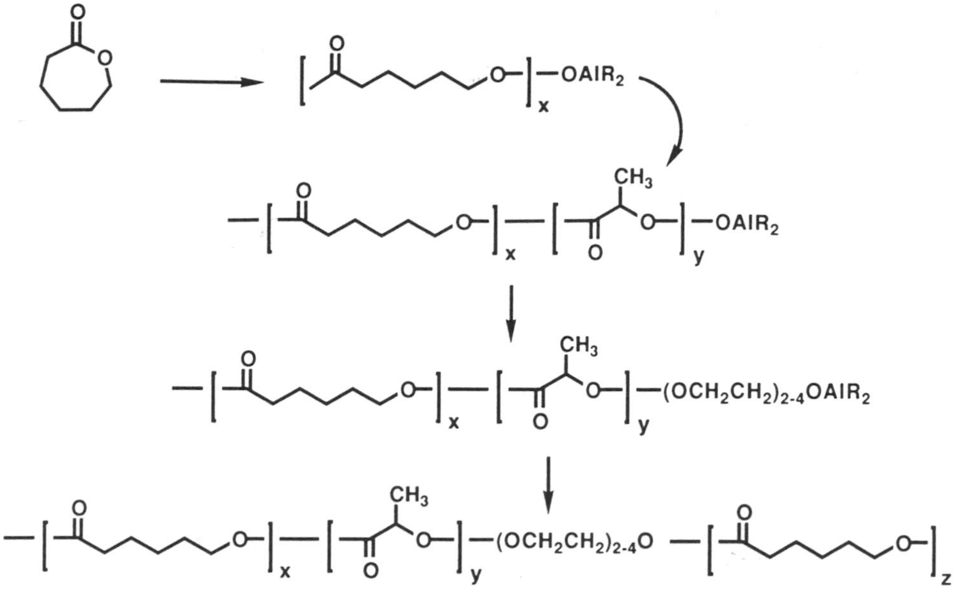

FIGURE 5   Stepwise synthesis of a triblock copolymer (PCL-PLA-PCL) of PCL and polylactic acid using aluminum coordination catalysts to minimize randomization of the block structure by transesterification. (From Ref. 43.)

order with respect to [Ph₃SnOAc].   Deliberately added water (0.5%) decreased the molecular weight from 50,000 to 30,000 at 100% conversion.   Added acetic acid did not affect the molecular weight.   It was concluded that the hydroxyl group is the active end in propagation while carboxylic groups are responsible for the induction period. The rate-determining step, responsible for zero-order kinetics, is proposed to be S_N1-type dissociation of the tin entity from the growing chain end, to recoordinate with monomer.

The choice of the alcohol permits manipulation of the structure of the polymer.   Water and monohydric alcohols afford linear chains with carboxylic acid and ester end groups, respectively.   Polyhydroxy initiators afford a route to ester end-blocked star and comb polymers (Fig. 4) (47).

Stannous octoate has the advantage of having been used to prepare polymers (Silastic, Capronor) for which substantial toxicological data are now available (6,48).   Stannous octoate-initiated polymerization has been used to prepare copolymers of ε-caprolactone with other lactones, including diglycolide, dilactide, δ-valerolactone, ε-decalactone, and other alkyl-substituted ε-caprolactones.   Conducting the polymerization at 120°C or higher results in random copolymers, even though the reactivities of the monomers differ significantly. The copolymerization parameters for copolymerization of dilactide and ε-caprolactone catalyzed by stannous octoate, stannous chloride, and tetrabutyl titinate have been determined (5).

Initiation of stannous octoate-catalyzed copolymerization of ε-caprolactone with glycerol was used to prepare a series of trifunctional hydroxy-end blocked oligomers, which were then treated with hexane-1,6-diisocyanate to form elastomeric polyesterurethanes with different crosslink densities (49). Initiation of ε-caprolactone polymerization with a hydroxypropyl-terminated polydimethylsiloxane in the presence of dibutyl tin dilaurate has been used to prepare a polyester-siloxane block copolymer (Fig. 4) (50).

4. Radical Polymerization

Ring-opening polymerization of 2-methylene-1,3-dioxepane (Fig. 6) represents the single example of a free radical polymerization route to PCL (51). Initiation with AIBN at 50°C afforded PCL with a M_V of 42,000 in 59% yield. While this monomer is not commercially available, the advantage of this method is that it may be used to obtain otherwise inaccessible copolymers. As an example, copolymerization with vinyl monomers has afforded copolymers of ε-caprolactone with styrene, 4-vinylanisole, methyl methacrylate, and vinyl acetate.

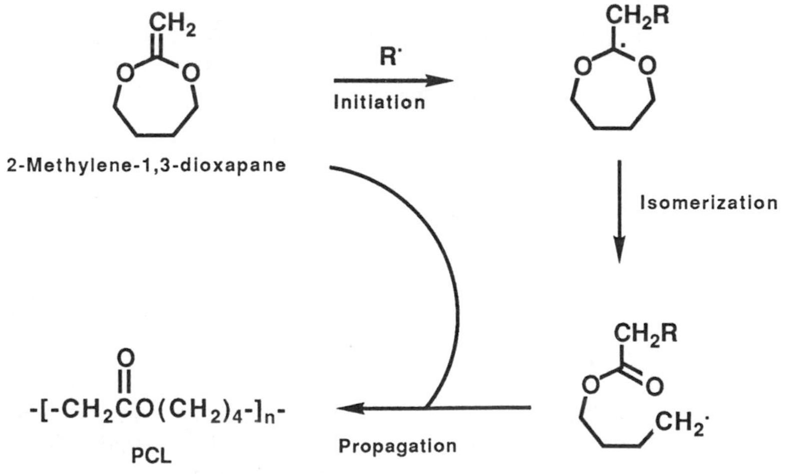

FIGURE 6 Synthesis of PCL by the free radical polymerization of 2-methylene-1,3-dioxepane. (From Ref. 51.)

C. Polymer Properties

Poly-ε-caprolactone is a semicrystalline polymer, melting in the range
of 59−64°C, depending on the crystallite size. Because of its low glass
transition temperature (T_g) of −60°C, the melt cannot be quenched to
a glass. The heat of fusion (ΔH_f) of 100% crystalline PCL is reported
to be 139.5 J/g (52), a value that has been used to estimate the crystallinity of PCL and its copolymers from differential scanning calorimetry
(DSC) traces. The crystallinity of PCL varies with its molecular weight
(53) and for molecular weights in excess of 100,000 the crystallinity is
about 40%, rising to 80% as the molecular weight decreases to 5000 (Fig.
7). Crystallinity is known to play an important role in determining both
permeability and biodegradability (vide infra) because of the generally
accepted fact that the bulk crystalline phase is inaccessible to water
and other permeants. That is, an increase in crystallinity reduces
the permeability by both reducing the solute solubility and increasing
the tortuosity of the diffusional pathway. The biodegradation rate
is reduced by the decrease in accessible ester bonds.

The T_g can be varied systematically by copolymerization [5].
For example, the T_g values of copolymers with dilactide increase in
proportion to the dilactide content according to the Fox equation:

$$(1/T_g)_{ab} = (W/T_g)_a + (W/T_g)_b \tag{1}$$

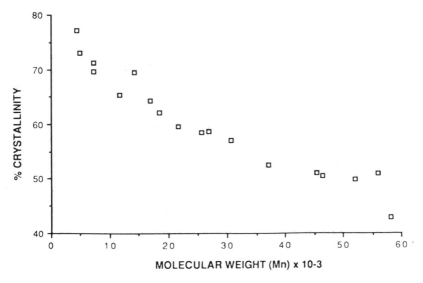

FIGURE 7 Relationship between the degree of crystallinity and the
molecular weight of PCL.

where the subscripts refer to the two constituent homopolymers and W is the weight fraction (Fig. 8). Only those polymers with a T_g below body temperature are in the rubbery state and exhibit a high permeability. This is evident from comparisons of the permeability of poly(lactic acid) with copolymers with increasing ε-caprolactone content (67). Some of the more important tensile properties of PCL are listed in Table 1 (7).

Studies of the hydrodynamic properties and unperturbed dimensions of fractionated PCL have shown that it is a flexible coil (54,55). The following Mark-Houwink equations have been reported:

$$[\eta] = 19.1 \times 10^{-5} \, M_w^{0.73} \text{ (DMF) (54)}$$

$$[\eta] = 9.94 \times 10^{-5} \, M_w^{0.82} \text{ (benzene) (54)}$$

$$[\eta] = 12.5 \times 10^{-5} \, M_n^{0.82} \text{ (benzene) (56)}$$

$$[\eta] = 14.1 \times 10^{-5} \, M_v^{0.79} \text{ (THF) (57)}$$

$$[\eta] = 13.0 \times 10^{-5} \, M_w^{0.83} \text{ (CHCl}_3\text{) (57)}$$

$$[\eta] = 4.59 \times 10^{-5} \, M_w^{0.91} \text{ (CHCl}_3\text{) (8)}$$

$$[\eta] = 5.15 \times 10^{-5} \, M_w^{0.86} \text{ (pyridine) (8)}$$

The solubility parameters of PCL are 20.8 and 20.4 $J^{1/2} \cdot cm^{-3/2}$ when calculated using the parameters of Fedors and Hoy, respectively (58). PCL is soluble in a number of solvents at room temperature, including THF, chloroform, methylene chloride, carbon tetrachloride, benzene, toluene, cyclohexanone, dihydropyran, and 2-nitropropane. It is poorly soluble in acetone, 2-butanone, ethyl acetate, acetonitrile, and DMF, and insoluble in alcohols, petroleum ether, and diethyl ether. The partition coefficients of a number of solutes between PCL and water have been measured and correlated with octanol-water partition coefficients (Fig. 9) (58,59). The linear correlation (Eq. 2) when combined with the water solubility of the solutes serves as a method of estimating the solubility of drugs in PCL from first principles.

$$\log P_{pcl/water} = 0.91(\pm 0.07)\log P_{oct/water} - 0.50(\pm 0.20)$$

$$\text{Solubility}_{pcl} = (P_{pcl/water}) \times (\text{solubility})_{water} \qquad (2)$$

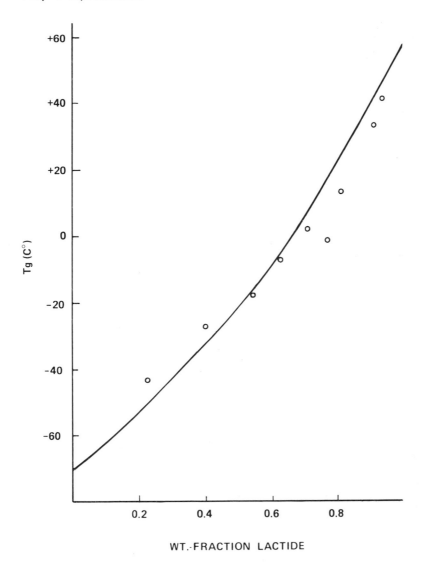

FIGURE 8   Relationship between the glass transition temperature and the composition of copolymers of ε-caprolactone and lactic acid. (From Ref. 5.)

TABLE 1 Some Important Physical Properties of Poly-ε-caprolactone[a]

1% Secant modulus (psi)	50,000
Elongation (%)	~750
Yield stress (psi)	1,600
Tensile strength (based on initial cross-sectional area) (psi)	3,500
Melting point (°C)	63
Glass transition temperature, 1 Hz	
Partially crystalline (°C)	−60
Amorphous (°C)	−71
Density (g/cm^3) 20°C	1.149
30°C	1.143
62°C	1.069
From 30 to −30°C (add)	5.6×10^{-4}
From 62 to 100°C (add)	-6.8×10^{-4}
Equilibrium moisture content (%)	
50% relative humidity	0.07
100% relative humidity	0.43

[a]Data for a polymer with M_W 40,000 (7).

D. Copolymers of ε-Caprolactone

The syntheses of block and random copolymers of PCL derived from
the following monomers and polymers have been described in Sec.
II.B: polyethylene glycol, ethylene oxide, polystyrene, diisocyanates
(urethanes), polyvinylchloride, chloroprene, THF, diglycolide, dilactide, δ-valerolactone, substituted ε-caprolactones, 4-vinyl anisole,
styrene, methyl methacrylate, and vinyl acetate. In addition to these
species, many copolymers have been prepared from oligomers of PCL.
In particular, a variety of polyester-urethanes have been synthesized
from hydroxy-terminated PCL, some of which have achieved commercial status (9). Graft copolymers with acrylic acid, acrylonitrile,
and styrene have been prepared using PCL as the backbone polymer
(60).

E. Blends of PCL with Other Polymers

The exceptional ability of PCL to form blends with many other polymers has stimulated a large amount of research. The subject has
been reviewed a number of times (7,8). To date, the potential of
such blends for drug delivery has been largely unexploited. The
permeability of blends of PCL with cellulose propionate, cellulose

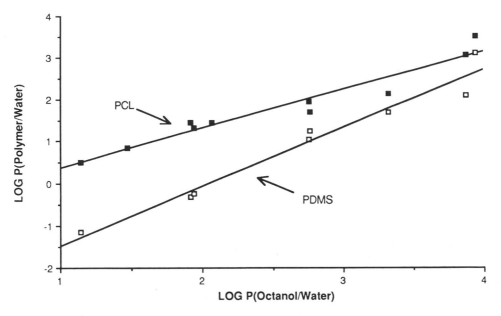

FIGURE 9 Linear correlation of octanol-water, PCL-water, and poly-dimethylsiloxane (PDMS)-water partition coefficients derived for a series of 10 solutes. (From Ref. 58.)

acetate butyrate, polylactic acid, and polylactic acid-co-glycolic acid have been studied and shown to be useful for manipulating the rate of release of drugs from microcapsules (61-63). Blends of poly(ε-caprolactone-co-lactic acid) and poly(lactic acid) have been used for the same purpose (64). Similarly, blending PCL with polylactic acid and polylactic acid-co-glycolic acid has been shown to be a useful method of modifying the rate of biodegradation of the composite (63, 65).

Polymer blends have been categorized as (1) compatible, exhibit-ing only a single T_g, (2) mechanically compatible, exhibiting the T_g values of each component but with superior mechanical properties, and (3) incompatible, exhibiting the unenhanced properties of phase-separated materials (8). Based on the mechanical properties, it has been suggested that PCL-cellulose acetate butyrate blends are com-patible (8). Dynamic mechanical measurements of the T_g of PCL-polylactic acid blends indicate that the compatability may depend on the ratios employed (65). Both of these blends have been used to control the permeability of delivery systems (vide infra).

Interpenetrating networks (IPNs) composed of different propor-tions of PCL and poly-2-hydroxyethyl methacrylate (pHEMA) have been prepared by radical-catalyzed polymerization of a solution of PCL in HEMA monomer containing 0.5 wt% ethylene bis(dimethacrylate) at 90°C (64,66). The size of the melting transition peak of PCL at 60°C (DSC) in the resulting IPN was proportional to the PCL content. The elongation at break remained constant, while the elastic modulus increased with the PCL content. Extraction of the PCL with ethyl acetate afforded a porous pHEMA film.

III. KINETICS OF DRUG RELEASE

A. Permeability Measurements

Early studies established that PCL and its copolymers have a high permeability to low molecular weight drugs (<400 D) (67). Diffusion coefficients (D) derived from diffusion cell and desorption studies are of the order of 10^{-9} cm$^2 \cdot$sec^{-1} for a series of steroids and narcotics (Table 2). These values are comparable to the diffusion coefficients of other organic polymers, e.g., polyethylene, but two orders less than silicone rubber. However, the solubility (C_S) of drugs is greater in PCL than in silicone rubber. As a consequence, the permeabilities (J) of the two polymers, the product of D and C_S, are not greatly different. This is illustrated with progesterone:

	C_S(g/cm^3)	D(cm^2/sec)	J(g/cm $\cdot$sec)
Silicone rubber	0.5×10^{-3}	4.5×10^{-7}	2.2×10^{-10}
PCL	16.9×10^{-3}	3.6×10^{-9}	0.6×10^{-10}

An analysis of partition coefficient data and drug solubilities in PCL and silicone rubber has been used to show how the relative permeabilities in PCL vary with the lipophilicity of the drug (58,59). The permeabilities of copolymers of ε-caprolactone and dl-lactic acid have also been measured and found to be relatively invariant for compositions up to 50% lactic acid (67). The permeability then decreases rapidly to that of the homopolymer of dl-lactic acid, which is 10^5 times smaller than the value of PCL. These results have been discussed in terms of the polymer morphologies.

The high permeability of PCL and its copolymers coupled with a controllable induction period prior to polymer weight loss (vide infra) lends itself to the development of delivery devices that are based on diffusion-controlled delivery of the drug during the induction period prior to weight loss. The subsequent biodegradation of the polymer serves the purpose of eliminating the need to recover the spent device.

TABLE 2   Diffusion Coefficients
($\times 10^9$ cm^2/sec) of Various Drugs
in Poly-ε-caprolactone

Progesterone	3.6
Testosterone	7.3
Norgestrel	4.1
Norethindrone	6.6
Ethynyl estradiol	3.3
Naltrexone	2.4
Codeine	3.8
Meperidine	2.5
Methadone	1.9
α-Acetylmethadol	2.6

Source: Data taken from
Ref. 106.

B.   Fabrication of Devices

Thin films of PCL and its copolymers have been prepared by casting
a common solution of the polymer and drug (68).   Thicknesses greater
than 0.1 mm were achieved by compression-molding an intimate drug-
polymer mixture at 100–130°C in a Carver press.   Tubing of PCL
with ID/OD values in the range of 1–3 mm has been prepared by
melt extrusion.   The conditions for extrusion are dependent on the
molecular weight or melt viscosity of the polymer but extrusion has
typically been conducted at about 160°C (57).   Fibers of PCL containing tetracycline hydrochloride, 0.5 mm O.D., have been prepared
by melt-spinning at 161°C using a Tinius Olsen extrusion plastometer
(69).

Microcapsules of PCL and its copolymers may be prepared by air-
coating (fluidized bed), mechanical, and, most commonly, solution
methods.   Typically, the solution method has involved emulsification
of the polymer and drug in a two-phase solvent-nonsolvent mixture
(e.g., CH_2Cl_2/water) in the presence of a surfactant such as polyvinyl alcohol.   Residual solvent is removed from the microcapsules
by evaporation or by extraction (70).   Alternatively, the solvent
combination can be miscible provided one of the solvents is high-
boiling (e.g., mineral spirits); phase separation is then achieved by
evaporation of the volatile solvent (71).   The products of solution
methods should more accurately be called microspheres, for they

generally consist of an intimate mixture or solution of the drug in
the polymer. Mechanical and fluidized bed methods are more likely
to produce polymer-coated drug crystals or droplets.

Porous membranes have been prepared by leaching an additive
from films and tubes of PCL (64,72). The procedure involves extrusion or casting blends of PCL and Pluronic F68, the latter being
an FDA-approved oxyethylene-co-oxypropylene triblock copolymer.
Treatment of the phase-separated blend with aqueous acetone or
aqueous alcohols causes both swelling of the polymer and extraction
of the Pluronic F68. The induced pore size and void volume may be
controlled by the time, temperature, and solvent composition.

C. Drug Release Rates

1. Monolithic Systems

The release of steroids such as progesterone from films of PCL and
its copolymers with lactic acid has been shown to be rapid (Fig. 10)
and to exhibit the expected $(time)^{1/2}$ kinetics when corrected for
the contribution of an aqueous boundary layer (68). The kinetics
were consistent with phase separation of the steroid in the polymer
and a Fickian diffusion process. The release rates, reflecting the
permeability coefficient, depended on the method of film preparation
and were greater with compression molded films than solution cast
films. In vivo release rates from films implanted in rabbits was very
rapid, being essentially identical to the rate of excretion of a bolus
injection of progesterone, i.e., the rate of excretion rather than the
rate of release from the polymer was rate determining.

The rate of release of levonorgestrel from films of block copolymers of ε-caprolactone and dl-lactic acid (drug load 30%) was shown
to be a function of the copolymer composition. The rate was unchanged for compositions of 100% and 88% ε-caprolactone, but decreased thereafter as the ε-caprolactone content decreased (42).

Release of tetracycline hydrochloride from PCL fibers was evaluated as a means of controlled administration to periodontal pockets
(69). Only small amounts of the drug were released rapidly in vitro
or in vivo, and poly(ethylene-co-vinyl acetate) gave superior results.
Because Fickian diffusion of an ionic hydrochloride salt in a lipophilic
polymer is unlikely, and because PCL and EVA have essentially identical Fickian permeabilities, we attribute this result to leaching of
the charged salt by a mechanism similar to release of proteins from
EVA (73). Poly-ε-caprolactone pellets have been found unsuitable
for the release of methylene blue, another ionic species (74,75). In
this case, blending PCL with polyvinyl alcohol (75% hydrolyzed) increased the release rate.

Poly-ε-caprolactone fibers have been used to control the release
of herbicides into an aquatic environment (76,77). A PCL blend with

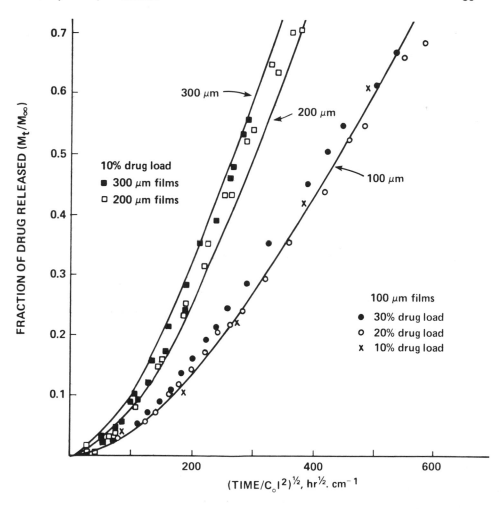

FIGURE 10 In vitro rates of release of progesterone from PCL films, illustrating their dependence on the film thickness and drug load. The deviation from (time)$^{1/2}$ kinetics reflects the contribution of an aqueous boundary layer. The solid lines were calculated assuming an aqueous boundary layer thickness of 19 μm. (From Ref. 68.)

fluridone (10–60% loading) was prepared by melt-spinning and, to minimize a burst of the agent, was then coated with an exterior layer of the herbicide-free polymer. The release kinetics were first order. Varying the fiber diameter made it possible to control the duration and rate of release, e.g., fibers with diameters of 0.8 and 1.2 mm released fluridone for 40 and 50 days, respectively. Effective control of hydrilla was achieved under conditions where the conventional herbicide formulation provided only marginal control.

In an extension of this work, pellets of a blend of PCL and hydroxypropylcellulose containing fluridone were prepared by grinding, blending, and then melt-spinning the mixture with a Berstorff twin screw extruder (78). The extruded rod was subsequently water-quenched and pelletized. Pellets were also prepared by coating bundles of extruded rods with the water-soluble excipients PEG 3350 and PEG 600 (95:5). In vitro release rate measurements were conducted in the simulant medium of 50% aqueous ethanol or hardened water. The observed release rates declined with time consistent with a monolithic formulation. Fluridone release for 20–40 days was achieved by adjustment of the pellet dimensions, the surface-to-volume ratio, and the hydroxypropylcellulose PCL ratio.

The in vitro rate of release of 5-fluorouracil from powders prepared by grinding a melt-dispersed PCL-drug mixture deviated somewhat from the expected time$^{1/2}$ kinetics, the extent depending on the drug loading (79). Drug release was complete within 100 hr. The in vitro rate of release of cytosine arabinoside (Ara C) from a PCL-Ara C rod prepared by dip-coating in a glacial acetic acid solution declined exponentially over an 8-day period (80).

Incorporation of a perfume into PCL or a PCL-polyolefin blend is reported to be a useful method of retaining the fragrance in soaps and detergent bars (81).

2. Microcapsules

Microencapsulation with PCL using the solvent evaporation method can be experimentally difficult. For example, PCL was the only polymer of five that failed to yield spherically shaped microcapsules using this technique (82). The insecticide Abate has been incorporated into PCL (21% loading) by the solvent separation method; in a comparative study, PCL afforded good-quality microspheres although poly(methyl methacrylate) microcapsules were smoother and had fewer defects (83).

Chloropromazine (8–34 wt% loading) has been microencapsulated in PCL-cellulose propionate blends by the emulsion solvent evaporation method (61). Phase separation for some ratios of the two polymers was detectable by SEM. The release rate from microcapsules in the size range of 180–250 µm in vitro (Fig. 11) was directly proportional to the PCL content of the blend, the half-life (50% drug release)

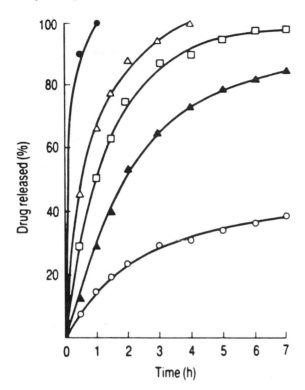

FIGURE 11 In vitro rates of release of chloropromazine from microspheres prepared from blends of PCL and cellulose propionate (CP),
illustrating the dependence of the rate on the blend composition.
100% PCL (•); 75% PCL/25% CP (△); 50% PCL/50% CP (◻); 25% PCL/
75% CP (▲); 100% CP (o). (From Ref. 61.)

varying from <1 hr (100% PCL) to >7 hr (100% cellulosic). The kinetics of release were accommodated by the standard Higuchi equation
for diffusion-controlled release from a sphere. An essentially identical result was achieved with PCL-cellulose acetate butyrate blends
(62). The rate of release of chloropromazine and progesterone from
microspheres of PCL alone was more rapid than the rate of dissolution
of the pure drug crystals of comparable dimensions (84).

Nitrofurantoin, an antibacterial agent, was incorporated into PCL
microcapsules by the solvent evaporation method (85). Capsule sizes
varied from 50 to 700 μm, depending on the stirring conditions. Drug
crystals were increasingly detectable on the surface as the loading
increased from 8.5 to 50 wt% (drug/polymer). The release rate depended on the microcapsule size and drug load, the half-life increasing

from >8 hr (10% load) to 1 hr (50% load). The results best conformed
to a (time)$^{1/2}$ rate law rather than first-order kinetics.

L-Methadone, a narcotic antagonist, was microencapsulated with
poly(ε-caprolactone-$\underline{co}$-lactic acid) using the solvent evaporation
method (63). At low drug loads (16%), L-methadone existed as a
solid solution in the polymer. Increasing the proportion of lactic
acid slowed the rate of drug release in vitro, and the mechanism of
release changed from diffusion to biodegradation control. By com-
bining different batches of microcapsules prepared with poly(ε-
caprolactone-$\underline{co}$-lactic acid), poly(L-lactic acid), and poly(L-lactic
acid-$\underline{co}$-glycolic acid), each with a different release rate profile, it
was possible to achieve zero-order kinetics for 6 days. A similar
control of release rates was achieved by preparing microcapsules
from polymer blends. For example, the L-methadone release rate
from a 1:1 blend of poly(L-lactic acid) and poly(L-lactic acid-$\underline{co}$-
glycolic acid) fell between the rates exhibited by the individual components (Fig. 12).

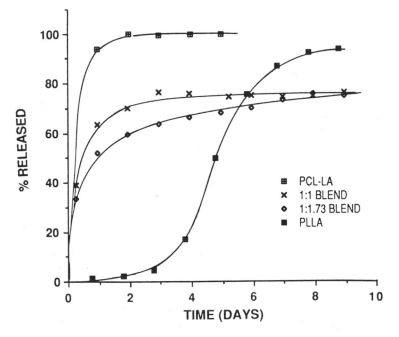

FIGURE 12 Control of the rate of release of methadone from microspheres by the use of blends of a ε-caprolactone-lactic acid copolymer and polylactic acid. (From Ref. 63.)

In the same study, poly(ε-caprolactone-<u>co</u>-lactic acid) micro-capsules with an 80% loading of L-methadone were prepared by an air-coating method. Subsequently, this technique was applied to the encapsulation of naltrexone and a 70% loading was achieved (86). Using 150- to 250-μm microcapsules, 90% of the drug was released at a constant rate over a 60-day period (Fig. 13). These micro-capsules blocked the action of morphine for the same time period when injected (i.m.) in five rats. Release of naltrexone from micro-spheres prepared from the homopolymer PCL was much more rapid, being complete within 2 days (87). This increase in release rate can be attributed to the greater permeability of the homopolymer relative to a copolymer containing a high proportion of lactic acid.

The microencapsulation and controlled release of nucleic acids, e.g., poly(I:C), for the stimulation of interferon production has been patented (87).

3. Reservoir Devices

Delivery systems for several steroids, LHRH analogs, and naltrexone have been described. A priori, a constant rate of release from a

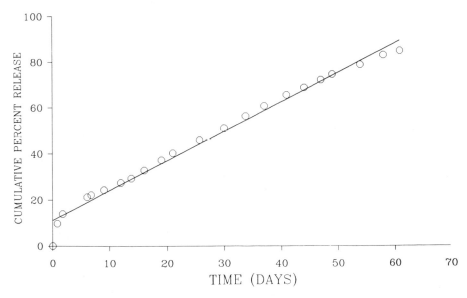

FIGURE 13 Rate of release of naltrexone from microspheres of a co-polymer of ε-caprolactone and lactic acid, in vitro at 37°C. (From Ref. 86.)

capsule is obtained provided solid drug remains within the reservoir, so maintaining a constant concentration gradient across the polymer membrane. In practice it can be quite difficult to achieve a constant rate and, unless the experimental variables are carefully selected, the dissolution of the drug in the polymer wall becomes rate limiting. This is illustrated by the in vitro rates of release of testosterone from PCL tubes of different wall thicknesses (Fig. 14) (68). The rate of testosterone release from the thinnest wall capsules fell continuously from 40 μg/day/cm capsule length to one-fourth of the value after 80 days. The rate from the thickest capsule was lower, as expected for the increased OD/ID ratio, but essentially constant over the same period. For the thinnest capsules, the initial rate was attributed to membrane-controlled release. However, this rate decreased as the smaller drug particles were dissipated, and the rate of dissolution of the larger particles decreased and became rate limiting. A constant rate of release was maintained only when the capsule wall thickness was sufficiently great that it remained rate determining.

With levonorgestrel, a contraceptive steroid, improved zero-order kinetics were obtained using an oil to suspend the solid drug within the PCL capsule (68). The oil apparently served to enhance the contact and dissolution of the drug in the capsule wall. The choice of suspending oil was found to be important, possibly because of plasticization of the polymer producing an increase in its permeability. After evaluating a number of suspending oils for levenorgestrel, ethyl oleate and tricaproin were identified as the optimum. Rates of release of levonorgestrel from PCL capsules for periods up to 1 year in vivo (rabbit) are illustrated in Fig. 15. Pharmacokinetic studies using pulsed intravenous injection of tritiated levonorgestrel and carbon-14-labeled levonorgestrel in PCL showed that the in vivo release rate was approximately 65% of the in vitro value (88). This discrepancy was attributed to boundary layer effects and less effective sink conditions in vivo. The contraceptive delivery system based on levonorgestrel in a PCL reservoir is now undergoing phase II clinical trials under the name of Capronor [vide infra (6)].

Increased permeability of reservoir devices has been achieved by using copolymers of ε-caprolactone (89). Naltrexone, a narcotic antagonist which blocks the action of morphine, requires a relatively high dose of 3–5 mg/day. Poly-ε-caprolactone is not permeable enough to achieve this rate of delivery. Furthermore, it biodegraded too slowly to be useful for the intended objective, a 1- to 2-month subdermal delivery system. An 80:20 copolymer of ε-caprolactone and dl-lactic acid was shown to possess the requisite permeability and biodegradability (89). Using capsules of this copolymer with dimensions 2.35 mm OD, 2.04 mm ID, 2.5 cm length, an in vitro release rate of 0.75–1.0 mg/day for 40 days was achieved (Fig. 16A).

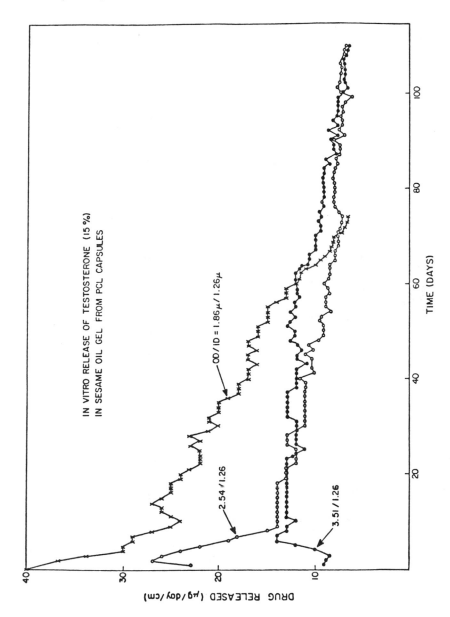

FIGURE 14 In vitro rate of release of testosterone from a PCL capsule (reservoir device), illustrating rate control by drug dissolution when the polymer membrane thickness is small. (From Ref. 68.)

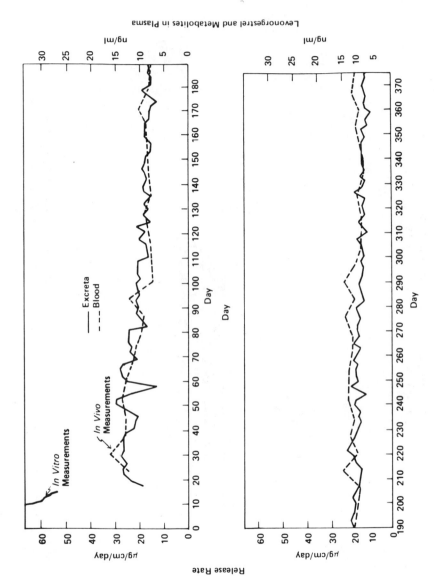

FIGURE 15 Long-term release of levonorgestrel from a PCL capsule (reservoir device) implanted (s.c.) in rabbit, showing both blood and excreta levels of the drug and its metabolites. (From Ref. 6.)

Poly-ε-caprolactone

Increasing the wall thickness reduced the release rate pr[
By choosing a copolymer with an intrinsic viscosity of 1.5
zene), capsules retained their integrity for the duration o[
livery but began to disintegrate after the drug was exhausted.   When
capsules 3 cm length were implanted in monkeys, morphine self-
administration was suppressed for 30 days (Fig. 16B).   The use of
tritium-labeled naltrexone to monitor tritium blood levels showed that
release rates were sustained for 30 days before declining.   At the
end of the study (day 46), the capsules were recovered as partially
degraded fragments.

The possibility of controlling anthropod pests in livestock has
been evaluated by implanting PCL capsules, dimensions 2 or 4 cm,
containing 20-hydroxyecdysone or $3\beta,5\beta,14\alpha$-trihydroxy-5β-cholest-
7-en-6-one.   No in vitro or in vivo release rates were reported but
effective biological control was achieved in rabbits (90,91).

Whereas PCL is very permeable to low molecular weight drugs
($M_W < 500$), Fickian diffusion of peptides and proteins through PCL
membranes is too slow to have any practical merit.   This deficiency
was overcome by preparing PCL capsules with controlled porosity by
the leaching procedure described in Sec. III.B.   For example, a constant rate of 40–50 µg/day/cm length of [D-Trp^6,des-Gly^{10}]LHRH
diethylamide, a peptide with a molecular weight of 1282, was maintained for more than 60 days using this technique (Fig. 17) (64,72).

IV.   BIODEGRADATION

A.   Poly(ε-caprolactone)

Studies of PCL have provided a relatively complete picture of the
factors and mechanisms responsible for the degradation of polyesters,
and several reviews are available (92-94).   When films or rods of
PCL are implanted subdermally in rats, rabbits, or monkeys, degradation begins with random hydrolytic chain scission of the ester
linkages, manifested by a reduction in the viscosity and molecular
weight of the polymer (5,6,53,95).   The same rate of cleavage is observed in vitro in water at 40°C (Fig. 18).   The rate does not
change despite 10-fold changes in the surface-to-volume ratio of the
implant; this is indicative of a bulk process.   Both of these observations, and the improbability of enzyme diffusion into the polymer
bulk, rule out a significant enzymatic contribution during this stage
of degradation.   Semilogarithmic plots of the change in viscosity or
molecular weight of the PCL versus time are linear, regardless of
the initial molecular weight, until the M_n has decreased to approximately 5000.   There is no weight loss during this first stage of the
degradation process.

This behavior is consistent with an autocatalytic process, whereby
the liberated carboxylic end groups catalyze the cleavage of additional
ester groups (Eq. 3):

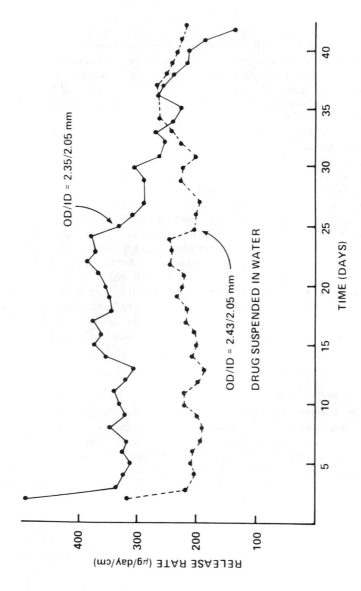

FIGURE 16A Daily rate of release of naltrexone from a poly-ε-caprolactone-co-lactic acid capsule (reservoir device) under in vitro conditions. (From Ref. 89.)

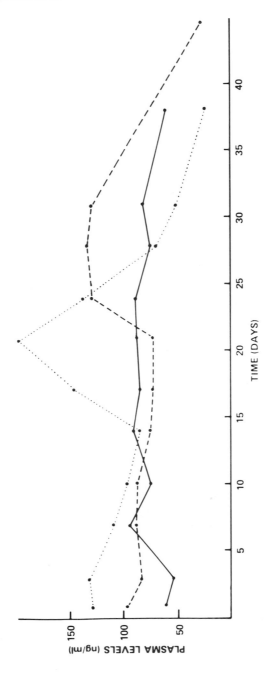

FIGURE 16B Release of naltrexone from the same reservoir device implanted in monkeys. (From Ref. 89.)

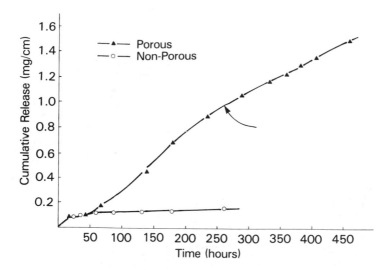

FIGURE 17 Use of a porous PCL membrane to achieve diffusion-controlled zero-order delivery of a LHRH analog from a reservoir device. (From Ref. 72.)

$$-COOR\ldots.HOOC-+ H_2O \longrightarrow 2-COOH + ROH \qquad (3)$$

Neglecting the contribution of uncatalyzed hydrolysis, the rate of chain scission is given by Eq. 4:

$$\frac{d[COOH]}{dt} = k_1[COOH][ester][H_2O] \qquad (4)$$

This equation assumes that the carboxylic acid group is not ionized in the polymer bulk and functions by hydrogen-bonding to the ester groups. For a small number of chain scissions, both the ester and water concentrations may be considered constants, and Eq. 4 simplifies to Eq. 5.

$$\frac{d[COOH]}{dt} = k_2[COOH] \qquad (5)$$

Integration results in the expression:

$$\frac{[COOH]}{[COOH]_0} = \exp(k_2 t) \qquad (6)$$

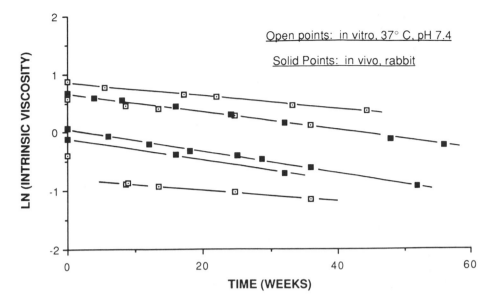

FIGURE 18 Rates of degradation of PCL in water at 40°C and in rabbit, demonstrating the kinetic equivalency of the two processes. (From Ref. 53.)

where [COOH] and [COOH]$_0$ are the carboxylic end-group concentrations at times t and zero, respectively. This expression holds until loss of oligomers begins to reduce the carboxylic end-group concentration of the polymer bulk. Provided [COOH] = M_n^{-1}, Eq. 6 can be rewritten as

$$\frac{1}{M_n} = \frac{1}{M_n^0} \exp(k_2 t) \tag{7}$$

or

$$Ln\left(\frac{M_n}{M_n^0}\right) = -k_2 t$$

If changes in the molecular weight distribution can be neglected, substitution of the Mark-Houwink equation into Eq. 7 leads to Eq. 8, where α is the Mark-Houwink exponent.

$$[\eta] = [\eta_0]\exp(-\alpha k_2 t) \tag{8}$$

If the hydrolysis is not autocatalyzed, the predicted kinetic expression is Eq. 9, where DP is the degree of polymerization.

$$\mathrm{Ln}\left(\frac{DP - 1}{DP}\right) = \mathrm{Ln}\left(\frac{DP_0 - 1}{DP_0}\right) - kt \tag{9}$$

Statistical evaluation shows that the experimental data are better correlated with Eqs. 7 and 8, i.e., an autocatalytic process, than with Eq. 9 although it is not possible to rule out the superposition of both mechanisms.

The second phase of polymer degradation is characterized by a decrease in the rate of chain scission (Fig. 19) and the onset of weight loss. Weight loss has been attributed to (1) the increased probability that chain scission of a low molecular weight polymer will produce a fragment small enough to diffuse out of the polymer bulk and (2) the breakup of the polymer mass to produce smaller particles with an increased probability of phagocytosis. The decrease in the rate of chain scission, as well as the increased brittleness of the polymer, is the result of an increase in the crystallinity of PCL,

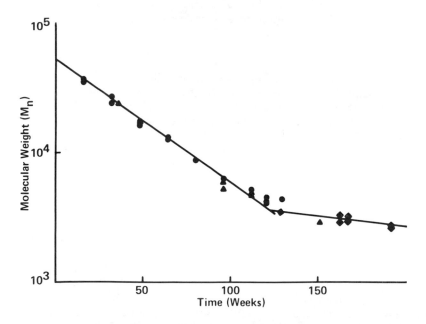

FIGURE 19 Decrease in the rate of chain scission of PCL in rabbit observed when the molecular weight (M_n) of the polymer has decreased to about 5000.

from 45 to 80% (Fig. 7). This is brought about by selective cleavage of chains in the amorphous phase followed by crystallization of the cleaved segments.

The phagocytosis of small particles and metabolism of oligomeric fragments have been studied in rats using PCL that had been pre-degraded under accelerated in vitro conditions to an M_n of 3000 D (96). Two different polymer particle sizes, 53–106 and 212–500 μm, were subdermally implanted under the ventral abdominal panniculus. Irrespective of particle size, bioabsorption occurred by intracellular degradation in phagosomes of macrophages, giant cells, and fibro-blasts, the process requiring less than 13 days for completion at some sites. The subsequent metabolism was studied by implanting low molecular weight carbon-14 and tritium-labeled PCL (53,97). Measurement of radioactivity in urine, feces, and expired air, and at the implant site of powdered PCL, demonstrated that absorption was complete in 60 days, although only 60% of the radioactivity was accounted for (Fig. 20). ε-Hydroxycaproic acid, derived from com-plete hydrolysis of the polymer, and (tritiated) water were the only metabolites detected (97).

B. Copolymers of ε-Caprolactone

Copolymers of ε-caprolactone with dilactide, diglycolide, δ-valerolac-tone, and ε-decalactone are degraded much more rapidly than PCL (Fig. 21) (95). A similar pattern of chain scission, with no weight loss until the M_n had declined to a limiting value of about 15,000, is observed. Weight loss, fragmentation of the implant, and deviation from the linearity of semilog plots of M_n or [η] versus time are coincident. The nonenzymatic nature of the degradation is again demonstrated by observation of the same processes in vitro.

The importance of the morphology of the polymer as well as the chemical structure is evident from the fact that both chain scission and weight loss of a random copolymer of dilactide and ε-caprolactone is more rapid than for either of the constituent homopolymers. In contrast, the rate of degradation of a block copolymer of dilactide and ε-caprolactone is intermediate in rate (42). A 1:1 blend of PCL and poly(glycolic acid-co-lactic acid) is similarly degraded in vitro (pH 7.4, 37°C) at a rate which is intermediate to the rates of de-gradation of the individual components (Fig. 22) (65). Some of the factors that are believed to control the relative rates of degradation of these copolymers and blends are crystallinity, glassy versus rub-bery state, permeability, and magnitude of water uptake.

Evidence of enzymatic degradation in vivo was first observed with a random copolymer of ε-caprolactone and ε-decalactone (93). The chain scission and weight loss of a semicrystalline copolymer with 8 mol % ε-decalactone was almost identical to that of PCL and was clearly nonenzymatic. When the proportion of ε-decalactone was

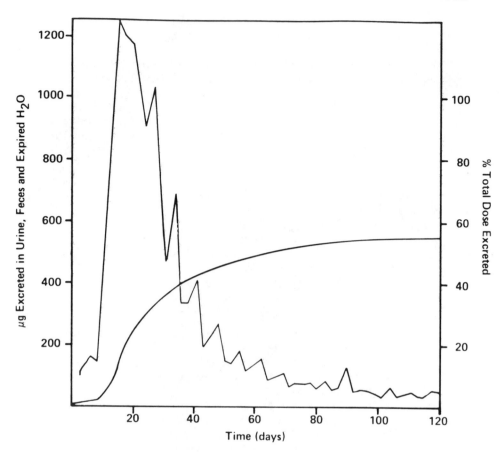

FIGURE 20 Rate of metabolism of PCL and its metabolites, determined from measurement of carbon-14 in the excreta and exhaled air of rats after implantation (s.c.) of the low molecular weight polymer. (From Ref. 53.)

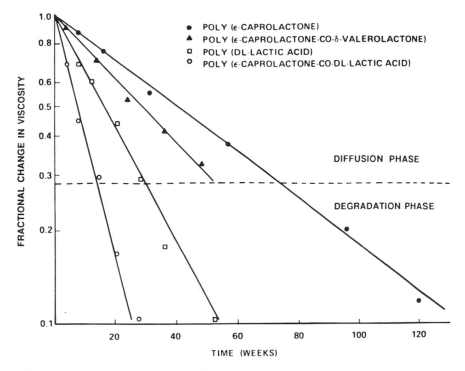

FIGURE 21 Semilog plot of the in vitro rate of hydrolytic chain scission of various copolymers of ε-caprolactone, measured under in vitro conditions. (From Ref. 95.)

increased to 13%, the resulting noncrystalline copolymer lost 80% weight after 8 weeks in rabbit. The large difference in the rate of weight loss for a small difference in copolymer composition suggested that the lack of crystallinity was a prerequisite for enzymatic degradation.

More definitive evidence of enzymatic attack was obtained with 1:1 copolymers of ε-caprolactone and δ-valerolactone crosslinked with varying amounts of a dilactone (98,99). The use of a 1:1 mixture of comonomers suppressed crystallization and, together with the crosslinks, resulted in a low-modulus elastomer. Under in vitro conditions, random hydrolytic chain cleavage, measured by the change in tensile properties, occurred throughout the bulk of the samples at a rate comparable to that experienced by the other polyesters; no weight loss was observed. However, when these elastomers were implanted in rabbits, the bulk hydrolytic process was accompanied by very rapid surface erosion. Weight loss was continuous, confined to the

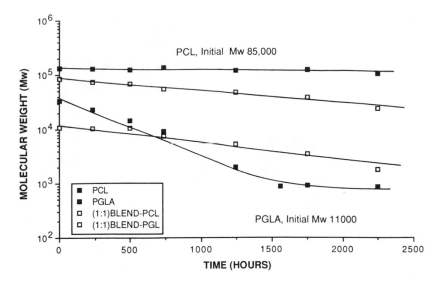

FIGURE 22 Semilog plot of the in vitro rate of hydrolytic chain scission of PCL, polyglycolic acid-co-lactic acid, and a 1:1 blend of the two polymers, demonstrating the use of blends to modify degradation rates. (From Refs. 64 and 65.)

surface, and was complete within 12–16 weeks (Fig. 23). Absorption was slightly more rapid in rats than in rabbits, being complete within 7 weeks. The rate of surface erosion was subject to control by modification of the chemical structure. Increasing the crosslink density reduced the rate of bioabsorption, as did the introduction of hydrophobic or sterically bulky substituents on the polyester chain (Fig. 23). It was not necessary to use a copolymer to observe enzymatic surface erosion. Crosslinking PCL with 12% of the bislactone served to eliminate the crystallinity of the homopolymer, and slow surface erosion of the resulting elastomer was observed (Fig. 24).

C. Manipulation of Polymer Degradation Rates

An implication of the kinetic analysis presented in Sec. IV.A is that the rate of chain scission of polyesters can be retarded by end-capping to reduce the initial carboxylic acid end-group concentration. Alternatively, the rate may be increased by acidic additives that supplement the effect of the carboxy end groups. The first expectation was confirmed by partial ethanolysis of high molecular weight

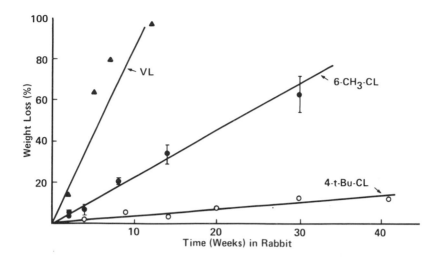

FIGURE 23   Rate of enzymatic surface erosion of a 1:1 copolymer of ε-caprolactone and δ-valerolactone, crosslinked with a dilactone to form an elastomer.   The effect of substitution of the ε-caprolactone nucleus is also shown.   (From Ref. 98).

PCL with ethanol-^{14}C (49).   The fraction of ethoxy end groups was derived from the carbon-14 introduced into the polymer:

$$HO[(CH_2)_5CO]_nOH + EtOH \longrightarrow HO[(CH_2)_5CO]_mOEt$$

$$+ HO[(CH_2)_5CO]_{n-m}OH$$

M_n 130,000 M_n 29,000

The rate of hydrolysis of the partially ethoxylated polymer was retarded, although not to the extent calculated from theory (Fig. 25), suggesting some contribution to the rate of chain scission by an uncatalyzed process.   End-capping poly(glycolic acid-co-lactic acid) has a similar effect on the rate of hydrolysis of this polyester (100).

Addition of oleic acid to PCL caused a significant increase in the rate of chain scission, the effect being proportional to the amount of acid added (Fig. 26).   The effect of added base, e.g., n-decylamine, was even more substantial, reducing the molecular weight of PCL from 60,000 to 20,000 in 20 days (Fig. 26).   This reduction in mo-

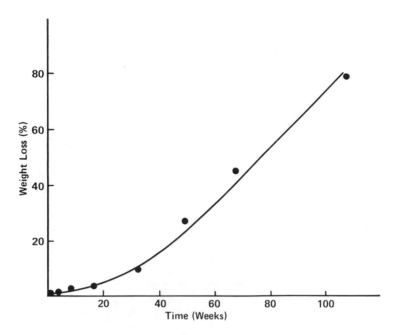

FIGURE 24   Rate of enzymatic surface erosion of ε-caprolactone crosslinked with 12 mol % of the dilactone, 2,2-bis(ε-caprolactone-4-yl)propane.   (From Ref. 98.)

lecular weight which is the result of aminolysis by the primary amine, would normally require >1 year in the absence of an acid or base. The effects of a series of added tertiary amines on the rate of chain scission of other polyesters, including poly(ε-caprolactone-co-lactic acid), has been studied and found to be equally great (65).   The mechanism with tertiary amines can only be general base catalysis for the effectiveness of the amines was not related to their pKa values or lipophilicities.   The acceleration of the hydrolysis of the polyesters was used as a strategy for controlling the drug release rate.

Blending of PCL and poly(glycolic acid-co-lactic acid) has been also used to control the rate of chain scission of the composite. Thus, while these two polymers differ greatly in their rate of hydrolytic chain cleavage, gel permeation chromatography (GPC) analysis of a 1:1 blend of PCL and poly(glycolic acid-co-lactic acid) in pH 7.4 buffer showed that both components of the blend were subject to the same rate of chain cleavage (65).

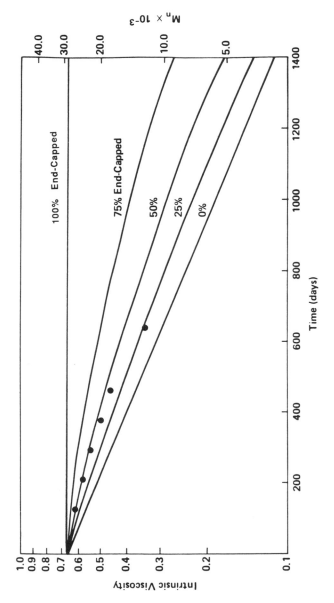

FIGURE 25 Reduction in the rate of hydrolytic chain scission of PCL achieved by ethoxylation of the carboxy end groups of the polymer. The experimental result is compared with calculated predictions of the effect of varying degrees of ethoxylation. (From Ref. 49.)

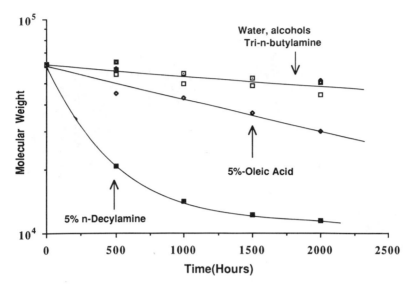

FIGURE 26 Enhancement of the rate of hydrolytic chain scission of PCL in the presence of oleic acid and n̲-decylamine. (From Ref. 49.)

V. TOXICOLOGY

The toxicology of PCL has mostly been conducted in conjunction with evaluations of Capronor, which is an implantable 1-year contraceptive delivery system composed of a levonorgestrel-ethyl oleate slurry within a PCL capsule (vide supra). An initial 90-day trial of Capronor in female rats and guinea pigs revealed no toxic effects (6). No systemic effects were observed. The capsules evoked a bland response at the implant site and a minimal tissue encapsulating response based on animal sacrifices at 90 and 187 days. The lack of an inflammatory response was confirmed by implanting polyvinyl alcohol sponges impregnated with the powdered polymer in rats. The latter technique has been used to study the toxicity of poly-α-cyanoacrylates (101). PCL and ε-caprolactone tested negatively in the Ames mutagenicity assay; the same result was obtained after their exposure to mammalian liver homogenates under conditions designed to effect metabolism (102).

In a second more extensive study, Capronor was implanted in rats and cynomolgus monkeys. Clinical chemistry observations, physical examinations, qualitative food consumption, urinalysis, and ophthalmoscopic analysis during a 2-year period, and histopathology at sacrifice, showed the only differences between the test and control groups were the typical effects of the progestational drug.

A method of assessing the toxicity of implants has been proposed based on the effects on cell ultrastructure in organ cultures, on cell surface characteristics, and cell population doubling times. The effects have been correlated with hemorrhage, fibrosis, and necrosis, respectively (103). Poly-ε-caprolactone was stated to give minimal tissue reaction and could not be scored in these tests.

VI. CLINICAL STUDIES

One-month clinical trials of Capronor were sponsored by the National Institutes of Health (NIH) at Duke University Medical Center and by

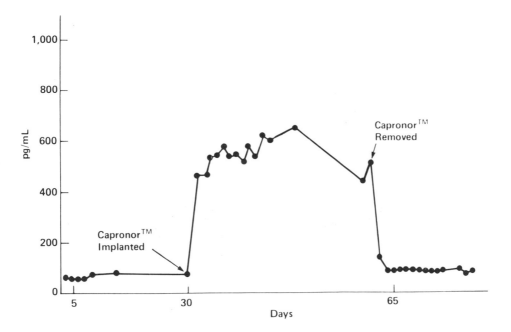

FIGURE 27   Mean levonorgestrel blood levels of eight subjects who received Capronor, implanted subdermally above the ischeal crest. The device was implanted at the beginning of the 4th cycle and withdrawn at the beginning of the 5th. The minimum sensitivity of the RIA was 100 ng/ml. (From Ref. 104.)

the World Health Organization (WHO) at Hammersmith Hospital, London, the University of Rome, and Chulalongkor Hospital, Bangkok (104,105). The trials generally spanned five menstrual cycles: three observation cycles to confirm ovulation, one cycle with Capronor implanted via a trocar above the hip (NIH) or in the upper arm (WHO), and one cycle after its removal. Basal body temperatures, and serum levels of levonorgestrel and endogenous LH, FSH, estradiol, and levonorgestrel were used as measures of efficacy. High serum levels of levonorgestrel (Fig. 27) together with the characteristic hormonal changes associated with minidoses of progestogens were observed during the test cycle.

A 1-year phase II clinical study of Capronor was recently completed at the University of California, San Francisco, and additional studies are in progress in Europe and Asia.

REFERENCES

1. Van Natta, F. J., Hill, J. W., and Carruthers, W. H., Studies of polymerization and ring formation. XXIII. ε-Caprolactone and its polymers, J. Am. Chem. Soc., 56, 455-457, 1934.
2. Potts, J. E., Clendinning, R. A., Ackart, W. B., and Niegisch, W. D., Biodegradability of synthetic polymers, Polym. Sci. Technol., 3, 61-79, 1973.
3. Potts, J. E., Clendinning, R. A., and Cohen, S., Biodegradable plastic containers for seedling transplants, Soc. Plast. Eng., Tech. Pap., 21, 567-569, 1975.
4. Huang, S., Biodegradable polymers, in Encyclopedia of Polymer Science and Engineering, Vol. 2 (H. F. Mark, N. M. Bikales, C. G. Overberger, G. Menges, and J. I. Kroschwitz, eds.), John Wiley and Sons, New York, 1985, pp. 220-243.
5. Schindler, A., Jeffcoat, R., Kimmel, G. L., Pitt, C. G., Wall, M. E., and Zweidinger, R., Biodegradable polymers for sustained drug delivery, Contemp. Top. Polym. Sci., 2, 251-289, 1977.
6. Pitt, C. G., and Schindler, A., Capronor: A biodegradable delivery system for levonorgestrel, in Long-Acting Contraceptive Delivery Systems (G. I. Zatuchni, A. Goldsmith, J. D. Shelton, and J. J. Sciarra, eds.), Harper and Row, Philadelphia, 1984, pp. 48-63.
7. Brode, G. L., and Koleske, J. V., Lactone polymerization and properties, J. Macromol. Sci., A6, 1109-1144, 1972.
8. Koleske, J. V., Blends containing poly(ε-caprolactone) and related polymers, in Polymer Blends, Vol. 2 (D. R. Paul and S. Newman, eds.), Academic Press, New York, 1978, pp. 369-389.

9. Jenkins, V. F., Caprolactone and its polymers, Polym. Paint Colour J., 167, 622, 624, 626-627, 1977.
10. Cox, E. F., Hostettler, F., and Kiser, R. R., Poly-ε-caprolactone, Macromol. Syn., 3, 111-113, 1969.
11. Morton, M., and Wu, M., Organolithium polymerization of ε-caprolactone, ACS Symp. Ser., 286, 175-182, 1985.
12. Velichkova, R. S., Toncheva, V. D., and Panayotov, I. M., Macromonomers from vinyl and cyclic monomers prepared by initiation with stable carbenium salts, J. Polym. Sci., Part A: Polym. Chem., 25, 3283-3292, 1987.
13. Dreyfuss, P., Adaway, T., and Kennedy, J. P., Polymerization and grafting of heterocyclic monomers from reactive mono- and macrohalides, Appl. Polym. Symp., 30, 183-192, 1977.
14. Lardelli, G., Lamberti, V., Weller, W. T., and de Jonge, A. P., The synthesis of lactones. Part I, Saturated δ- and ε-lactones, Rec. Trav. Chim., 86, 481-503, 1967.
15. Hassall, C. W., The Baeyer-Villiger oxidation of aldehydes and ketones, Organic Reactions, 9, 73-106, 1957.
16. Pitt, C. G., Gu, Z.-W., and Hendren, R. W., The synthesis of biodegradable polymers with functional side chains, J. Polym. Sci., Polym. Chem. Ed., 25, 955-956, 1987.
17. Starcher, P. S., Tinsley, S. W., and Phillips, B., Union Carbide Corp., Bis- ε-caprolactones, U.S. Patent 3,072,680, 1963.
18. Kruizinga, W. H., and Kellogg, R. M., Simple and high yield synthesis of macrocyclic lactones by ring-closure of caesium salts of ω-halogenoaliphatic acids, J. Chem. Soc., Chem. Commun., 286-288, 1979.
19. Johns, D. B., Lenz, R. W., and Leucke, A., Lactones, in Ring-Opening Polymerization (K. I. Ivin and T. Saegusa, eds.), Elsevier, New York, 1984, pp. 461-521.
20. Balas, A., Palka, G., Foks, J., and Janik, H., Properties of cast urethane elastomers prepared from poly(ε-caprolactone)s, J. Appl. Polym. Sci., 29, 2261-2670, 1984.
21. Slomkowski, S., and Penczek, S., Influence of dibenzo-18-crown-6 ether on the kinetics of anionic polymerization of β-propiolactone, Macromolecules, 9, 367-369, 1976.
22. Deffieux, A., and Boileau, S., Use of cryptates in anionic polymerization of lactones, Macromolecules, 9, 369-371, 1976.
23. Ito, K., and Yamashita, Y., Propagation and depropagation rates in the anionic polymerization of ε-caprolactone cyclic oligomers, Macromolecules, 11, 68-72, 1978.
24. Ito, K., Hashizuka, Y., and Yamashita, Y., Equilibrium cyclic oligomer formation in the anionic polymerization of ε-caprolactone, Macromolecules, 10, 821-824, 1977.
25. Rashkov, I. B., Gitsov, I., Panayotov, I. M., and Pascault,

J. P., Anionic polymerization of lactones initiated by alkali
graphitides. I Polymerization of ε-caprolactone initiated by
KC$_{24}$, J. Polym. Sci., Part A: Polym. Chem., 21, 923-936,
1983.

26. Perret, R., and Skoulios, A., Synthese et caracterisation de
     copolymeres sequences polyoxyethylene/poly-ε-caprolactone,
     Die Makromolekulare Chemie, 156, 143-156, 1972.

27. Nobutoki, K., and Sumitomo, H., Preparation of block copoly-
     mer of ε-caprolactone with living polystyrene, Bull. Chem.
     Soc. Jap., 40, 1741-1745, 1967.

28. Huang, S. J., Edelman, P. G., and Cameron, J. A., Cross-
     linkable polyesters for biomedical composites, Polym. Mater.
     Sci. Eng., 53, 515-519, 1985.

29. Dreyfuss, P., and Kennedy, J. P., Alkyl halides in conjunc-
     tion with inorganic salt for the initiation of the polymerization
     and graft copolymerization of heterocycles, J. Polym. Sci.,
     Polym. Symp., 56, 129-137, 1976.

30. Horlbeck, G., Siesler, H. W., Tittle, B., and Trafara, G.,
     Characterization of polyether and polyester homo- and copoly-
     mers prepared by ring-opening polymerization with a new cat-
     alytic system, Macromolecules, 10, 284-287, 1977.

31. Crivello, J. V., Lockhart, T. P., and Lee, J. L., Diaryl-
     iodonium salts as thermal initiators of cationic polymerization,
     J. Polym. Sci., Polym. Chem. Ed., 21, 97-109, 1983.

32. Rashkov, I. B., and Gitsov, I., Cationic polymerization initi-
     ated by intercalation compounds of Lewis acids. II. Initiating
     ability and mechanism of action of the initiators, J. Polym.
     Sci., Polym. Chem. Ed., 24, 155-165, 1986.

33. Kricheldorf, H. R., Mang, T., and Jonte, J. M., Polylactones.
     1. Copolymerizations of glycolide and ε-caprolactone, Macro-
     molecules, 17, 2173-2181, 1984.

34. Jonte, J. M., Dunsing, R., and Kricheldorf, H. R., Polylac-
     tones. 4. Cationic polymerization of lactones by means of al-
     kylsulfonates, J. Macromol. Sci., Chem., A23, 495-514, 1986.

35. Dunsing, R., and Kricheldorf, H. R., Polylactones. 14. Poly-
     merization of δ-valerolactone and ε-caprolactone by means of
     trimethylsilyl triflate, Eur. Polym. J., 24, 145-150, 1988.

36. Kricheldorf, H. R., Berl, M., and Scharnagl, N., Poly(lac-
     tones). 9. Polymerization mechanism of metal alkoxide initiated
     polymerizations of lactide and various lactones, Macromolecules,
     21, 286-293, 1988.

37. Yamashita, Y., Tsuda, T., Ishida, H., and Hasegawa, M.,
     Polymerization and Copolymerization of ε-Caprolactone, Kogyo
     Kagaku Zasshi, 71, 755-757, 1968.

38. Teyssie, P., Bioul, J. P., Hocks, L., and Ouhadi, T., Catal-

ysis with soluble M-O-M'-O-M bimetallic oxides, Chemtech, 7, 192-194, 1977.

39. Hamitou, A., Jerome, R., Hubert, A. J., and Teyssie, P., New catalyst for the living polymerization of lactones to polyesters, Macromolecules, 6, 651-652, 1973.

40. Hamitou, A., Ouhadi, T., Jerome, R., and Teyssie, P., J. Polym. Sci., Part A: Polym. Chem., 15, 865-887, 1983.

41. Ouhadi, T., Stevens, C., and Teyssie, P., Study of poly-ε-caprolactone bulk degradation, J. Appl. Polym. Sci., 20, 2963-2970, 1976.

42. Feng, X. D., Song, C. X., and Chen, W. Y., Synthesis and evaluation of biodegradable block copolymers of ε-caprolactone and DL-lactide, J. Polym. Sci., Polym. Lett. Ed., 21, 593-600, 1983.

43. Song, C. X., and Feng, X. D., Synthesis of ABA triblock copolymers of ε-caprolactone and DL-lactide, Macromolecules, 17, 2764-2767, 1984.

44. Endo, M., Aida, T., and Inoue, S., Immortal polymerization of ε-caprolactone initiated by aluminum porphyrin in the presence of alcohol, Macromolecules, 20, 2982-2988, 1987.

45. Yasuda, T., Aida, T., and Inoue, S., Reactivity of (porphinato)aluminum phenoxide and alkoxide as active initiators for polymerization of epoxide and lactone, Bull. Chem. Soc. Jap., 59, 3931-3934, 1986.

46. Bassi, M. B., Padias, A. B., and Hall, H. K., Jr., Kinetics of the hydrolytic polymerization of ε-caprolactone, Private communication.

47. Schindler, A., Hibionada, Y. M., and Pitt, C. G., Aliphatic polyesters III. Molecular weight and molecular weight distribution in alcohol-initiated polymerization of ε-caprolactone, J. Polym. Sci., Part A: Polym. Chem., 20, 319-326, 1982.

48. Segal, S., Contraceptive subdermal implants, in Advances in Fertility Research (D. R. Mishell, ed.), Raven Press, New York, 1982, pp. 117-137.

49. Pitt, C. G., and Gu, Z. W., Modification of the rates of chain cleavage of poly(ε-caprolactone) and related polyesters in the solid state, J. Control. Rel., 4, 283-292, 1987.

50. Haubennestel, K., and Bubat, A., Polyester-siloxane for coatings and moldings, Byk-Chemie G.m.b.H., Ger. Offen. DE 3535283 A1, April 9, 1987.

51. Bailey, W. J., Ni, Z., and Wu, S.-R., Synthesis of poly-ε-caprolactone via a free radical mechanism. Free radical ring-opening polymerization of 2-methylene-1,3-dioxepane, J. Polym. Sci., Polym. Chem. Ed., 20, 3021-3030, 1982.

52. Crescenzi, V., Manzini, G., Calzolari, G., and Borri, C.,

Thermodynamics of fusion of poly-β-propiolactone and poly-ε-caprolactone. Comparative analysis of the melting of aliphatic polylactone and polyester chains, Eur. Polym. J., 8, 449-463, 1972.

53. Pitt, C. G., Chasalow, F. I., Hibionada, Y. M., Klimas, D. M., and Schindler, A., Aliphatic polyesters. I. The degradation of poly(ε-caprolactone) in vivo, J. Appl. Polym. Sci., 26, 3779-3787, 1981.

54. Koleske, J. V., and Lundberg, R. D., Lactone polymers. II. Hydrodynamic properties and unperturbed dimensions of poly-ε-caprolactone, J. Polym. Sci., Part A-2, 7, 897-907, 1969.

55. Knecht, M. R., and Elias, H., Zur flexibilitat aliphatischer polyester, Makromol. Chem., 157, 1-2, 1972.

56. Sekiguchi, H., and Clarisse, C., Polymerization of ε-caprolactone with boron derivatives as catalysts, Makromol. Chem., 177, 591-606, 1976.

57. Schindler, A., Unpublished results.

58. Pitt, C. G., Bao, Y. T., Andrady, A. L., and Samuel, P. N. K., The correlation of polymer-water and octanol-water coefficients: Estimation of drug solubilities in polymers, Int. J. Pharm., 45, 1-11, 1988.

59. Bao, Y. T., Samuel, N. K. P., and Pitt, C. G., The prediction of drug solubilities in polymers, J. Polym. Sci., Part C, Polym. Lett., 26, 41-46, 1988.

60. Critchfield, F. E., and Koleske, J. V., Union Carbide Corp., Lactone graft copolymers, U.S. Patent 3,760,034, 1973.

61. Chang, R. K., Price, J. C., and Whitworth, C. W., Control of drug release rates through the use of mixtures of poly-caprolactone and cellulose propionate polymers, Pharm. Technol., 10, 24, 26, 29, 32-33, 1986.

62. Chang, R. K., Price, J., and Whitworth, C. W., Control of drug release rate by use of mixtures of polycaprolactone and cellulose acetate butyrate polymers, Drug Dev. Ind. Pharm., 13, 1119-1135, 1987.

63. Cha, Y., and Pitt, C. G., A one-week subdermal delivery system for L-methadone based on biodegradable microcapsules, J. Control. Rel., 7, 69-78, 1988.

64. Pitt, C. G., Cha, Y., Hendren, R. W., Holloman, M., and Schindler, A., Manipulation of the permeability and degradability of polymers, Proc. 14th Int. Symp. Control. Rel. Bioactive Materials, 75-76, 1987.

65. Cha, Y., and Pitt, C. G., Toronto Controlled Release Society, Biomaterials, 11, 108-112, 1990.

66. Davis, P. A. Nicolais, L., Ambrosio, L., and Huang, S. J., Synthesis and characterization of semi-interpenetrating polymer networks of poly(2-hydroxyethyl methacrylate) and poly(caprolactone), Polym. Mater. Sci. Eng., 56, 536-540, 1987.

67. Pitt, C. G., Jeffcoat, A. R., Zweidinger, R. A., and Schindler, A., Sustained drug delivery systems. I. The permeability of poly(ε-caprolactone), poly(DL-lactic acid), and their copolymers, J. Biomed. Mater. Res., 13, 497-507, 1979.

68. Pitt, C. G., Gratzl, M. M., Jeffcoat, A. R., Zweidinger, R., and Schindler, A., Sustained drug delivery systems. II. Factors affecting release rates from poly(ε-caprolactone) and related biodegradable polyesters, J. Pharm. Sci., 68, 1534-1538, 1979.

69. Goodson, J. M., Holborow, D., Dunn, R. L., Hogan, P., and Dunham, S., Monolithic tetracycline-containing fibers for controlled delivery to periodontal pockets, J. Periodontol., 54, 575-579, 1983.

70. Tice, T. R., Lewis, D. H., Cowsar, D. R., and Beck, L. R., Stolle Research and Development Co., Injectable, long-acting microparticle formulation for the delivery of antiinflammatory agents, Eur. Patent Appl. 102265 A2, March 7, 1984.

71. Gardner, D. L., Battelle Development Corp., Process of preparing microcapsules of lactides or lactide copolymers with glycolides and/or ε-caprolactones, U.S. Patent 4,637,905, A, Jan. 20, 1987.

72. Schindler, A., Research Triangle Institute, Porous bioabsorbable polyesters as controlled-release reservoirs for high molecular weight drugs, Eur. Patent Appl. EP 223708 A2, May 27, 1987.

73. Bawa, R., Siegel, R., Marasca, B., Karel, M., and Langer, R., An explanation for the controlled release of macromolecules from polymers., J. Control. Rel., 1, 259-267, 1985.

74. Wang, P. Y., Application of combined matrix concept in delivery of bioactive polypeptide, Polym. Mater. Sci. Eng., 57, 400-403, 1987.

75. Wang, P. Y., Insulin delivery by an implantable combined matrix system, Life Support Syst., 4, 380-382, 1986.

76. Lewis, D. H., and Dunn, R. L., Stolle Research and Development Corp., Controlled release aquatic biologically active agent formulations, Eur. Patent Appl. EP 126827 A1, Dec. 5, 1984.

77. Van, T. K., and Steward, K. K., The use of controlled-release fluoridone fibers for control of hydrilla (Hydrilla verticillata), Weed Sci., 34, 70-76, 1986.

78. Dagenhart, G. S., Dunn, R. L., Strobel, J. D., Perkins, B. M., Price, M. W., and Stoner, W. C., Jr., Polymeric-pellet delivery systems for controlled release of aquatic herbicides, Toronto Controlled Release Society, Proc. 14th Int. Symp. Control. Rel. Bioact. Materials, 291-292, 1987.

79. Gebelein, C. G., Mirza, T., and Chapman, M., The release of

5-fluorouracil from polycaprolactone matrixes, Polym. Mater. Sci. Eng., 57, 413-416, 1987.

80. Strobel, J. D., Laughlin, T. J., Austroy, F., Lilly, M. B., Perkins, B. H., and Dunn, R. L., Controlled-release systems for anticancer agents, Toronto Controlled Release Society, Proc. 14th Int. Symp. Control. Rel. Bioact. Materials, 261-262, 1987.

81. Munteanu, M. A., Oltarzewski, E. S., Shechter, L., and Warren, C. B., International Flavors and Fragrances Inc., Detergent bar containing poly(ϵ-caprolactone) and aromatizing agent, U.S. Patent 4,469,613, A, Sep. 4, 1984.

82. Bissery, M. C., Puisieux, F., and Thies, C., A study of the parameters in the making of microspheres by the solvent evaporation procedure, Expo. Congr. Int. Technol. Pharm. 3rd, 3, 233-239, 1983.

83. Jaffe, H., Hayes, D. K., Amarnath, N., and Chaney, N., Preparation and scanning electron microscopy of microcapsules of Abate, Pesticides, 20, 9-11, 1986.

84. Chang, R. K., Price, J. C., and Whitworth, C. W., Enhancement of dissolution rate by incorporation into a water insoluble polymer, polycaprolactone, Drug Dev. Ind. Pharm., 13, 249-256, 1987.

85. Dubernet, C., Benoit, J. P., Couarraze, G., and Duchene, D., Microencapsulation of nitrofurantoin in poly(ϵ-caprolactone): Tableting and in vitro release studies, Int. J. Pharm., 35, 145-156, 1987.

86. Nuwayser, E. S., Gay, M. H., Deroo, D. J., and Blaskovich, P. D., Sustained release injectable naltrexone microcapsules, Basel, Switzerland, Controlled Release Society, Proc. 15th Int. Symp. Control. Rel. Bioact. Materials, 201-202, 1988.

87. Tice, T. R., Pledger, K. L., and Gilley, R. M., Southern Research Institute, Microencapsulation of DNA and/or RNA, especially for stimulation of interferon production, Eur. Patent Appl. EP 248,531, A2, Dec. 9, 1987.

88. Hendren, R. W., Reel, J. R., and Pitt, G. C., Measurement of drug release rates from sustained delivery devices in vivo, Fort Lauderdale, Florida, Controlled Release Society, Proc. 11th Int. Symp. Control. Rel. Bioact. Materials, 110-111, 1984.

89. Pitt, C. G., Marks, T. A., and Schindler, A., Biodegradable drug delivery systems based on aliphatic polyesters: application to contraceptives and narcotic antagonists, in Controlled Release of Bioactive Materials (R. Baker, ed.), Academic Press, New York, 1980, pp. 19-43.

90. Jaffe, H., Sonenshine, D. E., Hayes, D. K., Dees, W. H., Beveridge, M., and Thompson, M. J., Effects of the controlled release of ecdysteroids on the development of sex pheremone

activity in ticks, Fort Lauderdale, Florida Controlled Release Society, Proc. 11th Symp. Control. Rel. Bioact. Materials, 118-119, 1984.

91. Jaffe, H., Hayes, D. K., Luthra, R. P., Shukla, P. G., Amarnath, N., and Chaney, N. A., Controlled-release microcapsules of insect hormone analogues, Geneva, Switzerland Controlled Release Society, Proc. 12th Int. Symp. Control. Rel. Bioact. Materials, 282-283, 1985.

92. Pitt, C. G., and Schindler, A., The design of controlled drug delivery systems based on biodegradable polymers, in Biodegradable and Delivery Systems for Contraception, Progress in Contraceptive Delivery Systems, Vol. 1 (E. S. E. Hafez and W. A. A. van Os, eds.), MTP Press Ltd., Lancaster, England, 1980, pp. 17-46.

93. Gilbert, R. D., Stannett, V., Pitt, C. G., and Schindler, A., Design of biodegradable polymers: Two approaches, in Development in Polymer Degradation, Vol. 4 (N. Grassie, ed.), Elsevier, New York, 1982, pp. 259-293.

94. Pitt, C. G., and Schindler, A., Biodegradation of polymers, in controlled drug delivery, Vol. 1 (S. D. Bruck, ed.), CRC Press, Boca Raton, 1983, pp. 53-80.

95. Pitt, C. G., Gratzl, M. M., Kimmel, G. L., Surles, J., and Schindler, A., Aliphatic polyesters. II. The degradation of poly (DL-lactide), poly (ε-caprolactone), and their copolymers in vivo, Biomaterials, 2, 215-220, 1981.

96. Woodward, S. C., Brewer, P. S., Moatamed, F., Schindler, A., and Pitt, C. G., The intracellular degradation of poly(ε-caprolactone), J. Biomed. Mater. Res., 19, 437-444, 1985.

97. Hendren, R. W., and Pitt, C. G., unpublished studies.

98. Pitt, C., Hendren, R. W., Schindler, A., and Woodward, S., The enzymatic surface erosion of aliphatic polyesters., J. Control. Rel., 1, 3-14, 1984.

99. Schindler, A., and Pitt, C. G., Biodegradable elastomeric polyesters, Am. Chem. Soc. Polym. Preprints, 23(2), 111, 1982.

100. Huffman, K. R., and Casey, D. J., Effect of Carboxylic end groups on hydrolysis of polyglycolic acid, J. Polym. Sci. Polym. Chem. Ed., 23, 1939-1954, 1985.

101. Hermann, J. B., and Woodward, S. C., An experimental study of wound healing accelerators, Am. Surg., 38, 26-28, 1972.

102. Research Triangle Institute, Report to NICHD, NIH, August 1977.

103. Wade, C. W. R., Hegyeli, A. F., and Kulkarni, R. K., Standards for in-vitro and in-vivo comparison and qualification of bioabsorbable polymers, J. Test. Eval., 5, 397-400, 1977.

104. Ory, S. J., Hammond, C. B., Yancy, S. G., Pitt, C. G.,

and Hendren, R. W., A clinical trial of a biodegradable contraceptive capsule containing levonorgestrel, Am. J. Obstet.
Gynecol., 145, 600-605, 1983.

105. Virutamasen, P., Elder, M., Benagiano, G., Pitt, C. G.,
Gabelnick, H. L., and Hall, P., Pharmacokinetic-pharmacody-
namic clinical trials of a bioerodible delivery system for levon-
orgestrel (Capronor implants), Task Force on Long-Acting
Agents for Fertility Regulation, World Health Organization.
In Press.

106. Pitt, C. G., Andrady, A. L., Bao, Y. T., and Samuel, N. K.
P., Estimation of rate of drug diffusion in polymers, in Con-
trolled-Release Technology, Pharmaceutical Applications, ACS
Symposium Series, Vol. 348 (P. I. Lee and W. R. Good, eds.),
American Chemical Society, Washington, D.C., 1987, pp. 49-
70.

4
Poly(ortho esters)

JORGE HELLER   SRI International, Menlo Park, California

RANDALL V. SPARER and GAYLEN M. ZENTNER   INTER$_X$ Research Corporation, Merck Sharp & Dohme Research Laboratories, Lawrence, Kansas

I. INTRODUCTION

The development of biodegradable polymers capable of releasing physically incorporated therapeutic agents by well-defined kinetics is a subject that has been receiving an increasing amount of attention (1). Major activity in this area has centered on aliphatic polyesters, principally due to the favorable toxicology of their degradation products. However, the dominant mode of drug release from such systems is diffusion with the consequent poor relationship between rate of polymer hydrolysis and rate of drug release. For this reason the development of new polymer systems where drug release is predominantly controlled by polymer hydrolysis is desirable. Further, the development of such polymers is becoming increasingly important due to the increased recognition that bioerodible polymers are an important means for the delivery of polypeptides and that these do not diffuse from dense, hydrophobic polymers at useful rates, particularly as their molecular weights increase.

II. POLYMER SYNTHESIS

One useful drug delivery system is derived from polymers that contain acid-labile linkages in their backbones because hydrolysis rates of such polymers can be readily manipulated by means of acidic or basic excipients physically incorporated into the matrix (2). Further, under certain conditions the hydrolysis of such polymers can be

predominantly confined to the outer surface and the resultant surface erosion allows excellent control of the release kinetics of incorporated therapeutic agents.

The required acid sensitivity can be achieved by preparing polymers with ortho ester linkages in their backbone. The first poly-
(ortho ester) has been described in a series of patents by Choi and
Heller (3-6), assigned to the ALZA corporation. These proprietary
materials were first designated with the trade name Chronomer and
later Alzamer. They are prepared by a transesterification as follows:

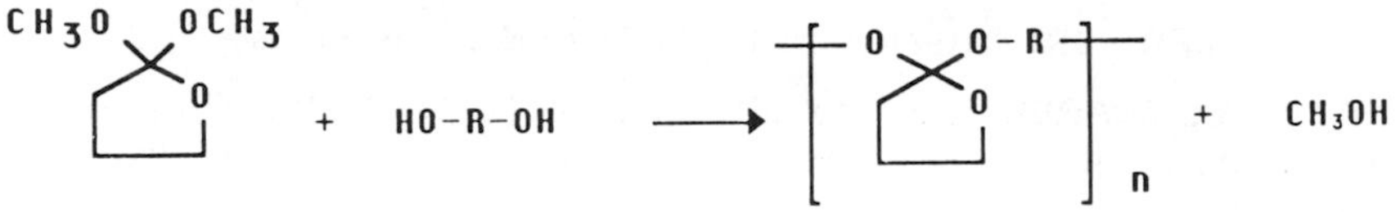

Hydrolysis of these polymers regenerates the diol and produces
γ-butyrolactone, which rapidly hydrolyzes to ω-hydroxybutyric acid.
Because poly(ortho esters) are acid-sensitive, a base is used to neu-
tralize the hydroxybutyric acid and to maintain the hydrolysis process
under control.

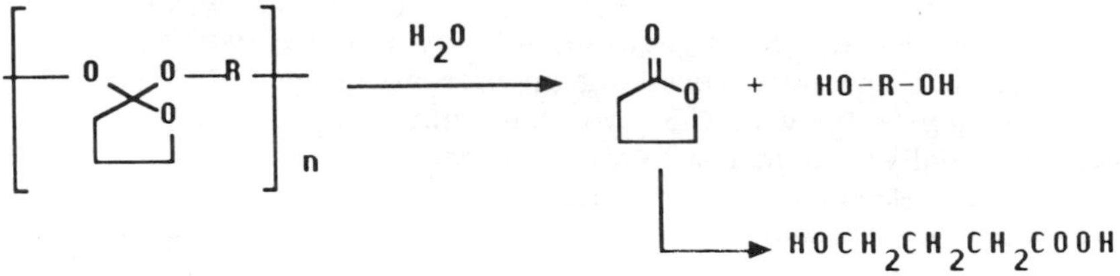

The preparation of such materials requires an ortho ester starting
material in which one alkoxy group has a greatly reduced reactivity
because materials with three reactive alkoxy groups only produce
crosslinked materials. The decreased reactivity of an alkoxy group
can only be achieved with a cyclic structure as shown. These Alzamer materials have been investigated as bioerodible inserts for the
delivery of the narcotic antagonist naltrexone (7) and for the delivery
of the contraceptive steroid norethisterone (8). Human trials of the
steroidal implant revealed local tissue irritation in two separate trials
and further work with this system was discontinued (9,10). The
reasons for the local irritation were never properly elucidated but it
is unlikely that it is due to polymer degradation products.

An entirely different class of poly(ortho esters) has also been
described (11). These polymers are prepared by the addition of
polyols to diketene acetals. Principally due to the relative ease of

synthesis, polymers were prepared by the addition of polyols to a diketene acetal derived from pentaerythritol.

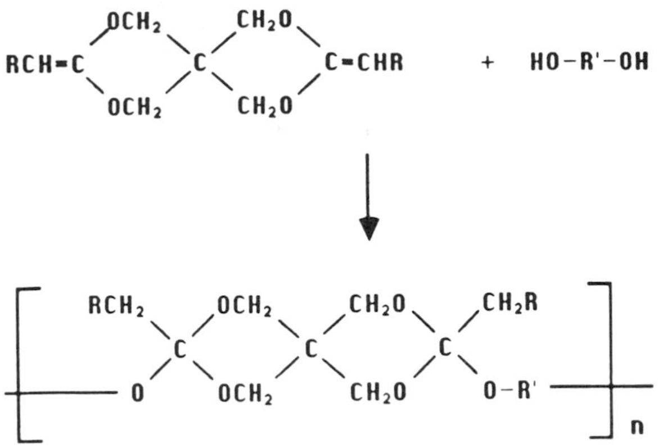

The reaction can proceed spontaneously but when catalyzed by small amounts of acid high molecular weight polymers are formed virtually instantaneously.

Initial work was carried out with 3,9-bis(methylene-2,4,8,10-tetraoxaspiro[5,5] undecane) where R = H (11). However, this monomer contains two electron donor alkoxy groups on one double bond which is thus highly susceptible to a cationic polymerization. For this reason, the monomer is extremely difficult to handle and cannot be analyzed by gas chromatography since it does not survive passage through the column. It is prepared by the dehydrohalogenation reaction of the reaction product of pentaerythritol and chloroacetaldehyde.

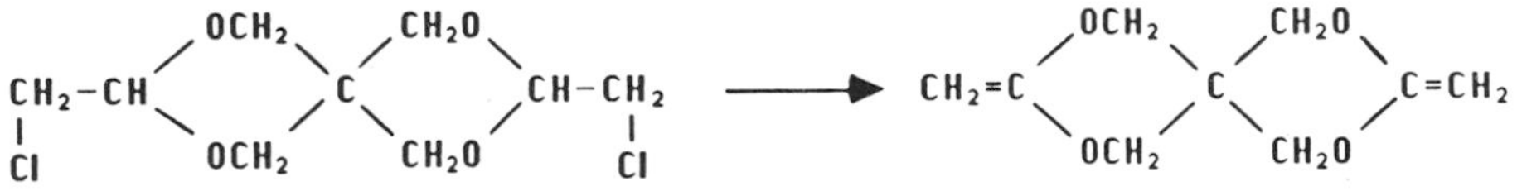

Because this diketene acetal is so susceptible to cationic polymerization, acids cannot be used to catalyze its condensation with diols because the competing cationic polymerization of the diketene acetal double bonds leads to a crosslinked product. Linear polymers can, however, be prepared by using iodine in pyridine (11). Polymer structure was verified by ^{13}C-NMR spectroscopy as shown in Fig. 1.

When one hydrogen in the diketene acetal is replaced with a methyl group, the resulting steric hinderance about the double bond

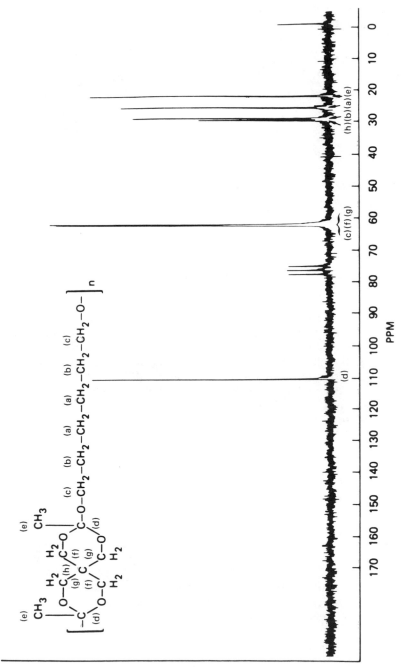

FIGURE 1 25.2 MHz ^{13}C-NMR spectrum of a polymer prepared from 3,9-bis(methylene-2,4,8,10-tetraoxaspiro[5,5]undecane) and 1,6-hexanediol in CDCl$_3$ at room temperature. (From Ref. 11.)

prevents the cationic polymerization without hindering addition of the diol to the double bond. The resulting monomer, 3,9-bis(ethylidene-2,4,8,10-tetraoxaspiro[5,5]undecane), where R = CH$_3$, is relatively stable and can be easily analyzed and purified. It is prepared by the rearrangement of diallyl pentaerythritol (12).

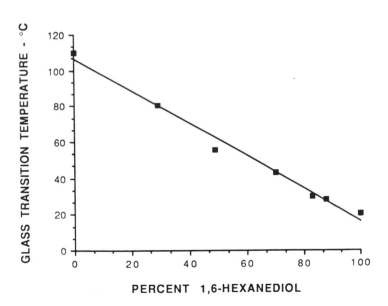

Mechanical properties of these linear polymers can be varied within wide limits by proper selection of the diol or mixture of diols. This is illustrated in Fig. 2 which shows variation in the glass transition temperature of the polymer as a function of diol composition

FIGURE 2   Glass transition temperature of a polymer prepared from 3,9-bis(ethylidene-2,4,8,10-tetraoxaspiro[5,5]undecane) and mixtures of trans-cyclohexane dimethanol and 1,6-hexanediol as a function of mol% 1,6-hexanediol.   (From Ref. 13.)

by using a rigid diol, <u>trans</u>-cyclohexanedimethanol and a flexible diol, 1,6-hexanediol (13).

As in all condensation polymerizations, molecular weight of the polymer is a function of stoichiometry and can be readily varied by appropriate skewing. This is important since an exact equivalence of functional groups can lead to polymers having molecular weights in excess of 100,000, which for some fabrication techniques is too high due to an undesirably high melt viscosity. Figure 3 shows a relationship between polymer molecular weight and stoichiometry (14). In carrying out these experiments it was assumed that both the diketene acetal and the diol were 100% pure even though gas chromatography analysis of the monomers indicated that while the diol was 100% pure, the diketene acetal was only 98.1% pure. The fact that the molecular weight peaks at about 98 mol% of <u>trans</u>-cyclohexanedimethanol verifies that the gas chromatographic analysis of the diketene acetal is accurate.

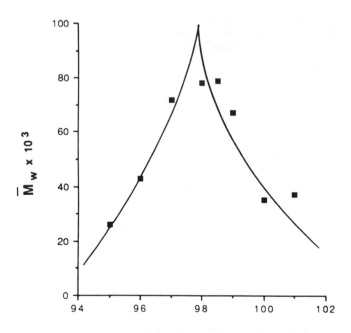

MOLE % trans-CYCLOHEXANEDIMETHANOL

FIGURE 3   Effect of stoichiometry on weight average molecular weight of a polymer prepared from 3,9-bis(ethylidene-2,4,8,10-tetraoxaspiro[5,5]undecane) and <u>trans</u>-cyclohexane dimethanol. (From Ref. 14.)

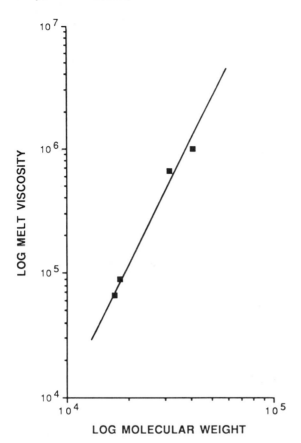

FIGURE 4   Melt viscosity versus weight average molecular weight of
a polymer prepared from 3,9-bis(ethylidene-2,4,8,10-tetraoxaspiro-
[5,5]undecane) and a 70:30 mole ratio of <u>trans</u>-cyclohexane dimethanol and 1,6-hexanediol measured at 150°C.   (From Ref. 14.)

Figure 4 shows a relationship between log (melt viscosity) and
log (weight average molecular weight) as determined by light scattering (14).   As shown, melt viscosity can be readily adjusted by controlling polymer molecular weight.

Because the condensation between a diketene acetal and a diol
proceeds without the evolution of volatile byproducts, this method
allows the preparation of dense, crosslinked materials by using reagents having a functionality greater than 2 (15).   Even though
either or both the ketene acetal and alcohol could have functionali-
ties greater than 2, only triols were investigated because the synthesis of trifunctional ketene acetals is extremely difficult.

To prepare crosslinked material, 2 eq of the diketene acetal is reacted with 1 eq of the diol and the resulting prepolymer is then reacted with a triol or a mixture of diols and triols.

$$CH_3CH=C\begin{smallmatrix}OCH_2\\\\OCH_2\end{smallmatrix}C\begin{smallmatrix}CH_2O\\\\CH_2O\end{smallmatrix}C=CHCH_3 \quad + \quad HO-R-OH$$

$$\downarrow$$

$$CH_3CH=C\begin{smallmatrix}OCH_2\\\\OCH_2\end{smallmatrix}C\begin{smallmatrix}CH_2O\\\\CH_2O\end{smallmatrix}C\begin{smallmatrix}C_2H_5\\\\O-R-O\end{smallmatrix}C\begin{smallmatrix}C_2H_5\\\\OCH_2\end{smallmatrix}C\begin{smallmatrix}OCH_2\\\\OCH_2\end{smallmatrix}C\begin{smallmatrix}CH_2O\\\\CH_2O\end{smallmatrix}C=CHCH_3$$

$$\downarrow R'(OH)_3$$

CROSSLINKED POLYMER

Because the ketene acetal-terminated prepolymer is a viscous liquid at room temperature, therapeutic agents and the triol can be mixed into the prepolymer at room temperature and the mixture crosslinked at temperatures as low as 40°C. This allows incorporation of heat-sensitive therapeutic agents into a solid polymer under very mild conditions of thermal stress. However, because the prepolymer contains reactive ketene acetal groups, any hydroxyl groups present in the therapeutic agent will result in the covalent attachment of the therapeutic agent to the matrix via ortho ester bonds (16).

III. POLYMER HYDROLYSIS

Initial polymer hydrolysis products are the diol or mixture of diols used in the reaction with the diketene acetal, and pentaerythritol dipropionate, or diacetate if 3,9-bis(methylene-2,4,8,10-tetraoxaspiro-[5,5]undecane) was used. These pentaerythritol esters hydrolyze at a slower rate to pentaerythritol and the corresponding aliphatic acid (13).

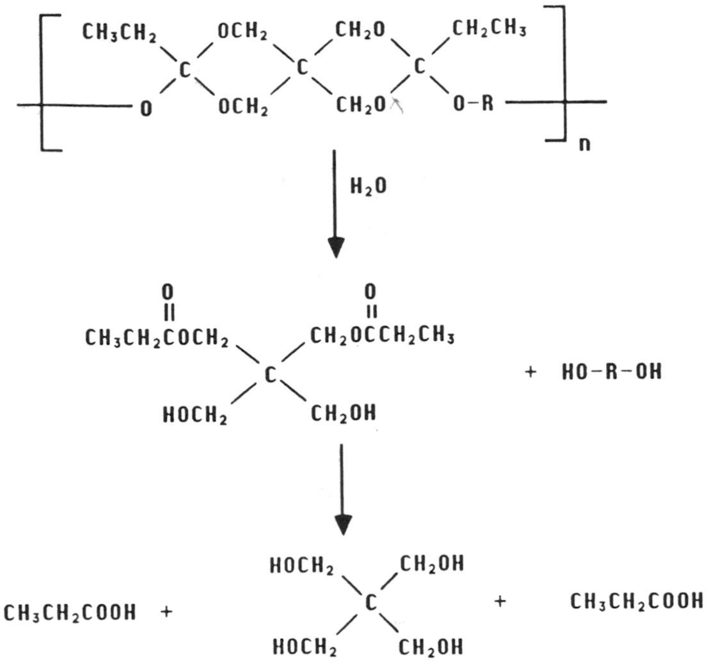

$$\left[\begin{array}{c} CH_3CH_2 \quad OCH_2 \quad CH_2O \quad CH_2CH_3 \\ \backslash C \diagup \quad \backslash C \diagup \quad \backslash C \diagup \\ O \quad OCH_2 \quad CH_2O \quad O\text{-}R \end{array}\right]_n$$

$\downarrow$ H_2O

$$\begin{array}{cc} O & O \\ \| & \| \\ CH_3CH_2COCH_2 \diagdown \quad \diagup CH_2OCCH_2CH_3 \\ C \\ HOCH_2 \diagup \quad \diagdown CH_2OH \end{array} \quad + \quad HO\text{-}R\text{-}OH$$

$\downarrow$

$$CH_3CH_2COOH \quad + \quad \begin{array}{c} HOCH_2 \diagdown \quad \diagup CH_2OH \\ C \\ HOCH_2 \diagup \quad \diagdown CH_2OH \end{array} \quad + \quad CH_3CH_2COOH$$

Details of the hydrolytic process are somewhat more complicated because the acid-catalyzed hydrolysis proceeds via the initial protonation of an alkoxy oxygen followed by bond cleavage. Because the protonation can involve the exocyclic or endocyclic alkoxy group, two different sets of initial products are possible. However, in both cases the ultimate degradation products remain the same. These two possible reaction paths are shown on page 130.

In order to ascertain the relative importance of these two paths, a careful analysis of the hydrolysis products from a linear polymer based on 1,6-hexanediol and 3,9-bis(ethylidene-2,4,8,10-tetraoxaspiro-[5,5]undecane) was carried out (16). Gas chromatographic analysis of the products is shown in Fig. 5. The chromatograph shows two major peaks and a minor peak that can be resolved into two components. The major peaks are 1,6-hexanediol and pentaerythritol dipropionate while the two minor peaks are 1,6-hexanediol monopropionate and pentaerythritol monopropionate. When the peak areas are corrected for unequal detector response, the area corresponding to the monopropionate of 1,6-hexanediol corresponds to about 4.5% of the total area. According to this study, 95.5% of the hydrolysis proceeds via cleavage of the exocyclic alkoxy bond and only 4.5% proceeds via the endocyclic alkoxy bond cleavage.

and

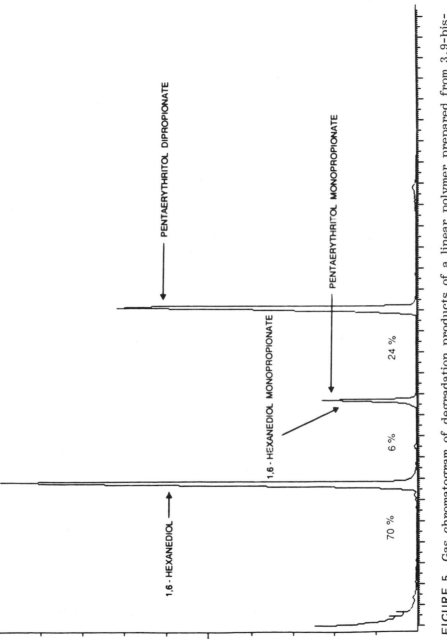

FIGURE 5 Gas chromatogram of degradation products of a linear polymer prepared from 3,9-bis-(ethylidene-2,4,8,10-tetraoxaspiro[5,5]undecane) and 1,6-hexanediol. (From Ref. 16.)

This study is in excellent agreement with hydrolysis studies of 2-methyl-1,3-dioxolane (17).

In these studies cleavage of the endocyclic alkoxy group produces ethylene glycol and since only about 5% was found, it was concluded that hydrolysis proceeds 95% via exocyclic cleavage, in excellent agreement with results found for poly(ortho esters).

IV.   CONTROL OF POLYMER HYDROLYSIS RATE

Because ortho ester linkages are acid-sensitive and stable in base, two fundamentally different methods for achieving control over erosion rate can be used.   In one method an acidic excipient is used to accelerate the rate of hydrolysis while in the other method a base is used to stabilize the interior of the device.

A.   Use of Acidic Excipients

When a hydrophobic polymer with a physically dispersed acidic excipient is placed into an aqueous environment, water will diffuse into the polymer, dissolving the acidic excipient, and consequently the lowered pH will accelerate hydrolysis of the ortho ester bonds.   The process is shown schematically in Fig. 6 (18).   It is clear that the erosional behavior of the device will be determined by the relative movements of the hydration front V_1 and that of the erosion front V_2.   If $V_1 > V_2$, the thickness of the reaction zone will gradually increase and at some point the matrix will be completely permeated with water, thus leading to an eventual bulk erosion process.   On the other hand, if $V_2 = V_1$, a surface erosion process will take place, and the rate of polymer erosion will be completely determined by the rate at which water intrudes into the matrix.

Because devices intended for long-term applications must erode at very slow rates, it is clear from the model shown in Fig. 6 that

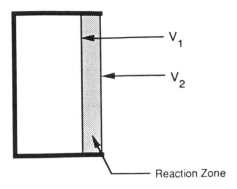

Reaction Zone

FIGURE 6   Schematic representation of water intrusion and erosion for one side of a bioerodible device.   (From Ref. 18.)

it is not possible to maintain long-term surface erosion because even the most hydrophobic polymer will be completely permeated by water in a matter of at most a week (19).   However, useful devices having a lifetime of up to 1 month have been prepared by using latent acid catalysts such as anhydrides (20).   In the presence of water the anhydrides will hydrolyze to the corresponding diacids.   Recent work (21) identified free acids with high potential as erosion catalysts. Physical features including high polymer solubility, low water solubility, and a pK_a of about 3 were critical to achieving the apparently conflicting goals of rapid erosion (12–24 hr duration) and long-term stability (6 months at 25°C).

The rate of polymer erosion in the presence of incorporated anhydride and release of an incorporated drug depends on the pK_a of the diacid formed by hydrolysis of the anhydride and its concentration in the matrix (20).   This dependence is shown in Fig. 7 for 2,3-pyridine dicarboxylic anhydride and for phthalic anhydride.   In this study, methylene blue was used as a marker.   The methylene blue release rate depends both on the pK_a and on the concentration of diacid hydrolysis product in the matrix.   However, at anhydride concentrations greater than 2 wt%, the erosion rate reaches a limiting value and further increases in anhydride concentration have no effect on the rate of polymer hydrolysis.   Presumably at that point V_1, the rate of water intrusion into the matrix, becomes rate limiting.

Convincing evidence for a surface erosion process is shown in Fig. 8, which shows the concomitant release of the incorporated marker, methylene blue, release of the anhydride excipient hydrolysis product, succinic acid, and total weight loss of the device.   According to these data, the release of an incorporated drug from an anhydride-catalyzed erosion of poly(ortho esters) can be unambiguously described by a polymer surface erosion mechanism.

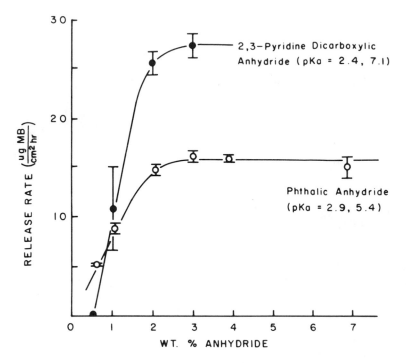

FIGURE 7   Effect of nature and amount of anhydride on methylene
blue release rate from a polymer prepared from 3,9-bis(ethylidene-
2,4,8,10-tetraoxaspiro[5,5]undecane) and a 35:65 mole ratio of trans-
cyclohexane dimethanol and 1,6-hexanediol.   (From Ref. 20.)

Release kinetics are characterized by an initial lag phase, a zero-
order release phase, and a depletion phase (22).   During the lag
phase water beings to intrude into the polymer and activates the
latent catalyst.   During the zero-order release phase, an equilibrium
between water intrusion and polymer erosion has been established
and an eroding front that penetrates the device is formed.   Because
thin discs, typically 8 × 0.8 mm, were used, device geometry remains
essentially unchanged and zero-order drug release is possible.   However, zero-order release can only be achieved if water penetration
or the hydrolysis of the anhydride catalyst or the polymer is the
rate-limiting step.   The depletion phase characterizes a decrease in
device size and depletion of the incorporated acidic excipient.

Surface erosion not only leads to zero-order drug release from
devices that maintain a constant surface area, but has other important consequences.   Among these are the following: (1) the rate of
drug release is directly proportional to drug loading, (2) the lifetime

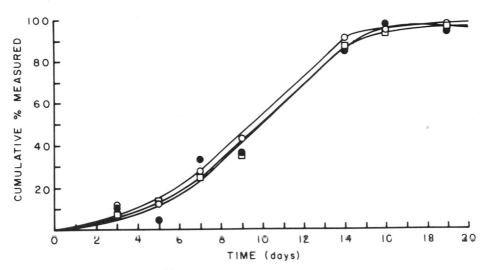

FIGURE 8   Cumulative release of methylene blue (o), [1,4 − [14]C]
succinic acid (□), and polymer weight loss (•) from polymer discs
prepared from 3,9-bis(ethylidene-2,4,8,10-tetraoxaspiro[5,5]undecane)
and a 50:50 mole ratio of <u>trans</u>-cyclohexane dimethanol and 1,6-
hexanediol at pH 7.4 and 37°C.   Polymer contains 0.1 wt% [1,4 −
[14]C]succinic anhydride and 0.3 wt% methylene blue.   (From Ref.
20.)

of the device is directly proportional to device thickness, and (3)
the rate of drug release is directly proportional to surface area.
These expectations have been experimentally verified as shown in
Figs. 9-11 (20).

B.   Use of Copolymerized Acidic Monomers

Because the use of excipients is not always desirable, other means
of accelerating polymer erosion have been explored.   One such approach is to use diols containing pendant carboxylic groups as comonomers in the synthesis of polymers (14).   The original intent of
this approach was to prepare polymers that contain pendant carboxylic acid groups and to rely on the ionization of these groups to
provide an acidic environment that would catalyze polymer erosion.
Indeed, it was found that the incorporation of less than 1 mol% of
9,10-dihydroxystearic acid into the polymer resulted in a significant
acceleration of polymer erosion as evidenced by the release of the
incorporated marker, p-nitroacetanilide (23).   It was further found
that the accelerating effect of 9,10-dihydroxystearic acid could be
greatly enhanced if polymer hydrophilicity is increased by the use

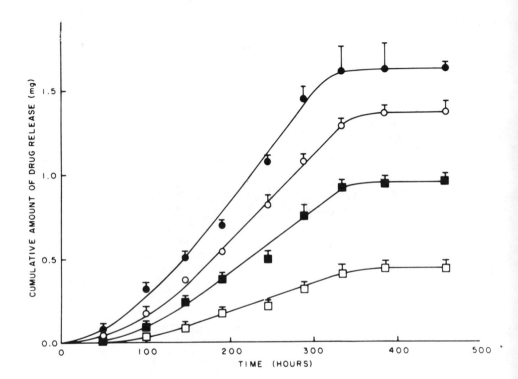

FIGURE 9   Effect of drug loading on cumulative drug release from
polymer discs prepared from 3,9-bis(ethylidene-2,4,8,10-tetraoxaspiro-
[5,5]undecane) and a 50:50 mole ratio of trans-cyclohexane dimethanol
and 1,6-hexanediol at pH 7.4 and 37°C.   Drug loading 8 wt% (o),
6 wt% (•), 4 wt% (■), and 2 wt% (□).   (From Ref. 20.)

of triethylene glycol as one of the monomers. Results of these studies are shown in Figs. 12 and 13.

However, in subsequent work it was found that carboxylic acid groups readily add to ketene acetals to form carboxyortho ester linkages (24). These are very labile linkages and on hydrolysis regenerate the carboxylic acid group which then exerts its catalytic function. Because carboxylic acids add so readily to ketene acetals, very labile polymers can be prepared by the addition of diacids to diketene acetals. The utilization of such polymers is currently under investigation.

C. Use of Basic Excipients

As already mentioned, acidic excipients can only be used to prepare surface-eroding devices that have a relatively short lifetime. If

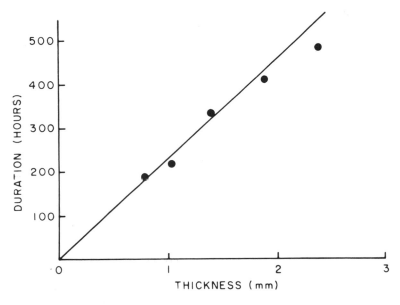

FIGURE 10 Effect of thickness on duration of drug release from polymer discs prepared from 3,9-bis(ethylidene-2,4,8,10-tetraoxaspiro-[5,5]undecane) and a 50:50 mole ratio of trans-cyclohexane dimethanol and 1,6-hexanediol at pH 7.4 and 37°C. Polymer contains 4 wt% drug and 0.2 wt% poly(sebasic anhydride). (From Ref. 20.)

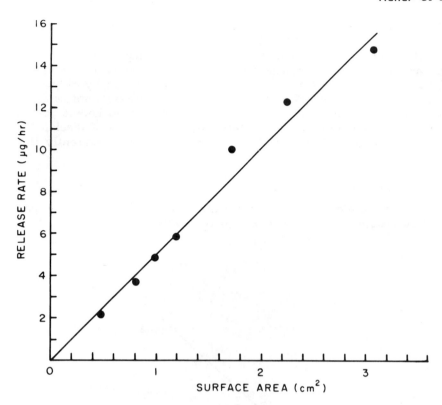

FIGURE 11 Effect of surface area on rate of drug release from polymer discs prepared from 3,9-bis(ethylidene-2,4,8,10-tetraoxaspiro-
[5,5]undecane) and a 50:50 mole ratio of trans-cyclohexane dimethanol
and 1,6-hexanediol at pH 7.4 and 37°C. Polymer contains 4 wt% drug
and 0.2 wt% poly(sebasic anhydride). (From Ref. 20.)

FIGURE 13 Cumulative release of p-nitroacetanilide (PNAC) from
polyester discs prepared from 3,9-bis(ethylidene-2,4,8,10-tetraoxas-
piro[5,5]undecane) and a 60:10:30 mole ratio of trans-cyclohexane
dimethanol, 1,6-hexanediol, and triethylene glycol at pH 7.4 and
37°C. PNAC content 2 wt%; (o) 0 mol%, (•) 0.25 mol%, (■) 0.50 mol%
9,10-dihydroxystearic acid. (From Ref. 23.)

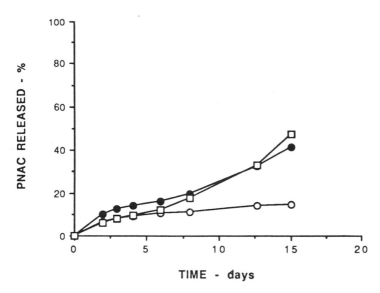

FIGURE 12 Cumulative release of p-nitroacetanilide (PNAC) from
polymer discs prepared from 3,9-bis(ethylidene-2,4,8,10-tetraoxaspiro-
[5,5]undecane) and a 60:40 mole ratio of trans-cyclohexane dimethanol
and 1,6-hexanediol at pH 7.4 and 37°C. PNAC content 2 wt%; (o)
0 mol%, (•) 0.25 mol%, (□) 0.50 mol% 9,10-dihydroxystearic acid.
(From Ref. 23.)

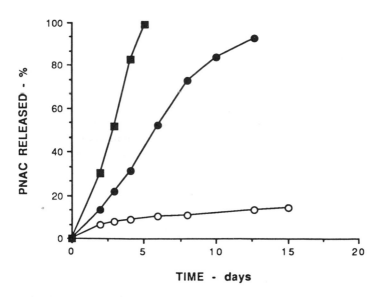

FIGURE 13

FIGURE 14   Schematic representation of erosion using $Mg(OH)_2$ excipient.   (From Ref. 2.)

long-term surface erosion is desired, the interior of the device must be stabilized so that water penetration into the matrix does not lead to hydrolysis.   Because ortho ester linkages are stable in base, such stabilization can be achieved by using basic excipients.

A plausible mechanism for the erosion of devices that contain $Mg(OH)_2$ is shown in Fig. 14 (2).   According to this mechanism, the base stabilizes the interior of the device and erosion can only occur in the surface layers where the base has been eluted or neutralized. This is believed to occur by water intrusion into the matrix and diffusion of the slightly water-soluble basic excipient out of the device where it is neutralized by the external buffer.   Polymer erosion then occurs in the base-depleted layer.

V.   DEVELOPMENT OF SPECIFIC DELIVERY SYSTEMS

A.   Implants

The development of a bioerodible implant capable of releasing controlled amounts of a contraceptive steriod from a subcutaneous implant for periods of time ranging from three months to about a year has been in progress for many years.   The three principal bioerodible polymers currently in use are copolymers of lactic and glycolic acid (25), poly(ε-caprolactone) (26), and poly(ortho esters) (14).   The desire to develop such a contraceptive system was the principal motivation for the initial development of the poly(ortho ester) polymer system.

Initial work with poly(ortho esters) focused on norethindrone and the use of water-soluble excipients such as Na_2CO_3, NaCl, and Na_2SO_4 (27).   As described by Fedors (28), the inclusion of such water-soluble salts leads to an osmotically driven water intake into the polymer.   This water intake leads to polymer swelling with consequent release of the incorporated norethindrone.   The effect of incorporated NaCl and Na_2CO_3 on erosion rate as compared to the

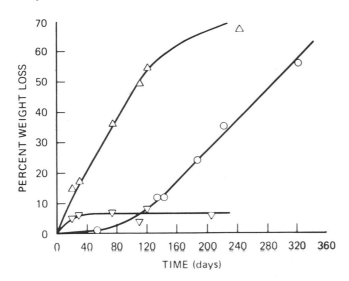

FIGURE 15 Weight loss of polymer discs prepared from 3,9-bis(meth-
ylene-2,4,8,10-tetraoxaspiro[5,5]undecane) and 1,6-hexanediol at
pH 7.4 and 37°C. (o) neat polymer, (∇) polymer with 10 wt% Na$_2$CO$_3$,
(Δ) polymer with 10 wt% NaCl. (From Ref. 27.)

pure polymer is shown in Fig. 15 (27). Because the pure polymer
is highly hydrophobic, it remained essentially unaffected by water
for about 2 months after which weight loss began and proceeded at
a relatively constant rate for about 1 year at which time the experiment was discontinued. Since NaCl is able to osmotically attract water
into the polymer, its incorporation serves to overcome the initial polymer hydrophobicity and thus eliminate the induction period. When
the basic salt Na$_2$CO$_3$ is used, essentially no polymer erosion takes
place because even though water is osmotically imbibed into the polymer, the alkaline environment within the matrix substantially reduces
the rate of hydrolysis of the ortho ester linkages.

When norethindrone is incorporated into the matrix along with a
water-soluble salt, the norethindrone release profiles shown in Fig.
16 were obtained (29). With devices that contain the basic Na$_2$CO$_3$
salt, release of norethindrone is caused by swelling of the device
and diffusion of the drug from the noneroding, swollen device. Because norethindrone is highly water-insoluble, the rate of release is
controlled by the dissolution rate. With devices that contain the
neutral NaCl salt release of norethindrone is controlled both by diffusion and polymer erosion. The nonlinearity of drug release is due
to an increase of the total surface area due to erosion of the polymer
in pockets around the incorporated salt particles and generation of

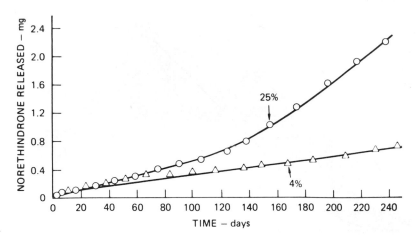

FIGURE 16   Norethindrone release from polymer discs prepared from 3,9-bis(methylene-2,4,8,10-tetraoxaspiro[5,5]undecane) and 1,6-hexanediol at pH 7.4 and 37°C.   Arrows indicate weight loss.   (o) 10 wt% NaCl and 4 mg drug; (Δ) 10 wt% Na₂CO₃ and 4 mg drug. (From Ref. 29.)

an open cell foam.   Formation of such a foam has been verified by scanning electron microscopy (13).

Because swelling and consequent bulk erosion induced by the water-soluble salt is not desirable, use of the low-water-solubility, slightly acidic salt calcium lactate was investigated (30).   By using this excipient it was hoped that a lowering of the pH within the surface layers of the device would take place and release of the drug would be controlled by polymer erosion confined to the surface layers of the device.   In these experiments norethindrone was replaced by the currently favored steroid levonorgestrel.

Although it was possible to achieve constant in vitro release of levonorgestrel for up to 410 days at which point the experiment was discontinued, release of the drug was not controlled by surface erosion of the polymer but instead the device underwent bulk erosion and release of levonorgestrel was completely controlled by its rate of dissolution.   Bulk erosion of the rod-shaped device was evident by scanning electron microscopy as shown in Fig. 17.

At the time these studies were conducted, the role of acidic excipients was not clearly understood, but the observed bulk erosion is of course consistent with the mechanism shown in Fig. 6.   Consequently, if gross bulk erosion is to be avoided and long-term erosion control of levonorgestrel achieved, it is necessary to stabilize the device interior.   To do so, devices with incorporated Mg(OH)₂

FIGURE 17   Scanning electron micrograph (30x) of polymer rods
(2.4 × 20 mm prepared from 3,9-bis(ethylidene-2,4,8,10-tetraoxaspiro-
[5,5]undecane) and a 70:30 mole ratio of <u>trans</u>-cyclohexane dimethanol
and 1,6-hexanediol after 10 weeks in rabbit.   Rods contain 30 wt%
levonorgestrel and 2 wt% calcium lactate.   (From Ref. 30.)

were investigated (15).   Because this salt has a water solubility of
only 0.8 mg/100 ml, osmotic effects should be minimal.

A ketene acetal-terminated prepolymer was first prepared from
2 eq of the diketene acetal 3,9-bis(ethylidene-2,4,8,10-tetraoxaspiro-
[5,5]undecane) and 1 eq of the diol 3-methyl-1,5-pentanediol and
then 30 wt% levonorgestrel, 7 wt% $Mg(OH)_2$, and a 30 mole% excess
of 1,2,6-hexanetriol mixed into the prepolymer.   This mixture was
then extruded into rods and cured.   Erosion and drug release from
these devices was studied by implanting the rod-shaped devices subcutaneously into rabbits, explanting at various time intervals, and
measuring weight loss and residual drug (15).

Weight loss and cumulative levonorgestrel release is shown in
Fig. 18.   Even though the data points, based on single devices,
show considerable scatter, polymer erosion and drug release appear
to occur concomitantly for at least 20 weeks.   Furthermore, scanning
electron microscopy examination of the explanted devices show none
of the voids indicative of bulk erosion that were apparent in the devices containing calcium lactate.   Thus, use of $Mg(OH)_2$ seems to
promote long-term surface erosion.   When these devices were implanted
into rabbits and levonorgestrel blood plasma level determined by radioimmunoassay, results shown in Fig. 19 were obtained (2).

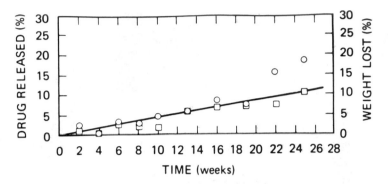

FIGURE 18 In vivo cumulative weight loss (□) and cumulative release of levonorgestrel (o) from a crosslinked polymer prepared from
a 3,9-bis(ethylidene-2,4,8,10-tetraoxaspiro[5,5]undecane)/3-methyl-
1,5-pentanediol prepolymer crosslinked with 1,2,6-hexane triol. Polymer rods, 2.4 × 20 mm, containing 30 wt% levonorgestrel and 7.1
mol% Mg(OH)$_2$. Devices implanted subcutaneously in rabbits. (From
Ref. 15.)

These data were encouraging since the blood plasma level was
reasonably constant once the initial burst subsided. However, the
steady state plasma level was too low and thus a more rapidly eroding polymer was needed. To achieve a more rapid erosion, a material containing 7 wt% Mg(OH)$_2$ and 1 mol% copolymerized 9,10-
dihydroxystearic acid was prepared. The devices were again implanted into rabbits and levonorgestrel blood plasma level determined.
Results of these studies are shown in Fig. 20 (2). Clearly, the
more rapidly eroding polymer produces a much higher drug plasma
level.

The explanted devices were also examined by scanning electron
microscopy and the results shown in Fig. 21 (18). The pictures
clearly show a progressive diminution of a central uneroded zone
and the development of voids around the periphery of the rod-shaped
device. The presence of voids suggest that once erosion starts,
generation of hydrophilic degradation products at that location ac-
celerates further polymer hydrolysis.

B. Oral Delivery Systems

Because acid excipients can be used to achieve rapid polymer erosion, the possibility of preparing devices useful for oral delivery
was investigated (31). In one such system, 2 wt% phthalic anhydride was incorporated into a polymer prepared from the diketene
acetal, trans-cyclohexanedimethanol and ^{14}C-labeled 1,6-hexanediol
and polymer erosion followed in a pH 7 buffer and in pH 1.5 canine

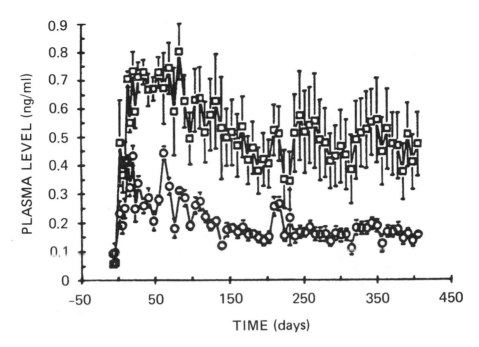

FIGURE 19 Daily rabbit blood plasma levels of levonorgestrel from
a crosslinked polymer prepared from a 3,9-bis(ethylidene-2,4,8,10-
tetraoxaspiro[5,5]undecane)/3-methyl-1,5-pentanediol prepolymer
crosslinked with 1,2,6-hexane triol. Polymer rods, 2.4 × 20 mm,
containing 30 wt% levonorgestrel and 7.1 mol% $Mg(OH)_2$. Devices
implanted subcutaneously in rabbits. (o) 1 device/rabbit, (□) 3
devices/rabbit. (From Ref. 2.)

gastric juice. Results of these studies are shown in Fig. 22. These
data show that within limits of experimental error, erosion rates are
nearly the same.

    In another system, 7.5 wt% cyclobenzaprine bydrochloride was
incorporated into a polymer prepared from 3,9-bis(ethylidene-2,4,8,
10-tetraoxaspiro[5,5]undecane) and an 80:20 mole ratio of 1,6-
hexanediol and trans-cyclohexanedimethanol containing 3 wt% phthalic
anhydride (32). In vitro drug release studies were carried out at
varying pH values. As shown in Fig. 23, release rate is not significantly affected by external pH. Drug release studies were also
carried out in fistulated beagle dogs. In these studies, discs were
tethered in the duodenum and removed at periodic intervals for
assay of residual drug. Results of this study, shown in Fig. 24,
demonstrate that the in vivo erosion was very similar to the in vitro
erosion at pH 7.4.

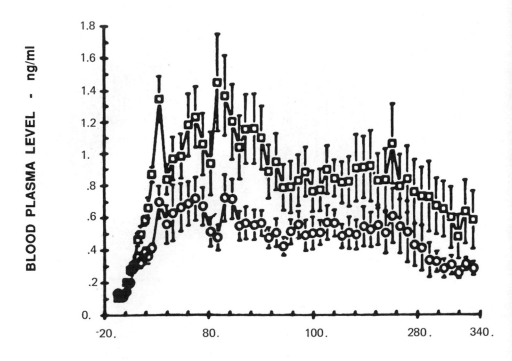

FIGURE 20 Daily rabbit blood plasma levels of levonorgestrel from a crosslinked polymer prepared from a 3,9-bis(ethylidene-2,4,8,10-tetraoxaspiro[5,5]undecane)/3-methyl-1,5-pentanediol prepolymer crosslinked with 1,2,6-hexane triol. Prepolymer contains 1 mol% copolymerized 9,10-dihydroxystearic acid. Polymer rods, 2.4 × 20 mm, containing 30 wt% levonorgestrel and 7.1 mol% Mg(OH)$_2$. Devices implanted subcutaneously in rabbits. (o) 1 device/rabbit, (□) 2 devices/rabbit.

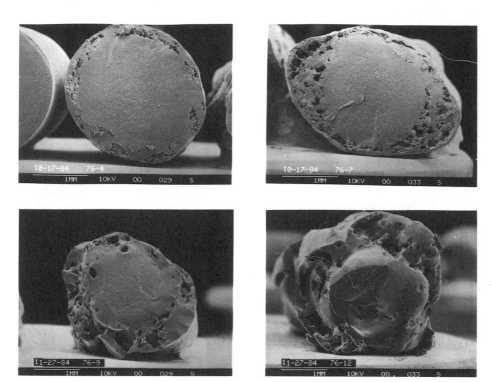

FIGURE 21 Scanning electron micrographs of crosslinked polymer
prepared from a 3,9-bis(ethylidene-2,4,8,10-tetraoxaspiro[5,5]un-
decane)/3-methyl-1,5-pentanediol prepolymer crosslinked with 1,2,6-
hexane triol. Prepolymer contains 1 mol% copolymerized 9,10-dihy-
droxystearic acid. Polymer rods, 2.4 × 20 mm, containing 30 wt%
levonorgestrel and 7.1 mol% $Mg(OH)_2$. Devices implanted subcutaneously in rabbits. (a) after 6 weeks, 30x; (b) after 9 weeks, 30x;
(c) after 12 weeks, 25x; (d) after 16 weeks, 25x. (From Ref. 18.)

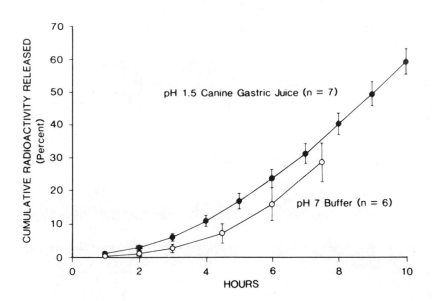

FIGURE 22 Release of radioactivity from polymer discs prepared
from 3,9-bis(ethylidene-2,4,8,10-tetraoxaspiro[5,5]undecane) and a
35:65 mole ratio of <u>trans</u>-cyclohexane dimethanol and [^{14}C]1,6-
hexanediol in phosphate buffer and canine gastric juice at 37°C.
Polymer contained 2.0% phthalic anhydride. (From Ref. 31.)

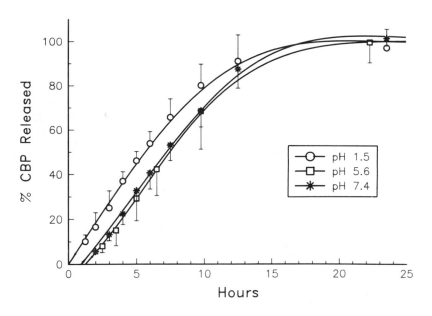

FIGURE 23

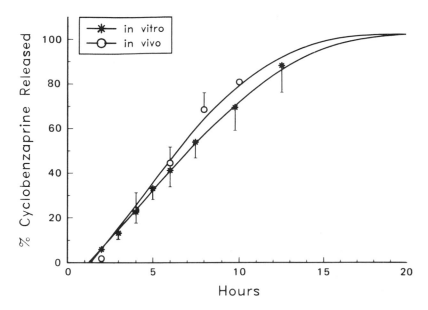

FIGURE 24 Release of cyclobenzaprine hydrochloride (CBP) from discs prepared from 3,9-bis(ethylidene-2,4,8,10-tetraoxaspiro[5,5]-undecane) and a 20:80 mole ratio of <u>trans</u>-cyclohexane dimethanol and 1,6-hexanediol in duodenally fistulated dogs and in a pH 7.4 buffer at 37°C. Polymer contained 3 wt% phthalic anhydride and 7.5 wt% CBP.

FIGURE 23 Release of cyclobenzaprine hydrochloride (CBP) from discs prepared from 3,9-bis(ethylidene-2,4,8,10-tetraoxaspiro[5,5]-undecane) and a 20:80 mole ratio of <u>trans</u>-cyclohexane dimethanol and 1,6-hexanediol at various pH values and 37°C. Polymer contained 3 wt% phthalic anhydride and 7.5 wt% CBP.

VI. POLYMER PROCESSING

Even though poly(ortho esters) contain hydrolytically labile linkages, they are highly hydrophobic materials and for this reason are very stable and can be stored without careful exclusion of moisture. However, the ortho ester linkage in the polymer is inherently thermally unstable and at elevated temperatures is believed to dissociate into an alcohol and a ketene acetal (33). A possible mechanism for the thermal degradation is shown below. This thermal degradation is similar to that observed with polyurethanes (34).

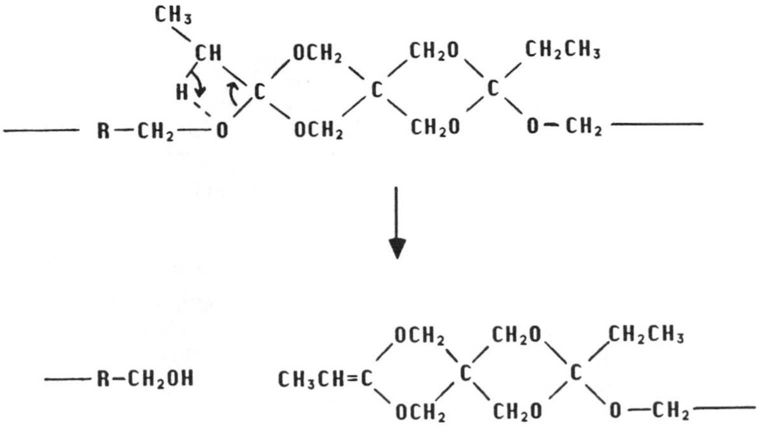

When acidic or latent acidic excipients (anhydrides) are incorporated into the polymer to control erosion rate, the polymers become quite sensitive to moisture and heat and must be processed in a dry environment. A rigorous exclusion of moisture is particularly important with materials that are designed to erode in less than 24 hr. Such materials may contain up to 5 wt% of an acidic catalyst and are analogous to a "loaded gun" in that even the slightest amount of moisture will initiate hydrolysis at the elevated processing temperatures.

Thus, catalyzed poly(ortho esters) must be processed and then packaged in a dry environment. Processing in an environment that has a relative humidity (RH) of less than 20% at 70°F is desirable with lower RH values leading to improved device properties. Prior to fabrication, the polymer, drug, and excipient must be dried and the final device packaged in a dry environment using a high moisture barrier aluminum-laminated strip packaging material. Care in maintaining a dry storage environment can not be overemphasized because significant hydrolysis can occur with catalyzed devices in contact with moisture even at storage temperatures below ambient.

The effect of the relative humidity during processing of poly-(ortho esters) is shown in Table 1. In these experiments a linear polymer based on 3,9-bis(ethylidene-2,4,8,10-tetraoxaspiro[5,5]undecane) and a 65:35 mole ratio of 1,6-hexanediol and trans-cyclohexanedimethanol was compounded with 2 wt% phthalic anhydride and 1 wt% hydrochlorothiazide. After compounding the polymer with the excipients in a high-intensity flux mixer (Brabender) at 175°C for 7 min, the melt was removed and cooled. After the compounding process gel permeation chromatrography indicated that the molecular weight decreased from 41,500 to 39,900. The compounded polymer was then injection molded (Custom Scientific Mini Max injection molder) at either 11 or 29% RH. As shown in Table 1, even though both RH values were quite low, a slightly higher reduction of molecular weight took place at the higher RH. Avoiding a reduction of molecular weight during thermal processing of linear polymers is important because it has been shown (20) that the erosion rate increases with decreasing polymer molecular weights. Therefore, the ability to prepare devices having reproducible behavior demands that moisture during fabrication be rigorously excluded.

Another important consideration in the thermal processing of linear poly(ortho esters) is the possibility of reactions between catalyst/polymer/drug. Because anhydrides are very good acylating

TABLE 1 Processing an Acid-Catalyzed Poly-(ortho ester) Under Different Relative Humidities (RH)

Sample	%RH[a]	Molecular weight[b]
Unprocessed[c]	—	41,500
Compounded[d]	11	39,900
Injection-molded	11	38,200
Injection-molded	29	34,300

[a]Room temperature 74°F.
[b]Determined by GPC.
[c]3,9-Bis(ethylidene-2,4,8,10-tetraoxaspiro[5,5]-undecane),(65:35) 1,6-hexanediol/trans-cyclohexanedimethanol.
[d]Same polymer as in note c but compounded with 2 wt% phthalic anhydride and 1 wt% hydrochlorothiazide.

agents they can acylate hydroxyl and amino groups under the elevated fabrication temperatures. The following study illustrates the
importance of these factors (32). A linear polymer was prepared
from 3,9-bis(ethylidene-2,4,8,10-tetraoxaspiro[5,5]undecane) and a
75:25 mole ratio of 1,6-hexanediol and trans-cyclohexanedimethanol.
The polymer was compounded in a 15% RH environment with 7.5 wt%
cyclobenzaprine hydrochloride (CPB) and 3 wt% phthalic anhydride
(PA) and then injection-molded (Custom Scientific Mini Max injection
molder) at either 130, 145, or 160°C. A control containing no CPB
was injection-molded at 145°C. Results of these studies presented
in Table 2 show mass balance recovery of CPB after processing at
each temperature; however, the recovery of PA decreased markedly
with increasing processing temperature. When samples molded at
145°C with and without CPB are compared, it is clear that addition
of CPB to the polymer markedly enhanced decreases in both polymer
molecular weight and amount of PA. The polymer molecular weight
of the sample injection-molded at 160°C was significantly lower than
that for samples molded at 130 and 145°C.

The activated carbonyl of anhydrides can acylate alcohols or
amines at the temperatures necessary for polymer processing. These
reactions have been verified by HPLC using the polymer system described in Table 2. An examination of the HPLC chromatograms in
Fig. 25 indicates that the phthalic anhydride peak (3.2 min) dimin-
ishes with increasing injection-molding temperatures and that two
new peaks (4.6 and 6.9 min) increase in intensity. These new peaks
corresponded to the half phthalate esters of 1,6-hexanediol and trans-

TABLE 2 Effect of Injection Molding Temperature on 3,9-Bis-
(ethylidene-2,4,8,10-tetraoxaspiro[5,5]undecane), (75:25) 1,6-
Hexanediol/trans-Cyclohexanedimethanol Polymer Compounded with
3 wt% Phthalic Anhydride and 7.5 wt% Cyclobenzaprine Hydrochloride (CBP)

Molding temp. (°C)	Molecular weight[a]	Cyclobenzaprine (wt%)	Phthalic anhydride (wt%)
130	18,500	7.4 ± 0.2	1.17 ± 0.13
145	30,500	None	2.80 ± 0.04
145	17,300	7.5 ± 1.1	1.01 ± 0.05
160	6,500	8.0 ± 0.3	0.02 ± 0.01

[a]Determined by GPC.
Source: Ref. 32.

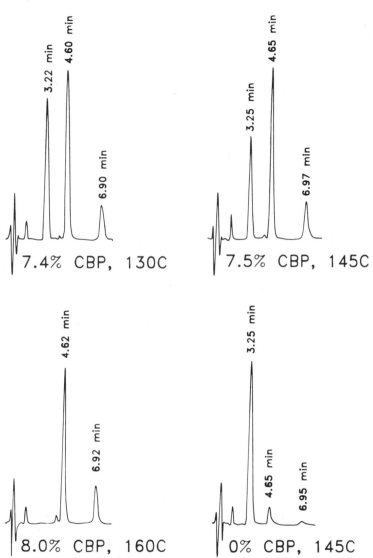

FIGURE 25   HPLC chromatograms of polymer samples hydrolyzed to cleave all ortho ester bonds.   Samples were prepared at either 130, 145, or 160°C.   Polymer prepared from 3,9-bis(ethylidene-2,4,8,10-tetraoxaspiro[5,5]undecane) and a 25:75 mole ratio of <u>trans</u>-cyclohexane dimethanol and 1,6-hexanediol and contained 3 wt% phthalic anhydride and 7.5 wt% cyclobenzaprine hydrochloride (CBP).

cyclohexanedimethanol formed by the reaction between phthalic anhydride and the hydroxyl end groups of the polymer. These catalyst
polymer interactions may be reduced through the incorporation of
free acids rather than anhydrides. The data of Table 3 compare the
recovery of phthalic anhydride and phthalic acid after processing.
The free acid devices were substantially free of acylation byproducts.
Infared spectra shown in Fig. 26 supports the conclusion that as the
injection-molding temperature increased, the phthalic anhydride content (1853, 1789, and 1777 cm^{-1}) decreased and the content of the
corresponding half ester (1735 cm^{-1}) increased.

Figure 27 shows that the release rate of CBP from the polymer
decreased with increasing injection-molding temperatures. Samples
with decreased levels of phthalic anhydride and increased levels of
the phthalate half ester erode at slower rates presumably because
the phthalic acid (pK_{a1} = 2.9 and pK_{a2} = 5.5) is a stronger acid
than its alkyl half ester (estimated pK_a = 3.6) and because the concentration of acidic groups is reduced by the esterification reaction.

A. Processing Methods for Linear Poly(ortho esters)

Linear poly(ortho esters) are processed by the same techniques as
nonerodible thermoplastic polymers (35). Drug, catalyst, and other
excipients are blended during a compounding procedure preceeding
a shaping/molding step. Compounding is typically a high-shear mixing of excipients and drug throughout the viscous polymer melt.
The excipients and drug must be homogeneously mixed to ensure
reproducible polymer erosion and drug release. Since relatively
high temperatures and high shear are attendant to melt mixing, these
conditions may result in unacceptable losses in polymer molecular
weight or chemical interactions between the polymer/excipient/drug.

TABLE 3 Acid Versus Anhydride Reactivity with Matrix[a]

Catalyst	Loading (moles/100 g)	% Recovered after processing[b]
Phthalic acid	0.02	100
Phthalic anhydride	0.02	34[c]

[a]3,9-Bis(ethylidene-2,4,8,10-tetraoxaspiro[5,5]undecane), (75:25)
1,6-hexanediol/trans-cyclohexanedimethanol polymer/cyclobenzaprine
HCl 7.5 wt%.
[b]Melt mix/injection mold.
[c]Hemiesters of diols formed.

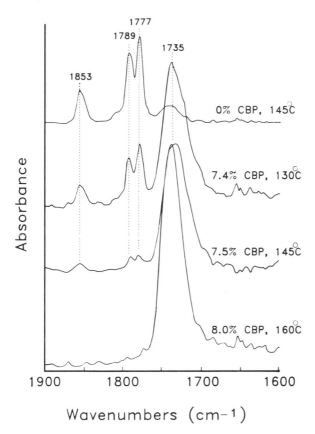

FIGURE 26 Fourier transform infrared spectroscopy of polymer
samples prepared at either 130, 145, or 160°C with or without cyclobenzaprine hydrochloride (CBP). Polymer prepared from 3,9-bis-
(ethylidene-2,4,8,10-tetraoxaspiro[5,5]undecane) and a 25:75 mole
ratio of trans-cyclohexane dimethanol and 1,6-hexanediol and contained 3 wt% phthalic anhydride and 7.5 wt% cyclobenzaprine hydrochloride (CBP).

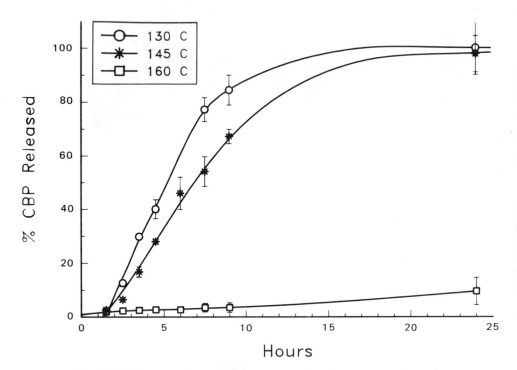

FIGURE 27   Release of cyclobenzaprine hydrochloride (CBP) at 37°C
and pH 7.4 from devices made at different injection-molding temperatures.   Polymer prepared from 3,9-bis(ethylidene-2,4,8,10-tetra-
oxaspiro[5,5]undecane) and a 25:75 mole ratio of trans-cyclohexane
dimethanol and 1,6-hexanediol and contained 3 wt% phthalic anhydride
and 7.5 wt% cyclobenzaprine hydrochloride (CBP).

    Alternatives to compounding in the melt are solution mixing or
powder blending of solid particles.   Mixing with the aid of solvents
can be performed at lower temperatures with minimal shear.   However,
difficulties in removal of the solvent results in plasticization of the
polymer matrix and altered erosion/drug release performance in addition to residual solvent toxicity concerns.   Powder blending at room
temperature minimizes thermal/shear stresses, but achieving intimate
mixtures is difficult.
    Stearic acid (5 wt%) was compounded with a linear polymer pre-
pared from 3,9-bis(ethylidene-2,4,8,10-tetraoxaspiro[5,5]undecane)
and a 30:70 mole ratio of 1,6-hexanediol and trans-cyclohexanedimethanol by thermal, solvent, or powder methods (Table 4).   The
thermal method (flux mixer and roller mill) resulted in good stearic

TABLE 4 Comparing the Compounding of Poly(ortho ester)[a] with 5 wt% Stearic Acid Using Different Compounding Methods

	Compounding method	Molecular weight after compounding[b]	Wt% stearic acid after compounding
Melt-mix	Brabender high-intensity flux mixer[c]	63,800 ± 900	4.3 ± 0.1
	Reliable roller mill[d]	70,800 ± 2600	4.7 ± 0.2
Solvent mix	Solvent casting[e]	80,200 ± 2800	4.5 ± 0.1
	Lyophilization[f]	95,500 ± 5400	4.7 ± 0.1
Powder mix	Dry powder blending[g]	99,600 ± 4000	4.4 ± 0.3

[a]3,9-Bis(ethylidene-2,4,8,10-tetraoxaspiro[5,5]undecane) (30:70) 1,6-hexanediol/trans-cyclohexanedimethanol polymer.
[b]By gel permeation chromatography. Molecular weight prior to compounding was 97,100 ± 3800.
[c]15 min at 130°C.
[d]30 min at 180°C.
[e]Dissolve to 10 wt% in methylene chloride; strip off the solvent in a rotary evaporator; place under vacuum at room temperature for 24 hr.
[f]Dissolve to 12.5 wt% in dry dioxane; freeze in dry ice/acetone, then place under vacuum for 24 hr.
[g]Blend powders at room temperature in a tumbling polyethylene jar for 1 hr.

acid content uniformity and the greatest losses in polymer molecular weight (approximately 30%) attributable to the high shear/temperature conditions. The solvent method (casting from methylene chloride and lyophilization from dioxane) resulted in good stearic acid uniformity with marked improvement in polymer molecular weight (not statistically different from initial). However, a residual solvent level of about 1 wt% accompanied these processes that could not be lowered further without prolonged vacuum treatment (several days). Powder blending also conserved polymer molecular weight; however, the content uniformity of stearic acid was lower than with the other methods.

The shaping/molding operation typically involves melting the compounded mixture and either molding in a cavity (injection or compression molding) or extrusion. All of these methods involve heat and shear, which may further degrade molecular weight. Data of Table 5 indicate that each of these methods resulted in decreases in polymer molecular weight with free stearic acid content also reduced by injection molding and extrusion. No loss of stearic acid was observed in compression-molded samples, presumably due to the less stressful conditions of heat and shear. The losses in stearic acid were due to esterification with hydroxyl polymer end groups. Clearly, processing methods that minimize heat/shear stress are preferred.

TABLE 5 Comparing the Shaping of a Poly(ortho ester)[a] Compound by Flux-Mixing with 5 wt% Stearic Acid

Shaping method	Molecular weight		Wt% stearic acid	
	Before	After	Before	After
Injection molding[b]	68,000 ± 500	59,300 ± 1300	4.8 ± 0.2	4.3 ± 0.1
Compression molding[c]	56,000 ± 1000	41,400 ± 900	4.5 ± 0.1	4.4 ± 0.1
Extrusion[d]	56,000 ± 1000	35,700 ± 2300	4.5 ± 0.1	3.7 ± 0.1

[a]3,9-Bis(ethylidene-2,4,8,10-tetraoxaspiro[5,5]undecane) (30:70) 1,6-hexanediol/trans-cyclohexanedimethanol polymer.
[b]140°C, 31 tons for 13 sec; Arburg Allrounder 250 hydraulic injection molder.
[c]80°C, 7 tons for 2 min. Quench to ambient temperature within 3 min; Carver Lab Press Model C (hydraulic drive).
[d]135°C screw, 150°C die, 20 rpm; Killion single-screw extruder.

REFERENCES

1. Heller, J., Biodegradable polymers in controlled drug delivery, CRC Crit. Rev. Ther. Drug Carrier Syst., 1, 39-90, 1984.
2. Heller, J., Control of polymer surface erosion by the use of excipients, in Polymers in Medicine II (E. Chielini, P. C. Migliaresi, Giusti, and L. Nicolais, eds.), Plenum Press, New York, 1986, pp. 357-368.
3. Choi, N. S., and Heller, J., Polycarbonates, U.S. Patent 4,079,038, March 14, 1978.
4. Choi, N. S., and Heller, J., Drug delivery devices manufactured from polyorthoesters and polyorthocarbonates, U.S. Patent 4,093,709, June 6, 1978.
5. Choi, N. S., and Heller, J., Structured orthoesters and ortho-carbonate drug delivery devices, U.S. Patent 4,131,648, December 26, 1978.
6. Choi, N. S., and Heller, J., Erodible agent releasing device comprising poly(ortho esters) and poly(ortho carbonates), U.S. Patent 4,138,344, February 6, 1979.
7. Capozza, R. C., Sendelbeck, L., and Balkenhol, W. J., Preparation and evaluation of a bioerodible naltrexone delivery system, in Polymeric Delivery Systems (R. J. Kostelnik, ed.), Gordon and Breach, New York, 1978, pp. 59-73.
8. Pharriss, B. B., Place, V. A., Sendelbeck, L., and Schmitt, E. E., Steroid systems for contraception, J. Reprod. Med., 17, 91-97, 1976.
9. World Health Organization 10th Annual Report, 1981, pp. 62-63.
10. World Health Organization 11th Annual Report, 1982, p. 61.
11. Heller, J., Penhale, D. W. H., and Helwing, R. F., Preparation of poly(ortho esters) by the reaction of ketene acetals and poly-ols, J. Polym. Sci., Polym. Lett. Ed., 18, 82-83, 1980.
12. Ng, S. Y., Penhale, D. W. H., and Heller, J., Preparation of poly(ortho esters) by the reaction of diketene acetals and al-cohols, Macromol. Synth. In Press.
13. Heller, J., Penhale, D. W. H., Fritzinger, B. K., Rose, J. E., and Helwing, R. F., Controlled release of contraceptive steroids from biodegradable poly(ortho esters), Contracept. Deliv. Syst., 4, 43-53, 1983.
14. Heller, J., Penhale, D. W. H., Fritzinger, B. K., and Ng, S. Y., Controlled release of contraceptive agents from poly(ortho ester), in Long Acting Contraceptive Delivery Systems (G. I. Zatuchni, A. Goldsmith, J. D. Shelton, and J. Sciarra, eds.), Harper and Row, Philadelphia, 1984, pp. 113-128.
15. Heller, J., Fritzinger, B. K., Ng, S. Y., and Penhale, D. W. H., In vitro and in vivo release of levonorgestrel from poly-(ortho esters). II. Crosslinked polymers, J. Control. Rel., 1, 233-238, 1985.

16. Heller, J., Ng, S. Y., Penhale, D. W. H., Fritzinger, B. K., Sanders, L. M., Burns, R. A., Gaynon, M. G., and Bhosale, S. S., The use of poly(ortho esters) for the controlled release of 5-fluorouracyl and a LHRH analogue, J. Control. Rel., 6, 217-224, 1987.

17. Chiang, Y., Krege, A. J., Salomaa, P., and Young, C. I., Effect of phenyl substitution on ortho ester hydrolysis, J. Am. Chem. Soc., 96, 4494-4499, 1974.

18. Heller, J., Controlled drug release from poly(ortho esters): A surface eroding polymer, J. Control. Rel., 2, 167-177, 1985.

19. Nguyen, T. H., Himmelstein, K. J., and Higuchi, T., Some equilibrium and kinetic aspects of water sorption in poly(ortho esters), Int. J. Pharm., 25, 1-12, 1985.

20. Sparer, R. V., Chung, S., Ringeisen, C. D., and Himmelstein, K. J., Controlled release from erodible poly(ortho ester) drug delivery systems, J. Control. Rel., 1, 23-32, 1984.

21. Zentner, G. M., Pogany, S. A., Sparer, R. V., Shih, C., and Kaul, F., The design, fabrication and performance of acid catalyzed poly(ortho ester) erodible devices, Abstracts of the Third Annual Meeting, American Association of Pharmaceutical Scientists, Orlando, FL, 1988.

22. Shih, C., Higuchi, T., and Himmelstein, K. J., Drug delivery from catalyzed erodible polymeric matrices of poly(ortho esters), Biomaterials, 5, 237-240, 1984.

23. Heller, J., Penhale, D. W. H., Fritzinger, B. K., and Ng, S. Y., The effect of copolymerized 9,10-dihydroxystearic acid on erosion rates of poly(ortho esters), J. Control. Rel., 5, 173-177, 1987.

24. Heller, J., Ng, S. Y., and Penhale, D. W. H., Preparation of poly(carboxy-ortho esters) by the reaction of diketene acetals and carboxylic acids. In Preparation.

25. Beck, L. R., and Tice, T. R., Poly(lactic acid) and poly(lactic-co-glycolic acid) contraceptive delivery system, in Long Acting Steroidal Contraception (D. R. Mishell, ed.), Raven Press, New York, 1983, pp. 175-199.

26. Ory, S. J., Hammond, C. B., Yancy, S. G., Hendren, R. W., and Pitt, C. G., The effect of a biodegradable contraceptive capsule (Capronor) containing levonorgestrel on gonadotropin, estrogen, and progesterone levels, Am. J. Obstet. Gynecol., 145, 600-605, 1983.

27. Heller, J., Penhale, D. W. H., Helwing, R. F., and Fritzinger, B. K., Release of norethindrone from poly(ortho esters), Polym. Eng. Sci., 21, 727-731, 1981.

28. Fedors, J., Osmotic effects. 1. Water absorption by polymers, Polymer, 21, 207-212, 1980.

29. Heller, J., and Himmelstein, K. J., Biodegradable poly(ortho esters) as drug delivery forms, in Directed Drug Delivery (R. T. Borchard, A. J. Repta, and V. J. Stella, eds.), Humana Press, NJ, 1985, pp. 171-188.

30. Heller, J., Fritzinger, B. K., Ng, S. Y., and Penhale, D. W. H., In vitro and in vivo release of levonorgestrel from poly-(ortho esters). I. Linear polymers, J. Control. Rel., 1, 225-232, 1985.

31. Fix, J. A., and Leppert, P. S., Erodible poly(ortho ester) drug delivery systems: In vitro erosion in canine and USP-stimulated gastric juice, J. Control. Rel., 4, 87-95, 1986.

32. Sparer, R. V., Physico-chemical characterization of phthalic anhydride catalyzed poly(ortho ester)/cyclobenzaprine devices, J. Control. Rel. In Press.

33. Heller, J., Penhale, D. W. H., Fritzinger, B. K., and Rose, J. E., Synthesis and release of contraceptive steroids from bio-erodible poly(ortho esters), in Polymers in Medicine (E. Chielini and P. Giusti, eds.), Plenum Publishing, New York, 1983, pp. 169-178.

34. Fabris, J. H., Thermal and oxidative stability of urethanes, in Advances in Urethane Science and Technology, Vol. 6 (K. C. Frisch and S. L. Reegen, eds.), Technomic Publishing, Westport, CT, 1978, pp. 173-196.

35. Tadmor, Z., and Gogos, C. G., Principles of Polymer Processing, John Wiley and Sons, New York, 1979.

5
Polyphosphazenes as New Biomedical and Bioactive Materials

HARRY R. ALLCOCK   The Pennsylvania State University, University Park, Pennsylvania

I. INTRODUCTION

Polymers used in medicine fall into two main categories: those that are sufficiently inert to fulfill a long-term structural function as biomaterials or membranes, and those that are sufficiently hydrolytically unstable to function as bioerodible materials, either in the form of sutures or as absorbable matrices for the controlled release of drugs. For the synthetic organic polymers widely used in biomedicine this often translates to a distinction between polymers that have a completely hydrocarbon backbone and those that have sites in the backbone that are hydrolytically sensitive. Ester, anhydride, amide, or urethane linkages in the backbone usually serve this function.

The purpose of this chapter is to introduce a new class of polymers for both types of biomedical uses: a polymer system in which the hydrolytic stability or instability is determined not by changes in the backbone structure, but by changes in the side groups attached to an unconventional macromolecular backbone. These polymers are polyphosphazenes, with the general molecular structure shown in structure 1.

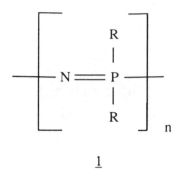

<u>1</u>

All polyphosphazenes contain a long-chain backbone of alternat-
ing phosphorus and nitrogen atoms, with two side groups attached
to each phosphorus. Typically, the chains may be 15,000 or more
repeating units long, which often corresponds to molecular weights
in the 2×10^5 to 5×10^6 region. As will be discussed, the synthesis routes used for the preparation of these polymers allow a wide
range of different side groups R to be attached to the backbone.
These range from hydrophobic groups that confer water insolubility
and protect the backbone against hydrolysis, through groups that
generate water solubility together with hydrolytic stability, to side
groups that provide a facile pathway for hydrolytic breakdown of
the polymer to innocuous, excretable, or metabolizable small molecules.
Several hundred different polyphosphazenes have been synthesized,
each with the same backbone but different side groups or combinations of side groups. The rest of this chapter will be devoted to a
review of this synthesis approach and the way in which specific side
groups generate different biomedical characteristics.

II. THE SYNTHESIS METHOD

Molecular structural changes in polyphosphazenes are achieved mainly
by macromolecular substitution reactions rather than by variations
in monomer types or monomer ratios (1-4). The method makes use
of a reactive macromolecular intermediate, poly(dichlorophosphazene)
structure (3), that allows the facile replacement of chloro side groups
by reactions of this macromolecule with a wide range of chemical reagents. The overall pathway is summarized in Scheme I.

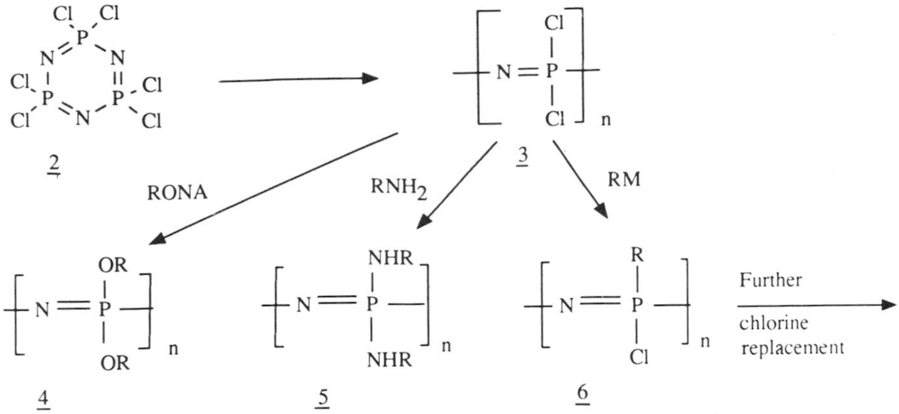

SCHEME I

Compound $\underline{2}$ is a white, crystalline, commercially available start-ing material, itself prepared from phosphorus pentachloride and am-monium chloride. Heating of compound $\underline{2}$ in the melt at 250°C brings about its conversion to the linear high polymer, compound $\underline{3}$. In the solid state, compound $\underline{3}$ is a rubbery elastomer that is sensitive to reaction with atmospheric moisture to give hydrochloric acid, am-monia, and phosphate. However, the principal limitation of this poly-mer is also its principal attribute. The reactivity of the P–Cl bonds in compound $\underline{3}$ is so high that, in solution in tetrahydrofuran or aromatic hydrocarbons, it reacts rapidly with nucleophiles such as alkoxides, aryloxides, amines, or organometallic reagents to give completely substituted products of structure $\underline{4}$ or $\underline{5}$. Mixed sub-stituent polymers can also be prepared by the sequential or simul-taneous reactions of compound $\underline{3}$ with two or more different reagents. Hence, the variations in side group arrangements are extremely broad. The properties of the final polymer depend critically on the type of side group introduced and on the disposition of different side groups. For example, single-substituent polymers are often microcrystalline materials. Random mixed-substituent polymers can be elastomers because of the lack of molecular symmetry and the lowered tendency to form ordered structures.

A variant of this method is used for the introduction of alkyl or aryl side groups into the polymer, and this is illustrated in Scheme II (5-8).

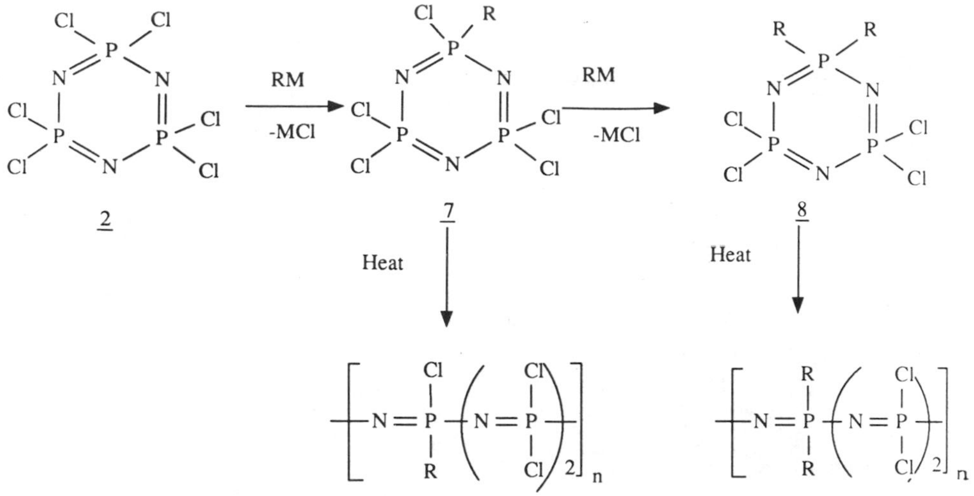

SCHEME II

Side groups such as alkyl, aryl, or organometallic units are easier to attach to the phosphazene skeleton at the cyclic trimer stage than at the macromolecular level. Thus, species such as compound 7 or 8 are readily prepared by the reaction of compound 2 (or its fluoro analog) with organolithium, organocopper, or Grignard reagents. Thermal polymerization to 9 or 10 followed by replacement of the chlorine atoms by alkoxy, aryloxy, or amino side units, yields a large number of additional polymer structures. Alkyl- or arylphosphazene polymers and their derivatives can also be prepared by a condensation reaction that starts from phosphinimines (9). Use of these synthesis methods yields polymers with a broad range of biomedically interesting properties. In the following sections the consequences of side group changes will be discussed.

III. SIDE GROUPS FOR BIOENERTNESS

The term "bioenertness" is a relative one since few if any synthetic polymers are totally biocompatible with living tissues. The term is used here on the basis of preliminary in vitro and in vivo tests, together with chemical evaluations based on analogies with other well-tested systems. Two different types of polyphosphazenes are of interest as bioinert materials: those with strongly hydrophobic surface characteristics and those with hydrophilic surfaces. These will be considered in turn.

A. Hydrophobic Side Groups

Some of the most hydrophobic synthetic polymers known are poly-phosphazenes that bear fluoroalkoxy side groups (1,2,10-12). Examples are shown in structures 11 and 12.

$$\left[N = P \begin{matrix} OCH_2 CF_3 \\ | \\ | \\ OCH_2 CF_3 \end{matrix} \right]_n \qquad \left[N = P \begin{matrix} OCH_2 CF_3 \\ | \\ | \\ OCH_2 (CF_2)_x CF_3 \end{matrix} \right]_n$$

11   12

Such polymers are as hydrophobic as poly(tetrafluoroethylene) (Teflon), with water contact angles in the range of 107°. But, unlike Teflon, polyphosphazenes of this type are easy to fabricate, are flexible or elastomeric, and can be used as coatings for other materials.

Subcutaneous in vivo testing of these polymers (13,14) has shown minimal tissue response—similar, in fact, to the response to poly(tetrafluoroethylene). These materials are candidates for use in heart valves, heart pumps, blood vessel prostheses, or as coating materials for pacemakers or other implantable devices.

Aryloxyphosphazene polymers, such as compound 13 or its mixed-substituent analogs, are also hydrophobic (contact angles in the region of 100°). These too show promise as inert biomaterials on the basis of preliminary in vivo tissue compatibility tests (13).

$$\left[N = P \begin{matrix} O\,C_6 H_5 \\ | \\ | \\ O\,C_6 H_5 \end{matrix} \right]_n \qquad \left[N = P \begin{matrix} CH_2 Si(CH_3)_3 \\ | \\ | \\ O\,CH_2 CF_3 \end{matrix} \right]_n$$

13   14

Hybrid organosilicon-organophosphazene polymers have also been synthesized (15-18) (structure 14) (the organosilicon groups were introduced via the chemistry shown in Scheme II). These are elastomers with surface contact angles in the region of 106°. Although no biocompatibility tests have been conducted on these polymers, the molecular structure and material properties would be expected to be similar to or an improvement over those of polysiloxane (silicone) polymers.

The connection between hydrophobicity and tissue compatibility has been noted for classical organic polymers (19). A key feature of the polyphosphazene substitutive synthesis method is the ease with which the surface hydrophobicity or hydrophilicity can be fine-tuned by variations in the ratios of two or more different side groups. It can also be varied by chemical reactions carried out on the organophosphazene polymer molecules themselves or on the surfaces of the solid materials.

B. Hydrophilic, Possibly Inert Polyphosphazenes

Intuitively it might be supposed that a polymer with a hydrophilic surface might be more biocompatible than one with a ("foreign") hydrophobic surface. This simple interpretation is not always correct. Nevertheless, the study of hydrophilic polymers, especially those that promote endothelial tissue overgrowth or ingrowth, is an area of great interest.

Several different classes of polyphosphazenes have hydrophilic surfaces. These include polymers with short-chain alkylamino side groups (3) ($-NHCH_3$, $-NHC_2H_5$, $-NHC_3H_7$, etc.), those with alkyl ether side units (20) ($OCH_2CH_2OCH_3$ and longer chain analogs), side groups that bear OH or COOH functional groups, and a varied class of derivatives that bear glyceryl or glucosyl side groups. Some of these polymers are so hydrophilic that they will dissolve in water. However, a judicious choice of cosubstituents, or the crosslinking of chains, can prevent dissolution. These materials will be considered in detail later under the heading of membranes and hydrogels.

C. Insoluble But Surface-Active Polymers

Biomaterials cover a broad range of properties from those that are designed to be inert to those that are intended to elicit a particular set of biological responses. In the latter category are materials with surfaces tailored to retard blood clotting, or surfaces that bear covalently bound bioactive agents such as enzymes or antigens. Four different types of surface-active polyphosphazenes have been synthesized. The biological responses have been studied in three of these.

First, at the simplest chemical level, aryloxyphosphazene polymers have been surface-sulfonated, as illustrated in Fig. 1 (21). This reaction results in a decrease in the contact angle from 101° for the surface of compound 15 to 0° for the surface of compound 16.

Second, bromination of methyl groups attached to arylphenoxyphosphazene polymers converted them to CH_2Br units (22). These were then quaternized with triethylamine, and the quaternary sites were used for anion exchange with sodium heparin (Fig. 2). The

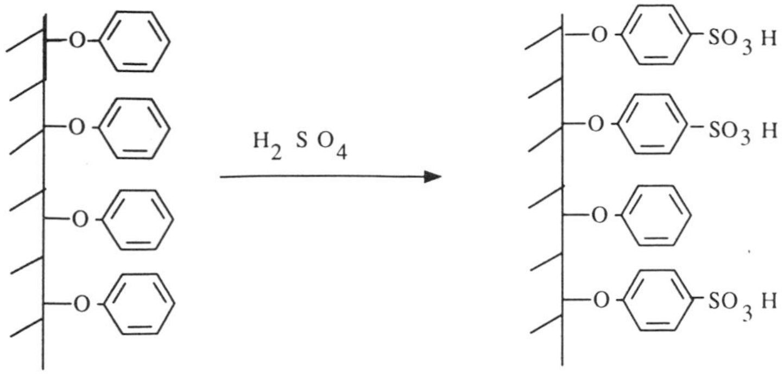

1 5   1 6

(From compound 13)

FIGURE 1   Sulfonation of the surface of poly(diphenoxyphosphazene).

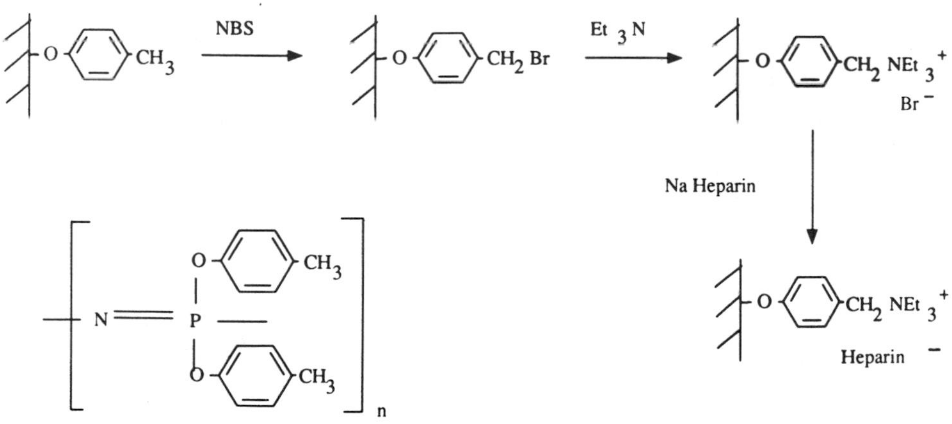

FIGURE 2   Bromination, quaternization, and binding of heparin to
the surface of poly[bis-(p-methylphenoxy)phosphazene].

heparinized surfaces lengthened the clotting time for bovine blood from 12 to 63 min (22). In these two examples, the surface chemistry is not unique to polyphosphazenes, but the ability to carry out such reactions on flexible polymers, and without extensive backbone cleavage, is a significant advantage.

The next two examples illustrate more complex surface reaction chemistry that brings about the covalent immobilization of bioactive species such as enzymes and catecholamines. Poly[bis(phenoxy)-phosphazene] (compound 15) can be used to coat particles of porous alumina with a high-surface-area film of the polymer (23). A scanning electron micrograph of the surface of a coated particle is shown in Fig. 3. The polymer surface is then nitrated and the arylnitro groups reduced to arylamino units. These then provided reactive sites for the immobilization of enzymes, as shown in Scheme III.

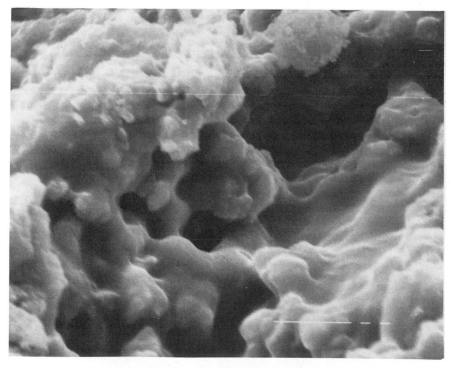

FIGURE 3 Scanning electron micrograph (1200x magnification) of the surface of a porous alumina particle coated with poly(diphenoxy-phosphazene). Surface nitration, reduction, and glutaric dialdehyde coupling immobilized enzyme molecules to the surface. (From Ref. 23.)

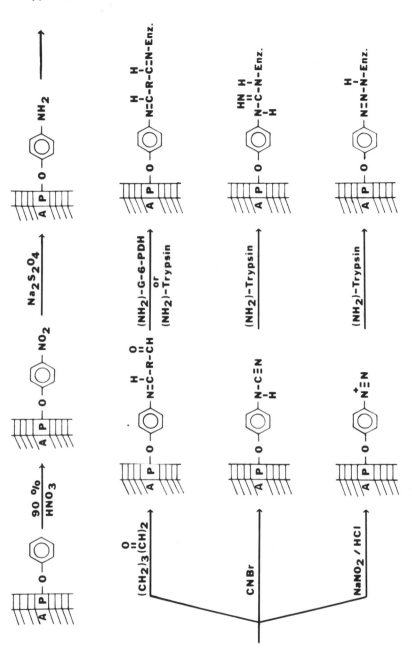

SCHEME III   A = Alumina, P = Polyphosphazene, R = $(CH_2)_3$

The use of glutaric dialdehyde as a coupling agent bound the enzymes trypsin or glucose-6-phosphate dehydrogenase to the surface. A large part of the enzymic activity was retained (Fig. 4), and the activity was such that the particle-enzyme conjugate could be used in laboratory scale continuous-flow reactors.

The final example involves the immobilization of catecholamines, such as dopamine, to a polyphosphazene surface (24). The chemistry is shown in Scheme IV. In this case, conversion of surface

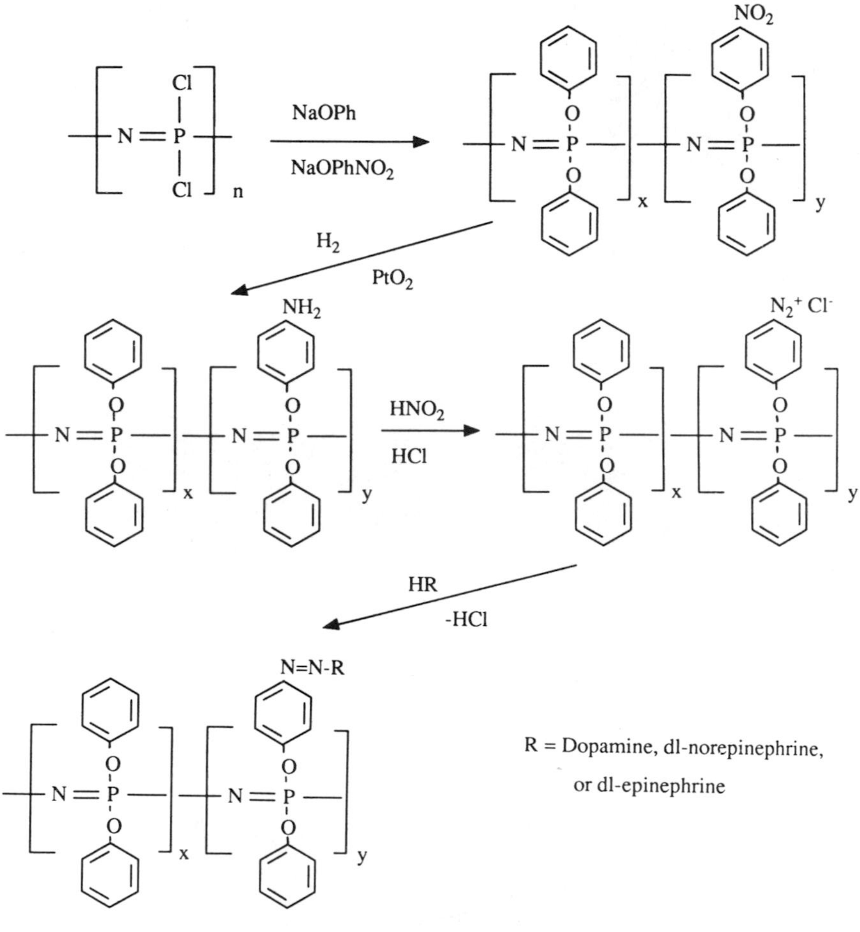

R = Dopamine, dl-norepinephrine, or dl-epinephrine

SCHEME IV

nitro groups to amino residues was followed by diazo coupling to the aromatic ring of dopamine (structure 17). Polymer films treated

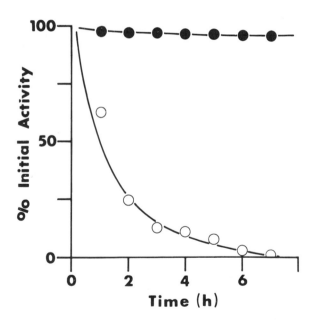

17

in this way elicited the same response from rat pituitary cells in culture as did free dopamine in solution. It appears that even the short spacer group employed allowed the immobilized catecholamine molecules to interact with the cell surface receptors. Utilization of this type of phenomenon with other bioactive agents offers many opportunities for the design of heterogeneous drug delivery devices.

FIGURE 4 Activity of glucose-6-phosphate dehydrogenase as a function of time for (•) the enzyme immobilized on the polyphosphazene/alumina support and (o) in the presence of free enzyme and non-activated support. (From Ref. 23.)

D. Insoluble Biodegradable Polyphosphazenes

First, it should be stated that most of the systems to be described
are erodible by nonenzymatic hydrolysis. With one exception, no
evidence exists yet that an in vivo degradation is different in mechanism from an uncatalyzed chemical hydrolysis.

The biomedical uses of polyphosphazenes mentioned earlier involve chemistry that could in principle be carried out on a classical
petrochemical-based polymer. However, in their bioerosion reactions,
polyphosphazenes display a uniqueness that sets them apart. This
uniqueness stems from the presence of the inorganic backbone, which
in the presence of appropriate side groups is capable of undergoing
facile hydrolysis to phosphate and ammonia. Phosphate can be me-
tabolized, and ammonia is excreted. If the side groups released in
this process are also metabolizable or excretable, the polymer can be
eroded under hydrolytic conditions without the danger of a toxic
response. Thus, polymers of this type are candidates for use as
erodible biostructural materials or sutures, or as matrices for the
controlled delivery of drugs. Four examples will be given to illustrate the opportunities that exist.

1. Amino Acid Ester Systems

The first bioerodible polyphosphazenes synthesized possessed amino
acid ester side groups (25). The structure and preparation of one
example is shown in Scheme V. The ethyl glycinato derivative shown

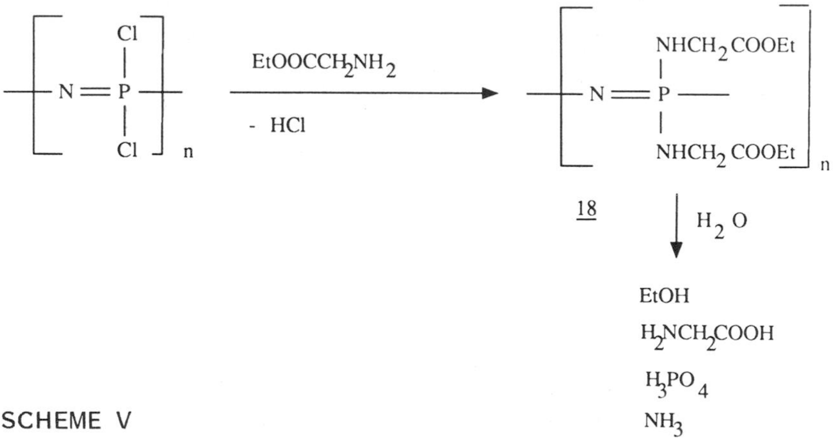

SCHEME V

(structure 18) is a leathery solid which erodes hydrolytically to
ethanol, glycine, phosphate, and ammonia (25). The hydrolysis of
this polymer in near-neutral media appears to proceed via an initial
hydrolysis of the ester function to yield the carboxylic acid (25,26).

This is followed by hydrolytic cleavage of the <u>side group</u> P–N bond
to release glycine and leave a hydroxyl group attached to phosphorus.
Hydroxyphosphazenes are themseleves hydrolytically unstable, de-
composing to phosphate and ammonia. In strongly basic aqueous
media the hydrolysis of the ester function proceeds rapidly, but the
subsequent breakdown pathways appear to be inhibited. In weak
base or weak acid the behavior is·similar to that in neutral media.
Polymer <u>18</u> has been tested in subcutaneous tissue response experiments, and the available evidence indicates an absence of cell toxicity, irritation, or giant-cell formation (13,27,28). The bioerosion
characteristics can be utilized either by use of the polymer as a matrix for release of a dissolved drug or by covalent attachment of the
drug as a cosubstituent. For example, the controlled release of the
antitumor agent, L-phenylalanine mustard (melphalan), from solid,
erodible matrices of polyphosphazenes that bear ethyl glycinate,
phenylalanine ethyl ester, or glutamic acid diethyl ester side groups
has shown considerable promise (27).

In addition, a series of papers has appeared on the use of amino
acid ester-substituted polyphosphazenes for the controlled release of
a covalently linked drug, naproxen [($^+$)-2-(6-methoxy-2-naphthyl)-
propionic acid] (28). The drug was connected to the polymer chain
through an amino acid residue. Again, no tissue inflammation was
detected during implantation experiments using rats. Enzymatic activity was found to increase the rate of drug release. However, the
long term release rate in vivo was lower than expected from in vitro
experiments, and optimization of the side group arrangement is clearly
needed.

2. Polymers with Steroidal and Amino Acid
    Ester Side Groups

Steroids that bear hydroxyl groups at the 3 position of an aromatic
A ring or at the 17 position can be linked covalently to a polyphosphazene chain by the use of the sodium salt of the hydroxyl unit
(29). Example polymers are shown in structures <u>19</u> and <u>20</u>.

<u>19</u>

20

The steric bulk of steroid structures prevents their use as the only organic side group present. However, mixed-substituent polymers that contain both steroidal side groups and amino acid ester or other cosubstituent units can be readily synthesized. If a saturated A ring is present in the steroid, linkage to the polymer chain is complicated by side reactions that result from dehydration of the steroid (chlorophosphazenes are powerful dehydrating agents).

3. Imidazolyl Derivatives

Imidazolyl groups linked to a phosphazene chain through the 1-nitrogen position are hydrolyzed especially easily (30). In fact, the imidazolyl single-substituent polymer shown as compound 21 is too sensitive to moisture to be handled in the atmosphere (30). However,

21   22

the rate of hydrolysis can be slowed by the incorporation of hydrophobic cosubstituent groups such as phenoxy or methylphenoxy units (31). These mixed-substituent derivatives are amphiphilic polymers that bioerode without swelling in water. For this reason, they are better candidates for zero-order drug release applications than their totally hydrophilic counterparts.

   A detailed study of the in vitro and in vivo release profile of
species such as progesterone from a matrix of polymer 22 has been
published (31).   Structure 22 is a schematic representation of the
polymer only.   In the polymers synthesized, 20–45% of the side
groups were imidazolyl units, with the remainder being p-methylphenoxy.   It was first demonstrated that the rate of erosion of these
polymers can be controlled by the ratio of imidazolyl to methylphenoxy
side groups.   Release profile studies, with the use of p-nitroaniline
as a model, indicated reproducible release rates that could be controlled by variations in the polymer side group ratios.   Polypeptide
release was demonstrated for polymer matrices that contained bovine
serum albumin.   Release profiles for the liberation of progesterone
in both in vitro and rat subcutaneous implant tests were also determined.   Typical data are shown in Fig. 5.   Good biocompatibility
was noted in all these tests.

4.   Schiff's Base-Linked Antibacterial Agents

Some exploratory work has been carried out on methods for the
linkage of antibacterial agents, such as sulfadiazine, to polyphosphazenes (32).   A terminal aldehydic group on an aryloxy side group
is capable of forming a Schiff's base linkage to the primary amino
group of the drug, as shown in structure 23.   The Schiff's base
linkage is hydrolytically unstable, and cleavage of this bond in aqueous media provides a mechanism for release of the drug in its small-
molecule form.   Dopamine and amino-4-picoline have been immobilized
in the same way.   The hydrolytic sensitivity of the system can be
varied by changes in the ratio of phenoxy to Schiff's base units.

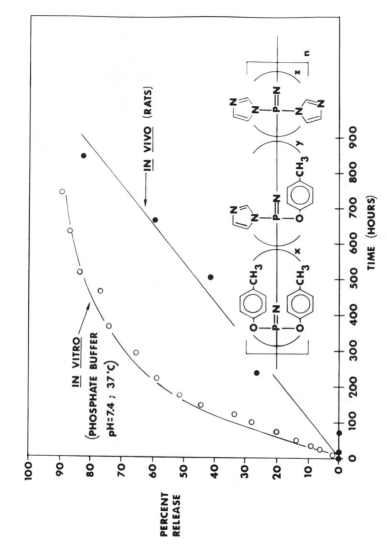

FIGURE 5   Erosion release profile for radioactive progesterone from a matrix of a mixed-substituent imidazolyl/p-methylphenoxyphosphazene polymer.   (From Ref. 31.)

5. Immobilized Local Anesthetics

A number of anesthetic molecules, including procaine, benzocaine, chloroprocaine, butyl p-aminobenzoate, and 2-aminopicoline, possess primary amino groups. These groups provide a means for the attachment of the bioactive molecules to a polyphosphazene skeleton through the chemistry shown in Scheme I (33).

Single-substituent polymers, such as the procaine derivative shown as compound 24, are insoluble in water. In principle, the bioerosion characteristics can be modified by the presence of amino acid ester, glyceryl, or glucosyl (see later) cosubstituents, and this offers the possibility of a long-term release of a local anesthetic at a targeted site in the body.

24

6. Linkage of Bioactive Molecules that Contain Carboxylic Acid Groups: Amide Linkages

A substantial number of bioactive molecules, such as polypeptides, N-acetyl-DL-penicillamine, p-(dipropylsulfamoyl)benzoic acid, and nicotinic acid, contain a carboxylic acid function, and this provides a site for linkage to a polyphosphazene chain. A number of prototype polymers have been synthesized in which pendent amino groups provide coupling sites for the carboxylic acid (34). The amide linkages so formed are potentially bioerodible, but the use of a hydrolytic sensitizing cosubstituent would be expected to accelerate the process.

Polymer 25 is prepared by the reduction of polyphosphazenes that bear p-cyanophenoxy side groups together with phenoxy cosubstituents. Dicyclohexylcarbodiimide-induced condensation with the carboxylic acids then yielded polymers of type 26. In principle, water-solubilizing cosubstituent groups could also be used to generate soluble polymeric drugs (see Section VI).

$$RCOOH = \quad \langle pyridyl \rangle COOH, \ CH_3C(O)NHCH_2COOH,$$

$$(CH_3)_2C(SH)CH(NHCOCH_3)COOH, \ or$$

$$(C_3H_7)_2NSO_2 \text{—} \langle C_6H_4 \rangle \text{—} COOH$$

E.   Glyceryl Derivatives

A recent development has been the synthesis of bioerodible polyphosphazenes that bear glyceryl side groups (35). The synthesis of these polymers requires a protection-deprotection sequence to reduce the functionality of the glycerol and prevent crosslinking. This is illustrated in Scheme VI. The protected glyceryl derivatives are insoluble in aqueous media and appear to be hydrolytically stable. The deprotected species (structure 27) is water-soluble and hydrolyzes in aqueous media at neutral pH at 37°C to give glycerol, phosphate, and ammonia. The free hydroxyl units of the deprotected polymer provide sites for the covalent attachment of drug molecules. Water insolubility can be imparted by the use of appropriate hydrophobic cosubstituent groups to generate solid, erodible materials.

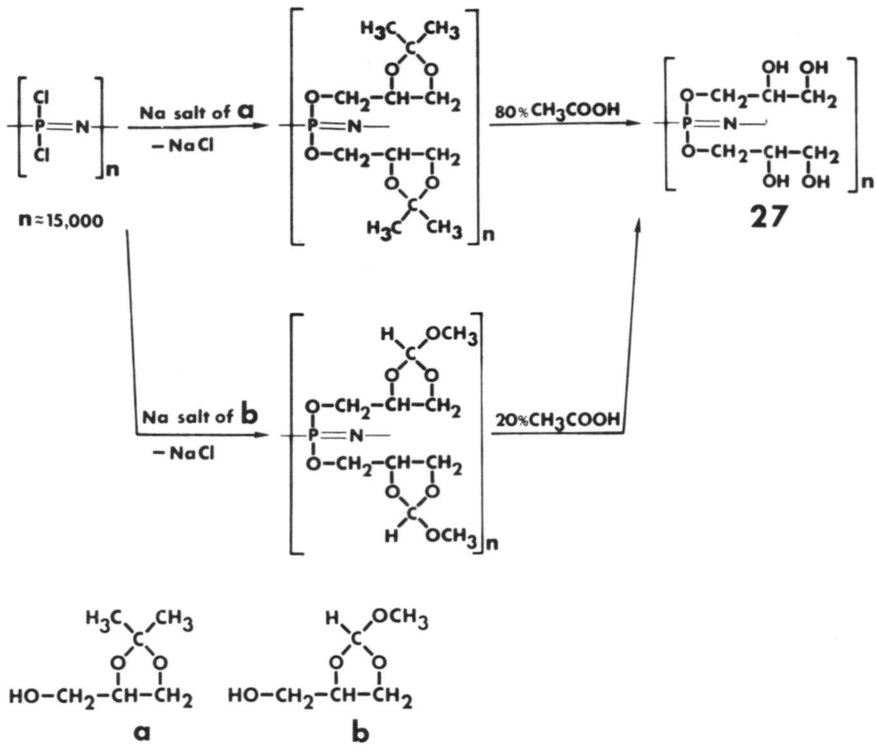

SCHEME VI

F. Glucosyl Derivatives

Polyphosphazenes that bear glucose residues as side groups (compound $\underline{28}$) have also been prepared. The synthesis involves a prior reaction of poly(dichlorophosphazene) with the sodium salt of diacetone glucose, together with cosubstitution by an amine or alkoxide or aryloxide. Deprotection of the sugar residue then yields polymers of type $\underline{28}$ (36,37). The glucosyl side groups can be oxidized, reduced, or acetylated in the manner well known for glucose or for cellulose. The most challenging parts of this synthesis procedure are (1) replacement of a large percentage of the chlorine atoms by the sterically hindered protected sugar residues and (2) the deprotection process itself (37). Failure to replace all the chlorine atoms leaves sites of extreme hydrolytic sensitivity, while excessive exposure to trifluoroacetic acid in the deprotection step brings about skeletal cleavage.

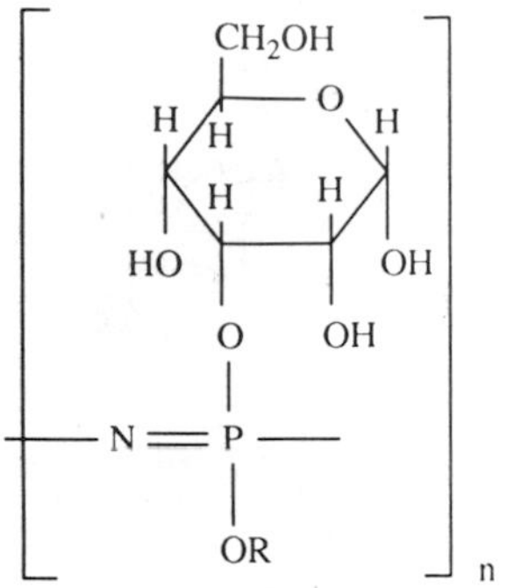

CH$_2$OH

28

Nevertheless, the potential uses of these polymers are intriguing, especially since preliminary evidence has been obtained that hydrolysis to glucose, phosphate, and ammonia takes place slowly in aqueous media. Mixed-substituent glucosyl polymers appear to have the greatest biomedical potential since the use of a second, less hindered, side group allows the steric hindrance and deprotection problems to be minimized (37).

IV. AMPHIPHILIC POLYMERS AND MEMBRANE MATERIALS

The attachment of hydrophobic and hydrophilic side groups to the same phosphazene polymer chain provides an opportunity to generate amphiphilic bulk and surface properties. The disposition of the two different groups along the chain can have a profound influence on the properties. Blocks of hydrophilic units would favor the formation of hydrophilic domains, whereas a uniform distribution of the two groups may lead to a more complex solid state behavior. These factors would strongly affect the behavior of a solid polymer as a semipermeable membrane material. Extensive domain formation could generate channels of hydrophobic or hydrophilic character that extend through the membrane.

The opportunities for biomedical membrane design and synthesis in the polyphosphazene system are wide ranging. So far, only preliminary explorations have been undertaken, and only a brief outline will be given here.

Trifluoroethoxy and phenoxy groups are hydrophobic. The methylamino group is hydrophilic. Studies have been made of the effect on surface properties and semipermeability of variations in

the ratios of trifluoroethoxy or phenoxy and methylamino units
(structures 29 and 30) over the range of 0–100% of each type of
group (38).   Structures 29 and 30 are schematic only since, for
most side group ratios, two identical side groups will be attached to
the same skeletal phosphorus atom.

29   30

    The change in surface character as the side group ratios are
changed can be monitored by contact angle measurements for water
droplets on the surface.   The contact angles varied over the range
of 100° for the trifluoroethoxy and phenoxy single-substituent polymers to 30° for the methylamino single-substituent polymer.   The
semipermeability behavior was investigated from the diffusion of small-
molecule dyes through films.   For example, a polymer with roughly
50% trifluoroethoxy and 50% methylamino side groups showed a faster
transmission of dye molecules than did standard regenerated cellulose
dialysis membranes (38).   The strength of membranes derived from
those polymers depended on the side group ratios.   In general, those
polymers with the highest ratios of methylamino side groups generated membranes that were difficult to handle because of brittleness
or solubility in aqueous media.   However, these same polymers were
the easiest to crosslink by γ-irradiation techniques, a process that
markedly enhanced the membrane- or hydrogel-forming properties.

V.   HYDROGELS

A hydrogel is formed by a water-soluble polymer that has been
lightly crosslinked.   Hydrogels swell as they absorb water but they
do not dissolve.   The volume expansion is limited by the degree of
crosslinking.   The minimum number of crosslinks needed to form a
three-dimensional matrix is approximately 1.5 crosslinks per chain,
and this yields the maximum expansion possible without separation
of the chains into a true solution.   Thus, a hydrogel may be more
than 95% water and, in that sense, has much in common with living
soft tissues.
    Hydrogels have many potential uses in biomedicine, ranging
from materials for the construction of soft-tissue prostheses or

tissue-like coatings or linings for heart valves or blood vessel prostheses, to matrices that imbibe drug molecules and release them by
diffusion. Additional prospective applications include their use as
substrates for enzyme or antigen immobilization and as materials that
favor tissue ingrowth. Soft contact lenses and intraocular lenses are
hydrogels. Hydrogels may be biostable or they may be bioerodible.
Erosion can occur by hydrolysis of the chains or by cleavage of the
crosslinks.

A number of polyphosphazenes are soluble in water. Examples
include polymers with methylamino (39), glucosyl (36), glyceryl (35),
or alkyl ether side groups (20). Here, we will consider one example,
a polymer that is also well suited to the "clean" method of radiation
crosslinking.

The polymer is poly[bis(methoxyethoxyethoxy)phosphazene]
(structure 31), also known as "MEEP" (20,40). In the solid state,

$$\left[\begin{array}{c} OCH_2CH_2OCH_2CH_2OCH_3 \\ | \\ N = P - \\ | \\ OCH_2CH_2OCH_2CH_2OCH_3 \end{array}\right]_n$$

31

at 25°C the molecular motions in this material are so facile that it
functions as a solid solvent for ionic species such as lithium triflate,
and the solid solution is a good ionic conductor of electricity. The
polymer-salt system is under consideration for use as an electrolyte
in solid, rechargeable, light-weight lithium batteries.

However, polymer 31 is exceedingly sensitive to γ-radiation-
induced crosslinking due to the presence of 11 carbon-hydrogen
bonds in each side group (41). Thus, relatively low doses of γ-
radiation (1.2 Mrad) convert it from a water-soluble polymer to a
material that imbibes water to form a hydrogel. The degree of water
imbibition depends on the degree of crosslinking which in turn depends on the radiation dose (Fig. 6). Hydrogels formed from this
polymer are stable to water and appear to be interesting candidates
for use as intraocular lenses, soft-tissue prostheses, or as hydrophilic coatings for biomedical devices.

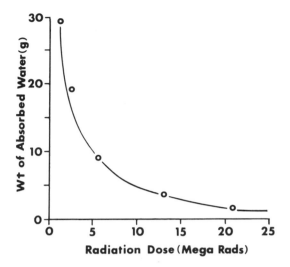

FIGURE 6 Degree of water absorption (and, therefore, inversely, the degree of crosslinking) by $[NP(OCH_2CH_2OCH_2CH_2OCH_3)_2]_n$ as a function of γ radiation dose. (From Ref. 41.)

VI. WATER-SOLUBLE, BIOACTIVE POLYMERS

In principle, water-soluble macromolecular drugs should be more effective for some forms of therapy than their small-molecule counterparts because of their restricted ability to escape through semipermeable membranes. Moreover, targeting groups can also be linked
to the macromolecule to further increase the effectiveness of the drug.
The macromolecular component of such drug-polymer conjugates should
be biodegradable in order to prevent eventual deposition of the polymer molecules at some site in the body, e.g., in the spleen. The
release of the small-molecule drug may precede, parallel, or follow
hydrolytic breakdown of the polymer skeleton.

We have already seen how water solubility and hydrolytic degradability can be built into the carrier macromolecule by the use
of specific side groups. Here we will review an additional way in
which drug molecules have been linked to polyphosphazenes—by
coordination.

A. Polymer-Bound Platinum Antitumor Agents

The platinum complex $(NH_3)_2PtCl_2$ is a well-known antitumor agent.
Because the drug is soluble in water, it is readily excreted through
the kidneys and can cause severe kidney damage. Various procedures are employed clinically to minimize these side effects, but the

linkage of the $PtCl_2$ component to a nonexcretable, water-soluble polymer offers an additional possibility for improving the effectiveness of the chemotherapy.

A polyphosphazene that is itself a base, and which can replace the ammonia ligands in $(NH_3)_2PtCl_2$, has been studied as a carrier macromolecule (42). The polymer is poly[bis(methylamino)phosphazene], $[NP(NHCH_3)_2]_n$. It forms a coordination complex with $PtCl_2$ through the backbone nitrogen atoms rather than through the methylamino side groups (43). A representation of the structure is shown in 32. Preliminary tissue culture testing has been carried

32

out using this polymer, and it appears that some of the antitumor activity is retained in the polymeric drug. This result is interesting and perhaps surprising since the active form of $(NH_3)_2PtCl_2$ is believed to be the small-molecule diaquo derivative $(H_2O)_2PtCl_2$, which must penetrate the tumor cell membrane in order to affect DNA replication. Thus, more work is needed to understand the behavior of macromolecular drugs of this type along with the role of dissociation or endocytosis in their biological activity.

B. Other Water-Soluble Systems

It was shown in earlier sections of this chapter that water solubility in polyphosphazenes can be generated not only by the presence of cosubstituent methylamino side groups, but also by glyceryl, glucosyl, or alkyl ether side groups. This provides a broad range of options for the tailoring of polymeric, water-soluble drugs. All of these options have not yet been explored in detail, but preliminary work has shown, for example, that water-soluble analogs of the anesthetic-bearing macromolecules discussed earlier are accessible (33). Water solubility is achieved if 50% if the side groups are methylamino units (structure 33).

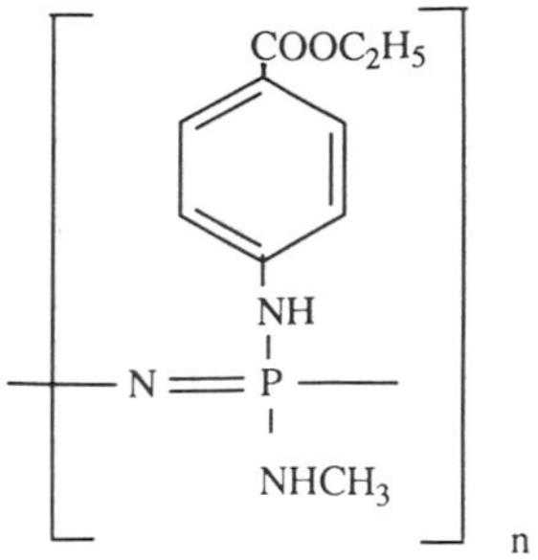

33

At another level, water-soluble polyphosphazenes are of interest as plasma extenders. In addition, specific polymers with pendent imidazolyl units have been studied as carrier macromolecules for heme and other iron porphyrins (structures 34 and 35) (44,45). (In structures 34 and 35 the ellipse and Fe symbol represent heme, hemin, or a synthetic heme analog.)

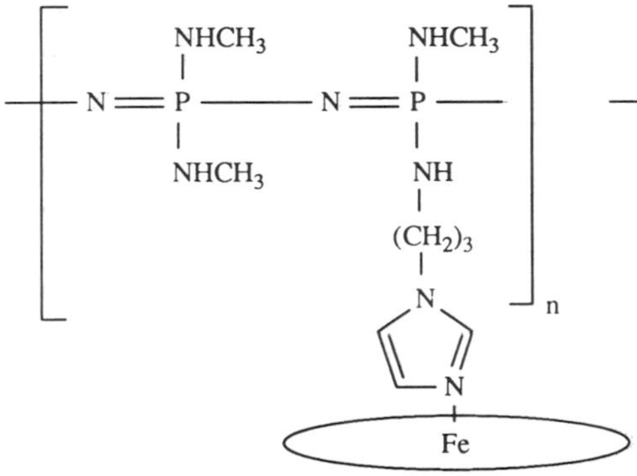

34 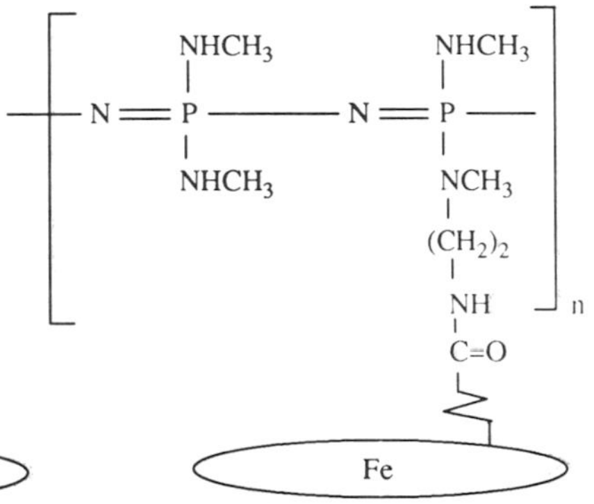   35

Finally, a new water-soluble polyphosphazene was recently synthesized that has the structure shown in 36 (46). This polymer has two attributes as a biomedical macromolecule. First, the pendent carboxylic acid groups are potential sites for condensation reactions with amines, alcohols, phenols, or other carboxylic acid units to generate amide, ester, or anhydride links to polypeptides or bioactive small molecules. Second, polymer 36 forms ionic crosslinks when brought into contact with di- or trivalent cations such as Ca^{2+} or Al^{3+}. The crosslinking process converts the water-soluble polymer to a hydrogel, a process that can be reversed when the system

$$\underline{36}$$

is infused with a monovalent cation such as Na^+ or K^+. This propetry opens pathways for (a) removal of ions from living systems, (b) temporary and reversible gelation of a solution as a soft-tissue substitute or for temporary sealing of a blood vessel, or (c) entrapment of drug molecules or living cells in the hydrogel and release of them by changes in pH or cation concentrations.

VII. SUMMARY AND COMMENTS

Understanding the relationship between molecular structure and materials properties or biological activity is one of the most important facets of biomaterials synthesis and new-drug design. This is especially true for polyphosphazenes, where the molecular structure and properties can be varied so widely by small modifications to the substitutive method of synthesis.

Table 1 is a summary of current knowledge of the relationship between side group structure in polyphosphazenes and biomedically important properties. Within rather broad limits two or more of these properties can be incorporated into the same polymer by a combination of different side groups attached to the same macromolecular chain.

It has been demonstrated that a variety of different polyphosphazenes can be developed as biomaterials, membranes or hydrogels, bioactive polymers, and bioerodible polymers. As with most new areas of polymer chemistry and biomaterials science, molecular design forms the basis of most new advances, but the rate-controlling step is the testing and evaluation of the materials in both in vitro and in vivo environments. This is particularly true for polyphosphazenes where the availability of research quantities only has limited the

TABLE 1 Summary of Side Group Structure-Property Relationships in Polyphosphazenes

$$\left[-N = \underset{\underset{R}{|}}{\overset{\overset{R}{|}}{P}} - \right]_n$$

Hydrophilic/Water Solubilization	Radiation Crosslinking
CH_3NH-	CH_3NH-
CH_3-	Protected glucosyl
Glucosyl	Glyceryl
Glyceryl	$CH_3OCH_2CH_2OCH_2CH_2O-$ etc.
$CH_3OCH_2CH_2OCH_2CH_2O-$	**Biological Activity (General)**
$HOOCC_6H_4O-$	Steroids (3- or 17-linked)
Water-Destabilization	Procaine, etc. (amino-linked)
$EtOOCCH_2NH-$ etc.	Dopamine (Schiff's base)
Imidazolyl	Sulfadiazine (Schiff's base)
Glyceryl	Peptides ⎫ amide-linked
(Glucosyl)	Penicillamines ⎭
Hydrophobic/Water Stabilization	$PtCl_2$ (coordination-linked)
CF_3CH_2O- etc.	**Surface Biological Activity**
C_6H_5O- etc.	Heparin (ionically bound)
C_6H_5-	Enzymes (glutaric aldehyde-linked)
$(CH_3)_3SiCH_2-$ etc.	Dopamine (Diazo-coupled)
	SO_3H

testing of many polymers. However, commercial quantities of the hydrophobic, bioinert polyphosphazene elastomers are now available, and the other polymers mentioned are accessible on a larger laboratory scale. The evaluation of those classical biomedical polymers that are currently in widespread use followed many years of testing and optimization. It seems likely that the development of polyphosphazenes as biomaterials or drug delivery systems will be a more concerted process because of the interrelated opportunities that exist for tailoring the properties and fine-tuning the biological responses.

ACKNOWLEDGMENTS

The work described here was carried out with the support of the
U.S. Public Health Service through the National Heart, Lung, and
Blood Institute. The biomedical evaluations were conducted through
collaborative projects with the research groups led by R. A. Langer
at MIT, W. S. Pierce and G. Nicholas at the Hershey Medical Center
of the Pennsylvania State University, W. C. Hymer at the University
Park campus of Pennsylvania State University, I. Hall at North Carolina State University, and C. W. R. Wade and R. E. Singler of the
U.S. army. It is also a pleasure to acknowledge the contributions
of my coworkers, whose names are given in the list of references.

REFERENCES

1. Allcock, H. R., and Kugel, R. L., Synthesis of high polymeric
    alkoxy- and aryloxyphosphonitriles, J. Am. Chem. Soc., 87,
    4216, 1965.
2. Allcock, H. R., Kugel, R. L., and Valan, K. J., High molecu-
    lar weight poly(alkoxy- and aryloxyphosphazenes), Inorg.
    Chem., 5, 1709, 1966.
3. Allcock, H. R., and Kugel, R. L., High molecular weight poly-
    (diaminophosphazenes), Inorg. Chem., 5, 1716, 1966.
4. Allcock, H. R., Inorganic macromolecules, Chem. Eng. News,
    63, 22, 1985.
5. Allcock, H. R., and Moore, G. Y., Polymerization and copoly-
    merization of phenyl-halogenocyclotriphosphazenes, Macromole-
    cules, 8, 377, 1975.
6. Allcock, H. R., Schmutz, J. L., and Kosydar, K. M., A new
    route for poly(organophosphazene) synthesis. Polymerization,
    copolymerization, and ring-ring equilibration of trifluoroethoxy-
    and chloro-substituted cyclotriphosphazenes, Macromolecules,
    11, 179, 1978.
7. Ritchie, R. J., Harris, P. J., and Allcock, H. R., Polymeriza-
    tion of monoalkyl-pentachlorocyclotriphosphazenes, Macromole-
    cules, 12, 1014, 1979.
8. Allcock, H. R., and Connolly, M. S., Polymerization and halogen
    scrambling behavior of phenyl-substituted cyclotriphosphazenes,
    Macromolecules, 18, 1330, 1985.
9. Neilson, R. H., and Wisian-Neilson, P., Poly(alkyl/arylphos-
    phazanes) and their precursors, Chem. Rev., 88, 541, 1988.
10. Rose, S. J. H., Synthesis of phosphonitrilic fluoroelastomers,
    J. Polym. Sci. B, 6, 837, 1968.
11. Singler, R. E., Sennett, M. S., and Willingham, R. A., Phos-
    phazene polymers: Synthesis, structure, and properties, in
    Inorganic and Organometallic Polymers (M. Zeldin, K. J. Wynne,

and H. R. Allcock, eds.), ACS Symp. Ser., 360, 1988, p. 360.

12. Penton, H. R., Polyphosphazenes: Performance polymers for specialty applications, in Inorganic and Organometallic Polymers (M. Zeldin, K. J. Wynne, and H. R. Allcock, eds.), ACS Symp. Ser., 360, 1988, p. 278.

13. Wade, C. W. R., Gourlay, S., Rice, R., Hegyeli, A., Singler, R., and White, J., Biocompatibility of eight poly(organophosphazenes), in Organometallic Polymers (C. E. Carraher, J. E. Sheats, and C. U. Pittman, eds.), Academic Press, New York, 1978, p. 289.

14. Nicholas, G. G., and Wright, S D.. Unpublished Results.

15. Allcock, H. R., Brennan, D. J., and Graaskamp, J. M., Ring-opening polymerization of methylsilane- and methylsiloxane-substituted cyclotriphosphazenes, Macromolecules, 21, 1, 1988.

16. Allcock, H. R., and Brennan, D. J., Organosilicon derivatives of cyclic and high polymeric phosphazenes, J. Organomet. Chem., 341, 231, 1988.

17. Allcock, H. R., Brennan, D. J., and Dunn, B. S., Synthesis of polyphosphazenes bearing geminal trimethylsilylmethylene and alkyl or phenyl side groups, Macromolecules, 22, 1534, 1989.

18. Allcock, H. R., Coggio, W. D., Archibald, R. S., and Brennan, D. J., Organosilylphosphazene oligomers and polymers: Synthesis via lithioaryloxyphosphazenes, Macromolecules, 22, 3571, 1989.

19. Lyman, D. J., and Knutson, K., Chemical, physical, and mechanical aspects of blood compatibility, in Biomedical Polymers (E. P. Goldberg and A. Nakajima, eds.), Academic Press, New York, 1980, p. 1.

20. Allcock, H. R., Austin, P. E., Neenan, T. X., Sisko, J. T., Blonsky, P. M., and Shriver, D. F., Polyphosphazenes with etheric side groups: Prospective biomedical and solid electrolyte polymers, Macromolecules, 19, 1508, 1986.

21. Allcock, H. R., and Fitzpatrick, R. J.. Unpublished Results.

22. Neenan, T. X., and Allcock, H. R., Synthesis of a heparinized poly(organophosphazene), Biomaterials, 3, 78, 1982.

23. Allcock, H. R., and Kwon, S., Covalent linkage of proteins to surface-modified poly(organophosphazenes): Immobilization of glucose-6-phosphate dehydrogenase and trypsin, Macromolecules, 19, 1502, 1986.

24. Allcock, H. R., Hymer, W. C., and Austin, P. E., Diazo coupling of catecholamines with poly(organophosphazenes), Macromolecules, 16, 1401, 1983.

25. Allcock, H. R., Fuller, T. J., Mack, D. P., Matsumura, K., and Smeltz, K. M., Synthesis of poly[(amino acid alkyl ester) phosphazenes], Macromolecules, 10, 824, 1977.

26. Allcock, H. R., Fuller, T. J., and Matsumura, K., Hydrolysis pathways for aminophosphazenes, Inorg. Chem., 21, 515, 1982.

27. Grolleman, C. W. J., de Visser, A. C., Wolke, J. G. C.,
     Klein, C. P. A. T., van der Goot, H., and Timmerman, H.,
     Studies on a bioerodible drug carrier system based on a poly-
     phosphazene, J. Control. Rel., 3, 143-154, 1986; 4, 119-131,
     1986; and 4, 133-142, 1986.
28. Goedemoed, J. H., and de Groot, K., Development of implant-
     able antitumor devices based on polyphosphazenes, Makromol.
     Chem., Macromol. Symp., 19, 341-365, 1988.
29. Allcock, H. R., and Fuller, T. J., Phosphazene high polymers
     with steroidal side groups, Macromolecules, 13, 1338, 1980.
30. Allcock, H. R., and Fuller, T. J., The synthesis and hydroly-
     sis of hexa(imidazolyl)cyclotriphosphazenes, J. Am. Chem. Soc.,
     103, 2250, 1981.
31. Laurencin, C., Koh, H. J., Neenan, T. X., Allcock, H. R.,
     and Langer, R. S., Controlled release using a new bioerodible
     polyphosphazene matrix system, J. Biomed. Mater. Res., 21,
     1231, 1987.
32. Allcock, H. R., and Austin, P. E., Schiff's-base coupling of
     cyclic and high polymeric phosphazenes to aldehydes and amines:
     Chemotherapeutic models, Macromolecules, 14, 1616, 1981.
33. Allcock, H. R., Austin, P. E., and Neenan, T. X., Phospha-
     zene high polymers with bioactive substituent groups: Pros-
     pective anesthetic aminophosphazenes, Macromolecules, 15, 689,
     1982.
34. Allcock, H. R., Neenan, T. X., and Kossa, W. C., Coupling
     of cyclic and high polymeric aminoaryloxyphosphazenes to car-
     boxylic acids: Prototypes for bioactive polymers, Macromole-
     cules, 15, 693, 1982.
35. Allcock, H. R., and Kwon, S., Glyceryl polyphosphazenes:
     Synthesis, properties, and hydrolysis, Macromolecules, 21,
     1980, 1988.
36. Allcock, H. R., and Scopelianos, A. G., Synthesis of sugar-
     substituted cyclic and polymeric phosphazenes, and their oxida-
     tion, reduction, and acetylation reactions, Macromolecules, 16,
     715, 1983.
37. Allcock, H. R., Kwon, S., and Pucher, S.   Results to be
     published.
38. Allcock, H. R., Gebura, M., Kwon, S., and Neenan, T. X.,
     Amphiphilic polyphosphazenes as membrane materials: Influence
     of side group on radiation crosslinking, semipermeability, and
     surface morphology, Biomaterials, 19, 500, 1988.
39. Allcock, H. R., Cook, W. J., and Mack, D. P., High molecular
     weight poly[bis(amino)phosphazenes] and mixed substituent
     poly(aminophosphazenes), Inorg. Chem., 11, 2584, 1972.
40. Blonsky, P. M., Shriver, D. F., Austin, P. E., and Allcock,
     H. R., Polyphosphazene solid electrolytes, J. Am. Chem. Soc.,
     106, 6854, 1984.

41. Allcock, H. R., Kwon, S., Riding, G. H., Fitzpatrick, R. J., and Bennett, J. L., Hydrophilic polyphosphazenes as hydrogels: Radiation crosslinking and hydrogel characteristics of poly[bis-(methoxyethoxyethoxy)phosphazene], Biomaterials, 19, 509, 1988.

42. Allcock, H. R., Allen, R. W., and O'Brien, J. P., Synthesis of platinum derivatives of polymeric and cyclic phosphazenes, J. Am. Chem. Soc., 99, 3984, 1977.

43. Allen, R. W., O'Brien, J. P., and Allcock, H. R., Crystal and molecular structure of a platinum-cyclophosphazene complex—cis-Dichloro[octa(methylamino)cyclotetraphosphazene-N,N']platinum(II), J. Am. Chem. Soc., 99, 3987, 1977.

44. Allcock, H. R., Greigger, P. P., Gardner, J. E., and Schmutz, J. L., Water-soluble polyphosphazenes as carrier molecules for iron(III) and iron(II) porphyrins, J. Am. Chem. Soc., 101, 606, 1979.

45. Allcock, H. R., Neenan, T. X., and Boso, B., Synthesis, oxygen-binding behavior, and Mossbauer spectroscopy of covalently-bound polyphosphaene heme complexes, Inorg. Chem., 24, 2656, 1985.

46. Allcock, H. R., and Kwon, S., An ionically-crosslinkable polyphosphazene: Poly[di(carboxylatophenoxy)phosphazene] and its hydrogels and membranes, Macromolecules, 22, 75, 1989.

6
Pseudopoly(amino acids)

JOACHIM KOHN   Rutgers, The State University of New Jersey, New Brunswick, New Jersey

I. INTRODUCTION

For a variety of medical applications (drug delivery, sutures, temporary vascular grafts or orthopedic implants) biodegradable polymers may offer significantly improved treatment options (1,2). However, the possible toxicity associated with a polymer that degrades slowly within the human body is a major concern. Attempts have been made to alleviate this problem by the exclusive use of naturally occurring nutrients or metabolites as monomeric starting materials. The development of polyesters derived from lactic and/or glycolic acid (3) would be a good example for the successful application of this approach. These polymers were the first synthetic biodegradable polymers approved for human use (4).

Based on the same rationale, poly(amino acids) should not give rise to toxic degradation products since they are derived from simple nutrients. Poly(amino acids) have therefore been extensively investigated as biomaterials (5). In spite of the large number of theoretically possible poly(amino acids), however, the number of promising biomaterials among them turned out to be surprisingly limited. One of the major limitations for the medical use of synthetic poly(amino acids) is the pronounced antigenicity of those poly(amino acids) that contain three or more different amino acids. For this reason alone, the search for biomaterials among the synthetic poly(amino acids) is confined to polymers derived from one or at most two different amino acids (6).

Another limitation is related to the fact that synthetic poly(amino acids) have rather unfavorable material properties. For instance, most synthetic poly(amino acids) derived from a single amino acid are insoluble, high-melting materials that cannot be processed into shaped objects by conventional fabrication techniques. The often undesirable tendency to absorb a significant amount of water when exposed to an aqueous environment is another common property of many poly(amino acids) (7). Finally, high molecular weight poly-(amino acids) are best prepared via $\underline{N}$-carboxyanhydrides which are expensive to make. Hence poly(amino acids) are comparatively costly polymers, even if they are derived from inexpensive amino acids (8).

Several attempts have been made to circumvent at least some of the limitations of poly(amino acids) as biomaterials. One commonly explored approach involves the modification of the amino acid side chains. For instance, poly(glutamic acid) has a chemically reactive side chain that can be modified in order to improve its material properties. The work of Sidman et al. was based on the rationale that poly(glutamic acid) can be made water-insoluble by attaching hydrophobic residues to the γ-carboxylic acid side chain (9). The resulting γ-alkyl-substituted derivatives of poly(glutamic acid) can be readily processed and have been suggested for a variety of medical uses, including sutures or drug delivery devices.

Unfortunately, the modification of the side chain is not a generally applicable approach. Among the major, naturally occurring amino acids, only L-lysine has a chemically reactive side chain that would be as readily available for chemical modification as the side chain of glutamic or aspartic acid. Since, however, poly(L-lysine) is known to be toxic (10), its derivatives cannot be candidates for generally applicable biomaterials. Thus, most of the poly(amino acids) that have so far been suggested as biomaterials are derivatives of glutamic or aspartic acid or copolymers of such derivatives with leucine, methionine, or a limited number of additional amino acids (11).

In view of these constraints, we recently suggested a different strategy for the improvement of the material properties of synthetic poly(amino acids) (12). Our approach is based on the replacement of the peptide bonds in the backbone of synthetic poly(amino acids) by a variety of "nonamide" linkages. "Backbone modification," as opposed to "side chain modification," represents a fundamentally different approach that has not yet been explored in detail and that can potentially be used to prepare a whole family of structurally new polymers.

In peptide chemistry, the term "pseudopeptide" is commonly used to denote a peptide in which some or all of the amino acids are linked together by bonds other than the conventional peptide linkage (13). Such pseudopeptides have found applications as specific structural

probes in biochemical research. Furthermore, pseudopeptide derivatives of several important peptides (luteinizing hormone-releasing
hormone, enkephalins, etc.) are presently investigated for their potential use as pharmacologically active drugs (13). In analogy to
the established practice in peptide chemistry, we have used the term
"pseudopoly(amino acid)" to denote a polymer in which naturally occurring amino acids are linked together by nonamide bonds.

Whereas conventional poly(amino acids) are probably best grouped
together with proteins, polysaccharides, and other endogenous polymeric materials, the pseudopoly(amino acids) can no longer be regarded as "natural polymers." Rather, they are synthetic polymers
derived from natural metabolites (e.g., α-L-amino acids) as monomers.
In this sense, pseudopoly(amino acids) are similar to polylactic acid,
which is also a synthetic polymer, derived exclusively from a natural
metabolite.

Surprisingly, so far very few attempts have been made to prepare backbone-modified poly(amino acids). For instance, Greenstein
tried to use the sulfhydryl group of cysteine for the synthesis of a
polysulfide, but no linear polymer was obtained (14). Later, Fasman
attempted to convert polyserine to poly(serine ester) by means of
the facile N⟶O acyl shift of serine. Since only a partial transfor-
mation of the backbone amide bonds to backbone ester linkages was
achieved, a structurally ill-defined random copolymer was obtained
(15). In 1977, Jarm and Fles prepared several poly(N-benzenesul-
fonamidoserine esters) of low molecular weight. Jarm's synthesis
was based on the ring-opening polymerization of N-benzenesulfona-
mido-L-serine-β-lactones. Except for molecular weight, optical rota-
tion, and melting range, no additional properties of the polymers
were reported (16). Since Jarm and Fles employed bioincompatible
benzenesulfonic acid derivatives for the protection of the serine N-
terminus, these polymers are not directly applicable as biomaterials.

The use of backbone-modified poly(amino acids) as biomaterials
was first suggested by Kohn and Langer (17) who prepared a polyester from N-protected trans-4-hydroxy-L-proline, and a poly(iminocarbonate) from tyrosine dipeptide as monomeric starting material
(12,18).

Our interest in the synthesis of poly(amino acids) with modified
backbones is based on the hypothesis that the replacement of conventional peptide bonds by nonamide linkages within the poly(amino
acid) backbone can significantly alter the physical, chemical, and
biological properties of the resulting polymer. Preliminary results
(see below) point to the possibility that the backbone modification
of poly(amino acids) circumvents many of the limitations of conventional poly(amino acids) as biomaterials. It seems that backbone-
modified poly(amino acids) tend to retain the nontoxicity and good
biocompatibility often associated with conventional poly(amino acids)

while at the same time exhibiting significantly improved material properties. Thus, the approach of backbone modification may eventually lead to a variety of useful biomaterials.

So far only a small number of backbone-modified poly(amino acids) have been prepared and carefully characterized. This chapter therefore represents an account of the initial investigations of this interesting and promising group of polymers.

II. SYNTHETIC CONCEPTS

Apparently there are no mild chemical reactions that can be used to transform the amide linkages in the backbone of conventional poly-(amino acids) into nonamide linkages such as ester, urethane, or carbonate bonds. Consequently, it is usually not possible to simply replace the backbone amide bonds of conventional poly(amino acids) by nonamide linkages. Pseudopoly(amino acids) must therefore be prepared from scratch by suitably designed polymerization reactions.

Our approach to the design of such polymerization reactions is based on the use of trifunctional amino acids as monomeric starting materials (12). In the simplest case, a trifunctional amino acid is polymerized by a reaction involving the side chain functional group. In this way one can envision the design of three structurally different types of polymers from each trifunctional amino acid: In Fig. 1a, the polymer backbone consists of linkages formed between the C terminus and the side chain functional group R; in Fig. 1b, the polymer backbone consists of linkages formed between the N terminus and the side chain functional group; and in Fig. 1c, the polymer backbone consists of linkages formed between the C terminus and the N terminus. If the linking bond in Fig. 1c is an amide bond, then obviously Fig. 1c is a schematic representation of a conventional poly(amino acid).

This approach is, among others, potentially applicable to serine, hydroxyproline, threonine, tyrosine, cysteine, glutamic acid, and lysine, and is only limited by the requirement that the nonamide backbone linkages give rise to polymers with desirable material properties. This concept can be further illustrated by a short description of the polymers that can possibly be derived from L-serine (Fig. 2): After protection of the side chain hydroxyl groups, polymerization via the N and C termini gives rise to conventional polyserine (Fig. 2c). This is a well-known polymer that so far has not found any significant industrial or medical application. After protection of the N terminus, polymerization via the side chain hydroxyl groups and the C terminus was recently shown to give rise to poly(serine ester) (19), a polymer in which serine residues are linked together by ester bonds instead of amide bonds (Fig. 2a). The structure of this polymer resembles that of poly(lactic acid), except that poly-

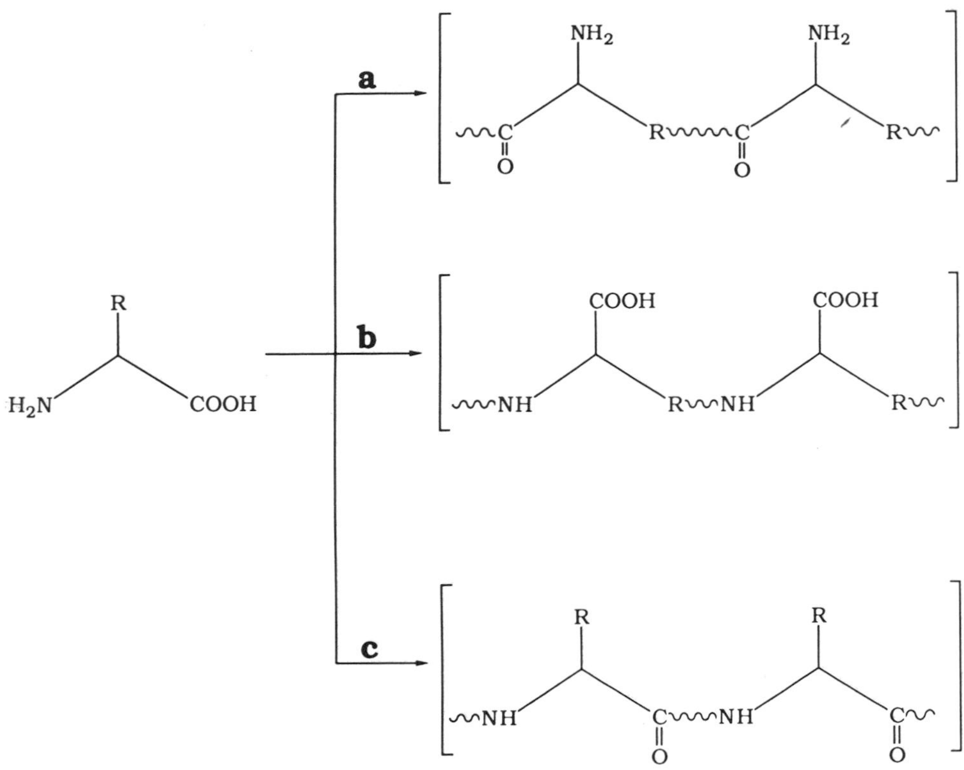

FIGURE 1   Schematic representation of the use of trifunctional amino acids as monomeric starting materials for the synthesis of pseudopoly-(amino acids).   (a) Polymerization via the C terminus and the side chain R.   (b)   Polymerization via the N terminus and the side chain R.   (c) Polymerization via the C terminus and the N terminus.   The wavy line symbolizes any suitable nonamide bond.   See text for details.

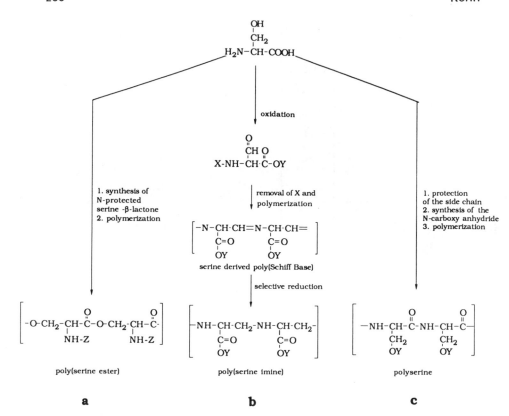

FIGURE 2   Polymers that can be derived from L-serine. (a) Poly-
(serine ester) was obtained by ring opening polymerization of N-
protected serine-β-lactones (19). (b) Poly(serine imine) has apparently not yet been prepared but can possibly be obtained by a reaction scheme that involves oxidation of the serine side chain, formation of the unstable poly(serine Schiff's base), followed by selective
reduction. (c) Conventional poly(L-serine).

(serine ester) possesses a pendant amino group that can be used
for the attachment of crosslinkers or drug molecules. Finally, at
least in theory one may envision a reaction scheme involving the
side chain hydroxyl group and the N terminus (Fig. 2b). After
oxidation of the side chain hydroxyl group to an aldehyde function,
this scheme would lead to a rapidly biodegradable poly(Schiff's base).
A nonbiodegradable polyimine could be obtained after suitable reduction of the Schiff's base linkages. Although poly(serine imine) itself
has apparently not yet been synthesized, the synthesis of the corresponding poly(threonine imine) has recently been accomplished by
a different synthetic route (20). With appropriate modifications, the
above principles can obviously be applied not only to the four major
hydroxyamino acids (Ser, Thr, Hpr, Tyr) but to virtually all trifunctional amino acids.

The structural versatility of pseudopoly(amino acids) can be increased further by considering dipeptides as monomeric starting materials as well. In this case polymerizations can be designed that
involve one of the amino acid side chains and the C terminus, one
of the amino acid side chains and the N terminus, or both of the
amino acid side chains as reactive groups. The use of dipeptides
as monomers in the manner described above results in the formation
of copolymers in which amide bonds and nonamide linkages strictly
alternate (Fig. 3). It is noteworthy that these polymers have both
an amino function and a carboxylic acid function as pendant chains.
This feature should facilitate the attachment of drug molecules or
crosslinkers.

Several such polymers have by now been prepared and were
found to possess a variety of interesting material properties. Tyrosine-derived poly(iminocarbonates) (see Sec. IV) would be a specific
example. These polymers were synthesized by means of a polymerization reaction involving the two phenolic hydroxyl groups located
on the side chains of a protected tyrosine dipeptide (12).

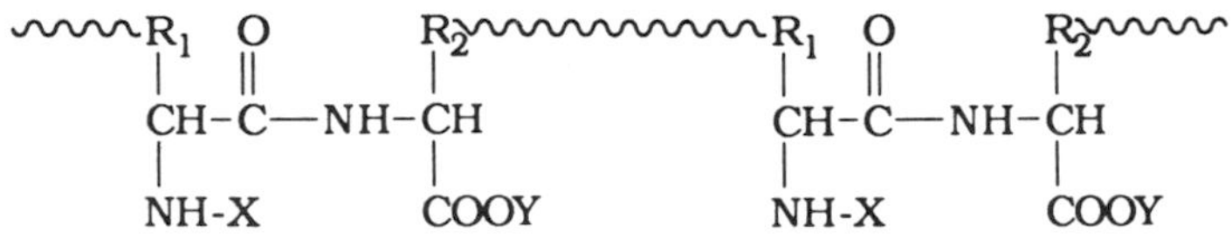

FIGURE 3 Schematic representation of a pseudopoly(amino acid)
derived from the side chain polymerization of a dipeptide carrying
protecting groups X and Y. The wavy line symbolizes a nonamide
bond. In this polymer, the amino acid side chains are an integral
part of the polymer backbone while the termini have become pendant
chains. In the backbone, amide and nonamide bonds strictly alternate.

III. POLYESTERS DERIVED FROM HYDROXYPROLINE

Hydroxyproline-derived polyesters represent a specific embodiment of the synthetic concepts described in Fig. 1a. The successful synthesis of a hydroxyproline-derived polyester was first reported by Kohn and Langer in 1987 (12). Thereafter, several of these new polyesters were investigated in detail by Yu and Langer (21-23).

A. Polymer Synthesis

In analogy to the transesterification of diethyl terephthalate used in the preparation of commercially important polyester fibers such as Dacron (24), a transesterification reaction was successfully employed for the preparation of poly($\underline{N}$-acylhydroxyproline esters) (Scheme 1).

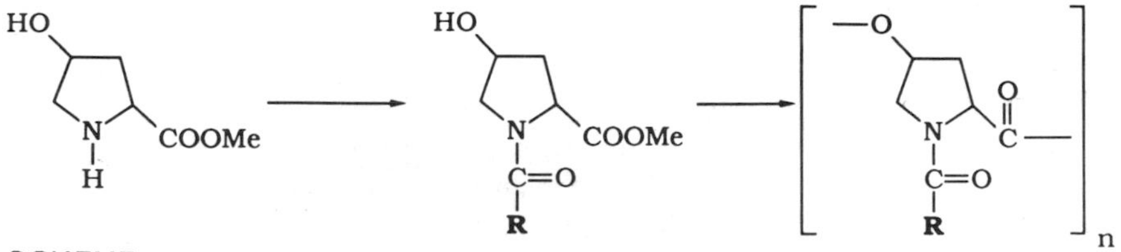

SCHEME 1

In order to avoid side reactions at the imino group of hydroxyproline, this functionality must obviously be protected prior to the actual polymerization step. The need to introduce a protecting group at the N terminus opens the possibility to systematically explore the effect of the size or chemical nature of the protecting group on the properties of the resulting polymer. Fortunately, a large variety of biocompatible carboxylic acids, ranging from acetic acid to stearic acid, are available for this purpose. Furthermore, the use of temporary blocking groups such as the benzyloxycarbonyl (Z) or the tert-butyloxycarbonyl group (Boc) facilitates the synthesis of unprotected poly(hydroxyproline ester).

So far, poly($\underline{N}$-acylhydroxyproline esters) containing six different N-protecting groups have been synthesized (Table 1). In order to optimize the reaction conditions poly($\underline{N}$-palmitoylhydroxyproline) was used as a model compound. First a variety of catalysts were tested for their ability to effectively catalyze the polymerization reaction. Using a limited selection of five model compounds ($\underline{p}$-toluenesulfonic acid, a strong acid; potassium tert-butoxide, a strong base; cadmium and zinc acetate, coordination catalysts; and aluminum isopropoxide, a Lewis acid), Kohn and Langer (12) initially identified aluminum isopropoxide as the best catalyst. In the presence of about 1 mol% of aluminum isopropoxide, poly($\underline{N}$-palmitoylhydroxyproline ester) was obtained with a weight average molecular weight of about 15,500 as

TABLE 1   Poly(N-acylhydroxyproline esters)

Polymer	Number of carbons in side chain	Highest molecular weight	Glass transition temperature $(C)°$
Poly(N-acetyl Hpr ester)	2	1,980	—
Poly(N-pivaloyl Hpr ester)	5	9,820	157
Poly(N-hexanoyl Hpr ester)	6	14,400	83
Poly(N-decanoyl Hpr ester)	10	41,500	71
Poly(N-tetradecanoyl Hpr ester)	14	24,600	96
Poly(N-palmitoyl Hpr ester)	16	42,400	97

determined by gel permeation chromatography (GPC) relative to poly-styrene standards.   Interestingly, p-toluenesulfonic acid was found to be virtually inactive, while the use of potassium tert-butoxide resulted in the formation of a strong brown color, indicative of mas-sive decomposition during the polymerization reaction.

These results were later confirmed by Yu and Langer (21,23) who screened 12 additional catalysts and identified titanium isopro-poxide (a coordination catalyst) as the best choice.   The reaction is not particularly sensitive to the concentration of the catalyst used: from 0.5 mol% up to 3 mol% of titanium isopropoxide no significant change of either the reaction rate or the maximum molecular weight of the resulting polymers was observed.   On the other hand, the reaction temperature is a far more critical parameter.   For the prep-aration of poly(N-palmitoylhydroxyproline) a reaction temperature of 180°C was found to be best.   Under the optimized conditions given in procedure 1, molecular weights (weight average, relative to poly-styrene standards) up to about 40,000 could be obtained.   At lower temperatures (e.g., 150°C) the polymerization was significantly slower and the maximum obtainable molecular weight was limited to below 20,000.   At temperatures above 180°C (e.g., 205°C), the molecular weight increased to about 35,000 in less than 10 hr, but thereafter it gradually decreased with time.   At 205°C, significant browning of the reaction mixture indicated the occurrence of decomposition re-actions.

An interesting observation was made in regard to the polymeriza-bility of N-acylhydroxyproline derivatives: The monomers with the smallest protecting groups such as N-acetyl and N-pivaloylhydroxy-proline gave rise to the polymers with the lowest molecular weights (Table 1).   When these monomers were polymerized, the initially molten reaction mixture rapidly solidified, even when the reaction temperature was raised to over 200°C. One possible explanation for the low molecular weight of these polymers is that their melting temperature was above the optimum reaction temperature. Thus, premature solidification of the reaction mixture (and not a chemical side reaction) seemed to have prevented the polymerization from going to completion. Based on this assessment, one may expect that the addition of a suitable plasticizer or high-boiling solvent to the reaction mixture should facilitate the formation of high molecular weight polymers. Unfortunately, this hypothesis has not yet been experimentally tested.

> Illustrative Procedure 1: Preparation of Poly(N-acylhydroxyproline esters) (21,23): A magnetic stir bar was placed into a melt polymerization tube. The tube was charged with purified N-acylhydroxyproline methyl ester. Under an atmosphere of argon, about 1 mol% of titanium isopropoxide was mixed with the monomer. The tube was then heated to 180°C for 24 hr under vacuum. Methanol evolved and was collected in a cold trap. The corresponding poly(N-acylhydroxyproline ester) was obtained as a slightly tinged, solid material.

B. Polymer Properties

The thermal properties of poly(N-acylhydroxyproline esters) were studied by differential scanning calorimetry (DSC) (22). All hydroxyproline-derived polyesters are thermally stable up to at least 300°C (open pan). Polymers carrying the acetyl, pivaloyl, hexyanoyl, and decanoyl protecting groups were apparently completely amorphous as demonstrated by a complete absence of melting transitions in the DSC curve. The glass transition temperatures (Table 1) varied from 71 to 157°C, confirming the general observation that the size of a pendant group strongly influences the solid state properties of a polymer.

The fact that the glass transition temperatures were fairly high in all polymers is probably a reflection of the stiffness of the polymer backbone imposed by the repeating hydroxyproline rings and the bulky side chains. Interestingly, among the polymers carrying unbranched side chains, the lowest glass transition temperature was found for poly(N-decanoylhydroxyproline ester). Initially, the pendant groups became more flexible with increasing length and T_g accordingly declined. Beyond a chain length of 10 carbons, however, the decline of the glass transition temperature was reversed due to the increased crystallinity of the side chains themselves. Actually, the side chain crystalline melting transitions could be clearly observed by DSC and occurred at 111 and 116°C for poly(N-tetradecanoylhydroxyproline ester) and poly(N-palmitoylhydroxyproline ester), respectively.

Hydroxyproline-derived polyesters are usually readily soluble in a variety of organic solvents (benzene, toluene, chloroform, dichloromethane, carbon tetrachloride, tetrahydrofuran, dimethylformamide, etc.). As expected, the solubility in hydrophobic solvents increased with increasing chain length of the N protecting group, while the solubility in polar solvents decreased. For example, poly(N-hexanoylhydroxyproline ester) is slightly soluble in ether but easily soluble in acetonitrile, while poly(N-palmitoylhydroxyproline ester) is readily soluble in ether but virtually insoluble in acetonitrile.

The combination of thermal stability in the molten state and the ready solubility in a variety of organic solvents greatly enhances the processibility of hydroxyproline-derived polyesters. Thus, films, porous membranes, and fibers have been prepared in our laboratory using solvent casting and wet spinning, while disclike objects and thin sheets were prepared by compression molding. Rods were obtained by extrusion from a heated syringe, a laboratory improvisation for conventional extrusion techniques.

The easy processibility of hydroxyproline-derived polyesters is in marked contrast to the unfavorable material properties of most conventional poly(amino acids) that cannot usually be processed into shaped objects by conventional polymer-processing techniques (7). Furthermore, since the synthesis of poly(N-acylhydroxyproline esters) does not require the expensive N-carboxyanhydrides as monomeric starting materials, poly(N-acylhydroxyproline esters) should be significantly less expensive than derivatives of conventional poly(hydroxyproline).

C. Stability and Biodegradability

The polyester backbone structure of poly(N-acylhydroxyproline esters) raises the obvious question of how the chemical stability and biodegradability of these polymers compares to the chemical properties of the widely used polyesters, derived from lactic or glycolic acid. Although the rates of hydrolysis of structurally diverse polyesters seem to be predominantly governed by the chemical nature of the ester linkage, additional factors such as polymer crystallinity, rigidity of the polymer backbone, packing density, and molecular weight can significantly influence the erosion rate of a polymeric device.

Considering the high hydrophobicity of the palmitoyl side chain and the rigidity of the polymer backbone, we assumed that poly(N-palmitoylhydroxyproline ester) would degrade somewhat more slowly than poly(lactic acid) or polycaprolactone. In order to confirm this hypothesis, a series of long-term stability and degradation studies have been performed over the last 2 years at MIT (22).

The polymer stability upon storage was tested by keeping samples of pure, powdered poly(N-palmitoylhydroxyproline ester) in

closed vials at temperatures ranging from −10 to 37°C. No measures
were taken to exclude ambient moisture. After storage for 1, 3, 6,
12, and 18 months, respectively, triplicate samples were examined
for any change in their molecular weight distribution by GPC. The
major conclusion of this study was that poly(N̲-palmitoylhydroxy-
proline ester) was stable for at least 18 months when stored at or
below 4°C. In samples stored at 25°C or 37°C increasingly large
peaks of a low molecular weight decomposition product were observed.
This decomposition product was tentatively proposed to be N̲-
palmitoylhydroxyproline, but no definitive assignment has so far
been made. Although the stability of poly(N̲-palmitoylhydroxyproline
ester) under absolutely anhydrous conditions (e.g., in sealed ampules under an inert atmosphere) still needs to be determined, it is
probably safe to predict that devices made of poly(N̲-palmitoyl-
hydroxyproline ester) will have adequate shelf lives when properly
packaged and stored below room temperature.

The degradation of poly(N̲-palmitoylhydroxyproline ester) in
aqueous media was tested at 37°C in solutions of 0.1 M HCl (pH 1),
0.1 M phosphate buffer (pH 7.4), and 0.1 M NaOH (pH 14). In
these test compression-molded polymer discs were employed. At intervals of 2−4 weeks the discs were removed, dried, and weighed.
The percent mass loss was recorded. Then the discs were dissolved
in chloroform and the polymer solution was analyzed by GPC. When
stored in buffer at 37°C, devices made of poly(N̲-palmitoylhydroxy-
proline ester) lost less than 5% of their weight over 200 days. Gel
permeation chromatographic analysis of the remaining polymer did
not reveal a significant reduction in the polymer molecular weight.
These observations clearly indicate that poly(N̲-palmitoylhydroxy-
proline ester) is significantly more stable than poly(lactic acid).

In addition, the surfaces of several discs were examined by
scanning electron microscopy (SEM). At the onset of the degradation studies, the device surfaces appeared to be smooth and struc-
tureless. Even after 200 days, no obvious change was observed in
the surface morphology of devices that had been exposed to buffer
at pH 7.4. Exposure to 0.1 N HCl at 37°C for 200 days resulted
in the formation of small, irregularly shaped cracks. Only devices
exposed to 0.1 N NaOH exhibited strong evidence for degradation
on their surfaces in the form of deep, irregular pits.

As intuitively expected, reducing the hydrophobicity of the side
chain resulted in an acceleration of the degradation process: whereas a device made of poly(palmitoylhydroxyproline ester) lost only
about 1% of its initial weight over 50 days in 0.1 N NaOH at 37°C,
a device made of poly(hexanoylhydroxyproline ester) lost over 5%
of its initial weight under identical degradation conditions. Surprisingly, however, the side chain effect on the degradation rate
is apparently quite small.

Based on these studies, it is obvious that poly(N-acylhydroxy-proline esters) are very slowly degrading polymers. These materials may therefore be useful for long-term applications, such as implant-able, multiyear contraceptive formulations. For such applications the degradation rates of poly(lactic acid)/poly(glycolic acid) devices would probably be too rapid.

D. Drug Release and Formulation

Several poly(N-acylhydroxyproline esters) with N-acyl protecting groups of varying length were tested as controlled drug release de-vices (22). So far, two different "model drugs," p-nitroaniline and acid orange, have been tested in vitro. p-Nitroaniline and acid or-ange were selected as models due to their different solubility pro-files: whereas p-nitroaniline is a hydrophobic compound with good solubility in methylene chloride and other nonpolar solvents, acid orange is a hydrophilic dye that is most readily soluble in water.

Compression-molded devices of poly(N-palmitoyl hydroxyproline ester) (side chain length: 16 carbons), poly(N-decanoylhydroxy-proline ester) (side chain length: 10 carbons), and poly(N-hexanoyl-hydroxyproline ester) (side chain length: 6 carbons) were prepared with dye contents of 1,5,10, and 20% of either p-nitroaniline or acid orange. Release curves were obtained by placing the loaded devices into phospate buffer (pH 7.4) at 37°C. The amount of released dye was followed spectrophotometrically in the usual fashion.

Completely different release profiles were obtained for the two model dyes. The release of acid orange was strongly dependent on the loading: at high loadings, the release of acid orange was com-plete within 20 hr. Prolonged release times were only obtained at loadings of 5% and below (Fig. 4a). On the other hand, the release of p-nitroaniline was a linear function of the square root of time, virtually independent of the loading (Fig. 4b). The dramatic dif-ference in the release profiles for acid orange and p-nitroaniline can be readily explained in terms of the different solubility of the two model dyes in the aqueous release medium.

Due to the high stability of poly(hydroxyproline esters), the degradation of the polymeric phase can, for all practical purposes, be neglected during the first 1000 hr of drug release. Thus, math-ematical models developed for the release of drugs from nondegrad-able polymers such as ethylene-vinyl acetate copolymer (EVA) should also be applicable to devices made of poly(hydroxyproline esters) (25,26). In compression-molded devices, particles of the model dye are heterogeneously dispersed within the polymeric phase. There-fore, as the dye particles are dissolved, channels are formed within the polymeric device that facilitate the penetration of water into the device and the release of dye from the device. The release of the highly water-soluble acid orange is governed by the rate of penetra-

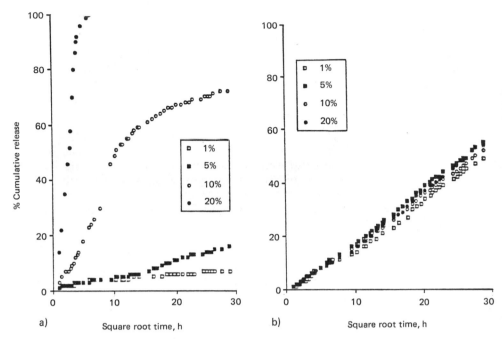

FIGURE 4 Release profiles of model dyes from compression-molded discs of poly(N-palmitoylhydroxyproline ester) (22). (a) Release of acid orange at loadings ranging from 1 to 20%. (b) Release of p-nitroaniline at loadings ranging from 1 to 20%. (From Ref. 22.)

tion of water into the device. Increasing levels of loading facilitate the penetration of water into the device and thus result in an increased rate of acid orange release. On the other hand, the release of p-nitroaniline was apparently "dissolution limited." If the diffusion of the dye out of the matrix is rapid relative to the rate of dissolution, additional channels available for the penetration of water into the device will have little effect on the observed release rates. Thus, increased levels of loading of p-nitroaniline did not translate into significantly increased release rates.

These studies indicate that in simple, compression-molded release devices, the observed release times may vary from several hours to many months, depending on the loading and the hydrophobicity of the released drug. However, in order to take full advantage of the long degradation time of poly(hydroxyproline esters) it is probably advantageous to consider the design of reservoir-type devices for which the rate of drug release is controlled by the rate of drug diffusion through the polymeric phase. In this configuration

it may be possible to significantly influence the rate of drug diffusion through the polymer by changing the protecting group attached
to the N terminus of hydroxyproline. Such devices should have active lifetimes well in excess of 1 year.

At present there is no reason evident why poly(N-acylhydroxyproline esters) should not be suitable for the formation of microcapsules or microspheres as well. For microencapsulated drug formulations the longer degradation times of poly(N-acylhydroxyproline
esters) as compared to poly(lactic acid) could again be a distinctive
advantage for long-term applications.

E. Biocompatibility

As part of the initial evaluation of hydroxyproline-derived polyesters,
poly(N-palmitoylhydroxyproline ester) was subjected to different biocompatibility screening tests (22). In the rabbit cornea bioassay
(27), small pieces of polymer are implanted into the cornea of white,
male New Zealand rabbits. Since the cornea is transparent, it is
possible to observe the local tissue response as a function of time
without having to gain surgical access to the implantation site. The
rabbit cornea bioassay has been used routinely for the initial screening of a variety of biomaterials. A significant pathological response
(reddening, formation of turbidity, ingrowth of blood vessels from
the periphery of the cornea) is a strong indication that the test material is unsuitable for implantation. For example, poly(vinyl pyr-
rolidone) results in a strong inflammatory response within 2 weeks
of implantation into the rabbit cornea. Local edema can cause the
cornea to appear two to three times the thickness of a normal cornea.
In cases of severe inflammation, histological examination typically re-
veals the presence of a large number of inflammatory cells and new
blood vessels at the implantation site (27).

In contrast, implantation of four small pieces of poly(N-palmitoylhydroxyproline ester) into four rabbit corneas elicited no pathological response in three corneas and a very mild inflammatory response
in one cornea. Histological examination of the corneas 4 weeks postimplantation showed no invading blood vessels or migrating inflammatory cells in the area around the implants.

Next, a 1 year subcutaneous implantation study in mice was performed (22). Small pieces of poly(N-palmitoylhydroxyproline ester)
(approximately 10 mg per implant) were implanted subcutaneously in
the dorsal area of the animals. The implants were placed between
the dermis and the adipose tissue layer. Groups of mice were sacrificed 4, 7, 14, 16, and 56 weeks postimplantation.

One major finding of this study was that the polymeric implants
remained intact throughout the 1-year course of the experiments, in
complete agreement with the predictions based on in vitro experiments.

This finding makes a significant contribution of enzymatic degradation to the overall degradation process highly unlikely.

No inflammatory response at the implantation site was seen in
any of the mice examined 4 or 7 weeks postimplantation. By week
14, a thin but distinct layer of fibrous connective tissue had formed
around the implant. Occasionally, a few multinucleated giant cells
were associated with the fibrous connective tissue. There was only
minimal evidence of any inflammatory cell infiltrate throughout the
1-year study. Thus, a typical "foreign body response" was seen,
resulting in the encapsulation of the foreign body without evidence
for any significant, pathological abnormalities at the implantation site.
In addition, necropsies were performed at the end of 16 weeks and
56 weeks postimplantation. A total of 24 organs (brain, eyes, nose,
adrenal glands, lacrimal glands, salivary glands, trachea, thyroid,
thymus, heart, lung, liver, spleen, kidney, pancreas, esophagus,
cecum, colon, duodenum, ileum, jejunum, stomach, accessory sex
glands, and testes) were examined and were found to be free of any
histopathological abnormalities.

In summary, preliminary results from two animal models (rabbit
and mouse) indicate that poly($\underline{N}$-palmitoylhydroxyproline ester) elicits
a very mild, local tissue response that compares favorably with the
responses observed for established biomaterials such as medical grade
stainless steel or poly(lactic acid)/poly(glycolic acid) implants. At
this point, additional assays need to be performed to evaluate possible allergic responses, as well as systemic toxic effects, carcino-
genic, teratogenic, or mutagenic activity, and adaptive responses.

IV. TYROSINE-DERIVED POLY(IMINOCARBONATES)

A. Rationale for the Development of Tyrosine-Derived Polymers

Although polymers with excellent material properties can be derived
from aliphatic components only, many of the most widely used industrial polymers contain aromatic components as integral parts of their
backbone structure (28). It has long been recognized that for many
different types of polymers, thermal stability, stiffness, and mechanical strength increase with an increasing proportion of aromatic
backbone components. In particular, noncrosslinked condensation
polymers must almost invariably contain cyclic or aromatic backbone
components in order to exhibit extremely high tensile strength and
Young's modulus.

For example, there is a dramatic improvement in modulus, tensile
strength, and thermal stability when the aliphatic components in
polyamides (nylons) are replaced by aromatic components, resulting
in polyaramides such as Kevlar (29). Likewise, poly(ether ether
ketone) (PEEK), one of the mechanically strongest condensation

polymers known so far (30), is exclusively derived from aromatic
backbone components (structure 1). Yet, the known toxicity of
many phenols has so far discouraged the incorporation of aromatic
monomers into degradable biomaterials. A noteworthy exception is
the polyanhydrides that contain bis(p-carboxyphenoxy)alkane units
as comonomers (31).

poly(ether-ether-ketone) (PEEK)

STRUCTURE 1

As the proposed applications for biodegradable polymers become
more sophisticated, one can envision an increasing need for polymers
that combine biodegradability with an exceptionally high level of
mechanical strength. The widely investigated use of biodegradable
polyesters for the fabrication of surgical ligating devices (32) or
orthopedic implants such as bone nails is a good example for this
general trend. Another example is the proposed use of biodegradable stents to prevent the restenosis of cardiac arteries after bal-
loon angioplasty (33,34). In both examples the availability of biodegradable yet mechanically strong implants would make it possible
to combine a mechanical support function with the site-specific delivery of pharmacologically active agents (e.g., bone growth factor
for bone nails, heparin derivatives or steroids for arterial stents).

Based on these considerations we felt that it would be a logical
extension of our studies on pseudopoly(amino acids) to investigate
whether biodegradable polymers that incorporate the aromatic side
chain of tyrosine in their backbone structure would indeed combine
a high degree of biocompatibility with a high degree of mechanical
strength.

Another important consideration for our interest in tyrosine-
derived polymers is related to the known immunological activity of
L-tyrosine. One of the first wholly synthetic antigens was a
branched polypeptide in which poly(D,L-alanine) side chains were
attached to the ε-amino groups of a poly(L-lysine) backbone. The
polyalanine chains in turn had to be elongated with tyrosine-
containing peptides in order to obtain a high degree of antigenicity
(35). Later, several tyrosine-containing poly(amino acids) were
used as hapten or antigen carriers in numerous immunological studies (36). In 1974, Miller and Tees (37) reported that tyrosine has
adjuvant properties as well. An immunological adjuvant is a material

that stimulates the immune response toward a given antigen without necessarily being antigenic by itself. Based on this report, we speculated that a tyrosine-derived, biodegradable polymer could provide sustained adjuvanticity while simultaneously serving as repository for antigen. Such implantable antigen delivery devices would obviously be of value for vaccinations, in animal breeding and possibly in human medicine.

Although tyrosine is often regarded as the most antigenic amino acid, it is interesting to note that no sensitization was observed in guinea pigs when a copolymer of L-tyrosine and L-aspartic acid was repeatedly injected intraabdominally (38). Other attempts to induce antibodies with derivatives of poly(L-tyrosine) also failed (39). These findings indicate that polymers that contain tyrosine as part of their backbone structure (as opposed to polymers with tyrosine-containing pendant chains) are not necessarily strongly antigenic materials. Evidently, the level of the immunological activity exhibited by tyrosine is strongly dependent on the overall structure of the tyrosine-containing polymer. Thus it is important to investigate the immunological properties of tyrosine-containing pseudopoly-(amino acids).

B. Poly(iminocarbonates) as a Model System

Using the general synthetic concepts described in Sec. II, we employed tyrosine dipeptides as the monomeric starting material. After protection of the N and C termini, the reactivity of a fully protected tyrosine dipeptide (structure 2) could be expected to resemble the

STRUCTURE 2

reactivity of simple diphenols such as Bisphenol A (BPA). In the past, diphenols have been used in the synthesis of a variety of polymers such as aromatic polyesters, polyurethanes, polyethers, polycarbonates (24,28), and, recently, poly(iminocarbonates) (40, 41). Since among these polymers, only the poly(iminocarbonates) are readily degradable under physiological conditions, we selected these as a model system for our further studies.

Poly(iminocarbonates) are little known polymers that, in a formal sense, are derived from polycarbonates by the replacement of the carbonyl oxygen by an imino group (Fig. 5). This backbone modification dramatically increases the hydrolytic lability of the backbone, without appreciably affecting the physicomechanical properties of the polymer: the mechanical strength and toughness of thin,

poly(BPA-carbonate)   poly(BPA-iminocarbonate)

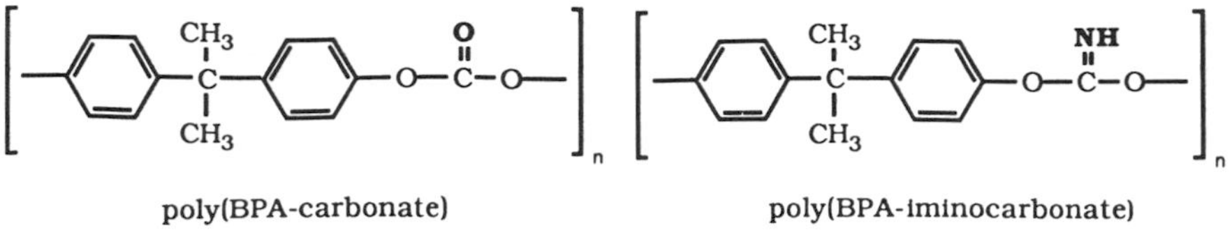

FIGURE 5 Molecular structures of poly(Bisphenol A carbonate) and poly(Bisphenol A iminocarbonate). The poly(iminocarbonates) are, in a formal sense, derived from polycarbonates by replacement of the carbonyl oxygen by an imino group.

solvent cast films of poly(BPA-iminocarbonate) is virtually identical to that of the corresponding poly(BPA-carbonate). On the other hand, poly(BPA-iminocarbonate) degraded within 9–12 months when implanted subcutaneously (41,42), while poly(BPA-carbonate) is stable for very long periods of time under physiological conditions (24).

Although the initially reported tissue compatibility tests for subcutaneous implants of poly(BPA-iminocarbonate) were encouraging (41,42), it is doubtful whether this polymer will pass more stringent biocompatibility tests. In correspondence with the properties of most synthetic phenols, BPA is a known irritant and most recent results indicate that BPA is cytotoxic toward chick embryo fibroblasts in vitro (43). Thus, initial results indicate that poly(BPA-iminocarbonate) is a polymer with highly promising material properties, whose ultimate applicability as a biomaterial is questionable due to the possible toxicity of its monomeric building blocks.

It was therefore particularly interesting to investiage whether it would be possible to replace BPA by various derivatives of L-tyrosine as monomeric building blocks for the synthesis of poly-(iminocarbonates). In order to be practically useful in drug delivery applications, the replacement of BPA by derivatives of tyrosine must give rise to mechanically strong yet fully biocompatible polymers.

C. Polymer Synthesis

Compared to polycarbonates, little work has so far been published on the synthesis of poly(iminocarbonates). The first attempted synthesis of a poly(iminocarbonate) was reported by Hedayatullah (44), who reacted aqueous solutions of various chlorinated diphenolate sodium salts with cyanogen bromide dissolved in methylene chloride. Unfortunately, Hedayatullah only reported the melting points and elemental analyses of the obtained products which, according to Schminke (40), were oligomers with molecular weights below 5000.

Only after pure, aromatic dicyanates had become available (45) a patent by Schminke et al. (40) described the synthesis of poly-(iminocarbonates) with molecular weights of about 50,000 by the solution polymerization of a diphenol and a dicyanate (Scheme 2). Bulk polymerization was also claimed to be possible.

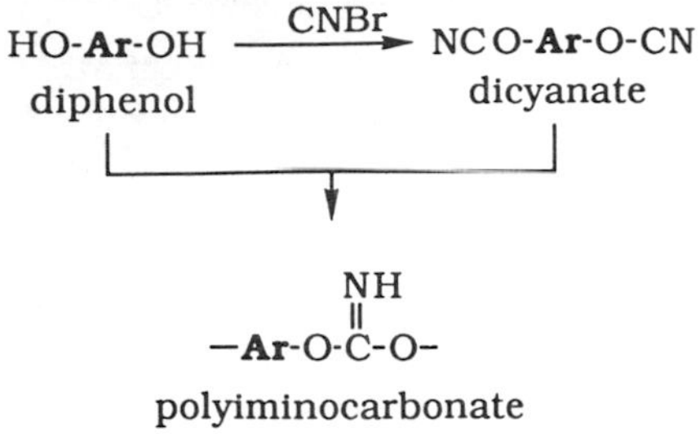

$$HO\text{-}\mathbf{Ar}\text{-}OH \xrightarrow{\text{CNBr}} NCO\text{-}\mathbf{Ar}\text{-}O\text{-}CN$$

diphenol   dicyanate

$$-\mathbf{Ar}\text{-}O\overset{NH}{\underset{}{\overset{\|}{C}}}O-$$

polyiminocarbonate

SCHEME 2

Initially we simply followed the published procedures for the synthesis of poly(iminocarbonates), replacing BPA by a variety of derivatives of tyrosine dipeptide (Figs. 6, 7 and Table 2). Although we were able to obtain the corresponding poly(iminocarbonates), the reactions were somewhat erratic and the polymer molecular weights were generally below 15,000. Since the reactions leading to poly-(iminocarbonates) had never been investigated in detail, we systematically optimized the reaction conditions (46).

Contrary to a literature report (40), we found that the bulk polymerization of diphenols with dicyanates is unsuitable for the

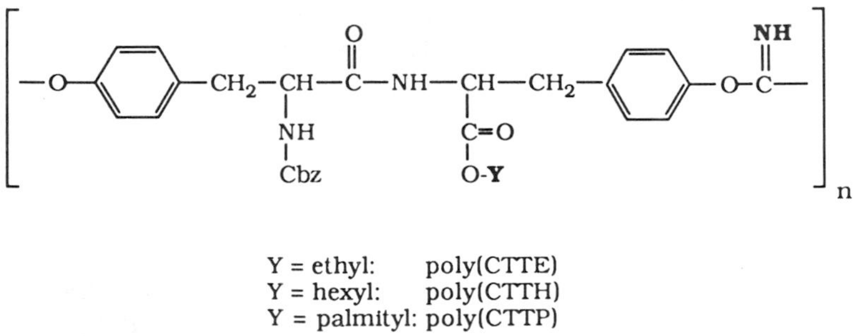

Y = ethyl: poly(CTTE)
Y = hexyl: poly(CTTH)
Y = palmityl: poly(CTTP)

FIGURE 6 Molecular structures of poly(CTTE), poly(CTTH), and poly(CTTP), a homologous series of tyrosine-derived polymers used in a study of the effect of the C-terminus protecting group on the materials properties of the resulting polymers. "Cbz" stands for the benzyloxycarbonyl group (47).

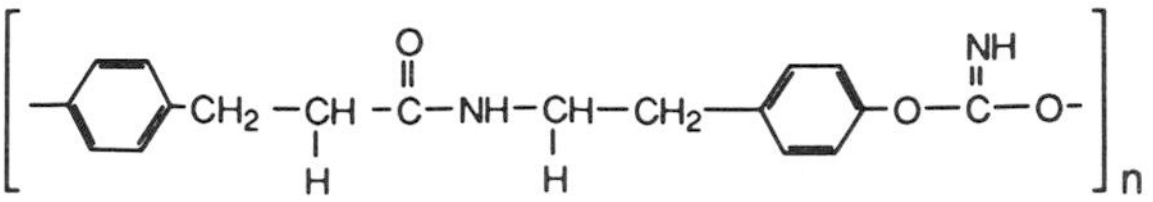

poly(desamino-tyrosyl-tyramine- iminocarbonate)

poly(Dat-Tym)

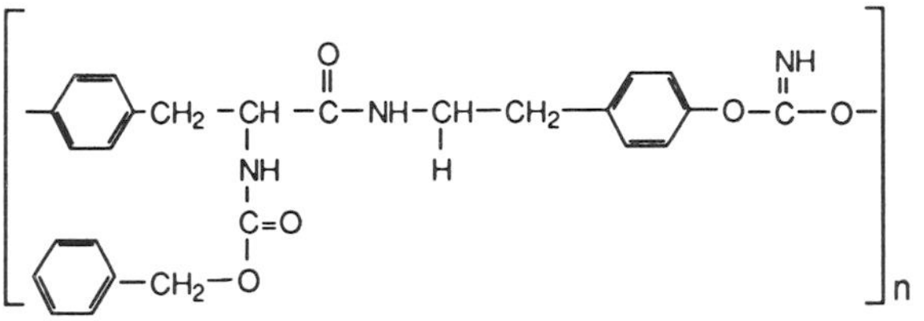

poly(N-benzyloxycarbonyl-tyrosyl-tyramine-iminocarbonate)

poly(Z-Tyr-Tym)

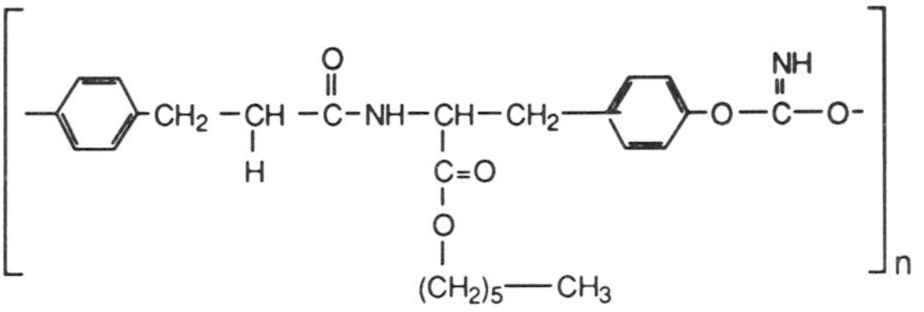

poly(desamino-tyrosyl-tyrosine hexyl ester- iminocarbonate)

poly(Dat-Tyr-Hex)

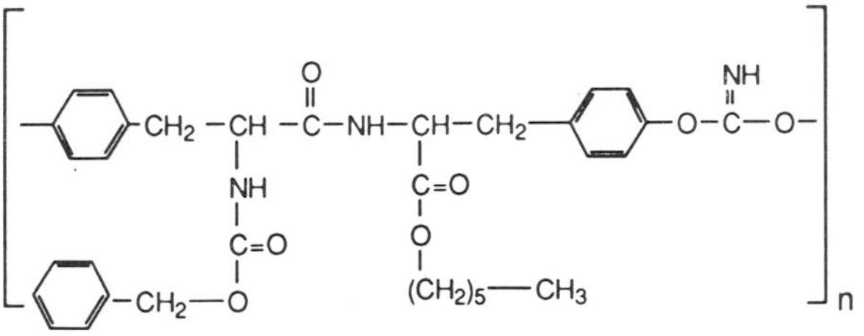

poly(N-benzyloxycarbonyl-tyrosyl-tyrosine hexyl ester-iminocarbonate)

poly CTTH

FIGURE 7 Tyrosine-derived poly(iminocarbonates) used to evaluate the effect of various side chain configurations on the physicomechanical properties of the resulting polymers.

TABLE 2   Mechanical Strength of Poly(iminocarbonates)[a]

Polymer	Tensile strength (kg/cm^2)	Tensile modulus (kg/cm^2)	Elongation (%)	
			yield	break
Poly(ether ether ketone)	820[a]	31,000[a]	6.5[a]	70[a]
	940[b]	36,500[b]	5.0[b]	50[b]
Poly(BPA-carbonate)	560[a]	20,000[a]	4.0[a]	100[a]
	620[b]	23,500[b]	n/a	80[b]
Poly(BPA-iminocarbonate)	510[a]	22,000[a]	3.5[a]	4.0[a]
Poly(CTTH-co-BPA iminocarbonate)	500[a]	18,600[a]	4.2[a]	4.4[a]
Poly(TDP-co-BPA iminocarbonate)[c]	515[a]	17,000[a]	4.1[a]	5.3[a]
Poly(L-lactic acid)[d]	115–390[b]	6,600[b]	6.4[b]	n/a
Poly(γ-methyl-D-glutamate)	40[b]	1,900[b]	5.0[b]	90[b]

[a]Tested in the author's laboratory according to ASTM D882-83.
[b]Literature values (see text for references).
[c]TDP, thiodiphenol.
[d]Low molecular weight. For poly(L-lactic acid) of high molecular weight [MW = 500,000], the tensile strength is about 500 kg/cm^2 and the tensile modulus is about 25,000 kg/cm^2.

preparation of structurally well-defined, linear poly(iminocarbonates).
Continuing the work by Schminke (40), we developed an optimized
solution polymerization procedure (see Illustrative Procedure 2) that
afforded tyrosine-derived poly(iminocarbonates) with molecular
weights above 50,000.

Illustrative Procedure 2: Poly(iminocarbonates) by Solution
Polymerization (46): Under argon, 1 g of a diphenol and an
exact stoichiometric equivalent of a dicyanate were dissolved in
5 ml of freshly distilled THF. 1 mol% of potassium tert-butoxide
was added, and the reaction was stirred for 4 hr at room temperature. Thereafter, the poly(iminocarbonate) was precipitated
as a gumlike material by the addition of acetone. The crude
poly(iminocarbonate) can be purified by extensive washings with
an excess of acetone. The molecular weight (in chloroform,
relative to polystyrene standards by GPC) is typically in the
range of 50,000–80,000.

Perhaps the most interesting finding of our synthetic studies
was that the interfacial preparation of poly(iminocarbonates) is possible in spite of the pronounced hydrolytic instability of the cyanate
moiety (see Illustrative Procedure 3). Hydrolysis of the chemically
reactive monomer is usually a highly undesirable side reaction during interfacial polymerizations. During the preparation of nylons,
for example, the hydrolysis of the acid chloride component to an
inert carboxylic acid represents a wasteful loss.

In contrast, during the interfacial preparation of poly(iminocarbonates), the hydrolysis of the dicyanate component regenerates the
diphenol component, which is a necessary reactant. Consequently,
it is possible to obtain poly(iminocarbonates) simply by the controlled
hydrolysis of a dicyanate under phase transfer conditions.

This feature of the interfacial preparation of poly(iminocarbonates) has an important consequence for the synthesis of copolymers:
if the dicyanate component is structurally different from the diphenol, partial hydrolysis of the dicyanate will lead to the presence
of two structurally different diphenol components that will compete
for the reaction with the remaining dicyanate. The interfacial copolymerization will therefore result in a random copolymer. On the
other hand, during solution polymerization no hydrolysis can occur.
Since the dicyanates can only react with diphenols and vice versa,
solution polymerization results in the formation of a strictly alternat-
ing copolymer.

Illustrative Procedure 3: Poly(iminocarbonates) by Interfacial
Polymerization (46): 4.5 mmol of a diphenol was dissolved in
45 ml of a 0.2 N aqueous solution of NaOH. Tetrabutylammonium
bromide (0.45 mmol) was added. The dicyanate (5.5 mmol) was

dissolved in 45 ml of a suitable chlorinated hydrocarbon such as carbon tetrachloride or methylene chloride. The two phases were vigorously mixed at 23°C by an overhead stirrer operating at 2000 rpm. If a polymeric product precipitated, it was collected on a Buchner funnel. Alternatively, the two phases were separated and the polymeric product was isolated by evaporation of the organic phase to dryness. The molecular weight (relative to polystyrene standards by GPC) is typically in the range of 80,000–150,000.

Finally, since dicyanates are prepared by the reaction of a diphenol with cyanogen chloride or cyanogen bromide (Scheme 2), we investigated whether it is possible to synthesize poly(iminocarbonates) directly from a diphenol and cyanogen halide. The use of cyanogen halide as an in situ cyanylating agent would eliminate the need to prepare and purify the reactive dicyanates in a separate reaction step and could significantly reduce the cost associated with the industrial production of poly(iminocarbonates). The reaction of diphenols with cyanogen halide would then be analogous to the preparation of polycarbonates from diphenols and phosgene.

Following the general procedure for the preparation of polycarbonates, Hedayatullah (44) had failed to obtain poly(iminocarbonates) of high molecular weight when he added an organic solution of cyanogen bromide into a strongly basic aqueous solution containing the diphenolate salt.

Since cyanogen bromide is freely soluble in water while phosgene is only sparingly soluble, the failure to obtain poly(iminocarbonates) from diphenols and cyanogen bromide can be explained by the marked tendency of cyanogen bromide to diffuse into the aqueous phase where it is rapidly hydrolyzed. Since Hedayatullah added an organic solution of cyanogen bromide into an aqueous solution containing BPA and sodium hydroxide, cyanogen bromide was by necessity exposed to an excess of base, leading to preferential hydrolysis rather than reaction with diphenol.

Based on this hypothesis we reversed the order in which the reactants are brought into contact. Consequently, we added an aqueous solution of BPA, sodium hydroxide, and phase transfer catalyst into a well-stirred solution of cyanogen bromide in carbon tetrachloride. Under these conditions poly(iminocarbonates) of high molecular weight were readily obtained (Fig. 8.)

The structure of poly(iminocarbonates) synthesized by the direct interfacial polymerization of BPA and cyanogen bromide was analyzed by NMR, Fourier transform infrared spectroscopy and elemental analysis and found to be identical in all aspects to authentic poly(iminocarbonates) obtained by solution polymerization (46).

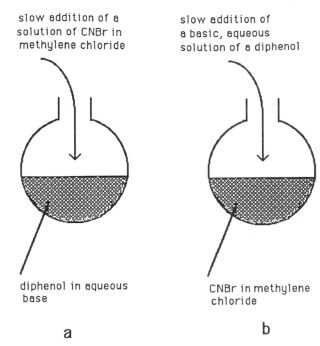

slow addition of a
solution of CNBr in
methylene chloride

slow addition of
a basic, aqueous
solution of a diphenol

diphenol in aqueous
base

CNBr in methylene
chloride

a   b

FIGURE 8   Interfacial synthesis of poly(iminocarbonates).   (a)
Hedayatulla (44) added a solution of cyanogen bromide into an aqueous solution of the diphenol.   Due to the excess of base present in
the reaction mixture at all times, most of the cyanogen bromide is
hydrolyzed, resulting in the formation of low molecular weight oligomers.   (b) In a recently published procedure, the order of addition
of the reagents was reversed.   In this way the excess of base in
the reaction mixture can be minimized and poly(iminocarbonates) of
high molecular weight were obtained (46).

    In summary, our synthetic studies led to the development of in-
terfacial and solution polymerization procedures for the preparation
of poly(iminocarbonates) of high molecular weight.   These procedures
have so far been employed for the synthesis of a small number of
structurally diverse poly(iminocarbonates).

D.   Physicomechanical Properties

In order to test the influence of the C-terminus protecting groups
on the properties of the resulting polymer, the ethyl, hexyl, and
palmityl esters of N-benzyloxycarbonyl-L-tyrosyl-L-tyrosine were
synthesized and the corresponding polymers [poly(CTTE),

poly(CTTH), and poly(CTTP); see Fig. 6] were prepared according to Scheme 2 (47). These polymers represent a homologous series differing only in the length of the alkyl group attached to the C terminus.

The first member of this series, poly(CTTE), was a high-melting and virtually insoluble polymer that could not be processed into shaped objects by any one of the conventional polymer-processing techniques. In this respect, poly(CTTE) was highly reminiscent of conventional poly(L-tyrosine).

A relative small increase in length of the C-terminus protecting group was found to have a significant influence on the polymer properties. Poly(CTTH) was readily soluble in many common organic solvents including methylene chloride, chloroform, and tetrahydrofuran. In addition, the melting range and glass transition temperature were significantly lowered by the replacement of the ethyl ester by a hexyl ester. Consequently, poly(CTTH) formed clear, transparent films and mechanically strong discs were obtained by compression molding.

As compared to poly(CTTH), the solubility and melting range of poly(CTTP) was not drastically changed in spite of a large increase in the length of the C-terminus protecting group. This seems to indicate that the C-terminus protecting group must have a certain minimum length in order for the tyrosine poly(iminocarbonate) to become soluble in organic solvents. Any further increase in the length of the C-terminus alkyl ester chain does not seem to significantly alter the melting range or solubility of the corresponding polymer.

In a related series of experiments, the amino group and/or the carboxylic acid group of tyrosine were replaced by hydrogen atoms. The corresponding tyrosine derivatives are 3-(4'-hydroxyphenyl)-propionic acid, commonly known as desaminotyrosine (Dat), and tyramine (Tym) (structures 3–5).

tyrosine	3-(4'-hydroxyphenyl) propionic acid	tyramine
(Tyr)	(Dat)	(Tym)
3	**4**	**5**

STRUCTURES 3, 4, and 5

Based on these monomeric building blocks a series of four structurally related poly(iminocarbonates) were synthesized carrying either no pendant chains at all [poly(Dat-Tym)], a N-benzyloxycarbonyl group as pendant chain [poly(Z-Tyr-Tym)], a hexyl ester group as pendant chain [poly(Dat-Tyr-Hex)], or both types of pendant chains simultaneously [poly(CTTH)] (Fig. 7).

This series of polymers made it possible to investigate the contribution of each type of pendant chain separately. Poly(Dat-Tym), the polymer carrying no pendant chains at all, was virtually insoluble in all common organic solvents. Due to its thermal instability in the molten state, poly(Dat-Tym) could be neither compression- nor injection-molded. In this respect, the processibility of poly(Dat-Tym) was not improved as compared to conventional poly(L-tyrosine).

Usually, the addition of pendant chains to the polymer backbone increases the solubility of the polymer in organic solvents by decreasing its crystallinity. We found, however, that there are significant differences in the magnitude of this effect between the N-terminus and the C-terminus pendant chains. In our test system, the incorporation of a pendant benzyloxycarbonyl group into the polymer structure did not significantly improve the polymer solubility. Like poly(Dat-Tym), poly(Z-Tyr-Tym) was virtually insoluble in all common organic solvents. On the other hand, the attachment of a pendant hexyl ester chain to the C terminus increased the solubility of the corresponding polymer significantly: Poly(Dat-Tyr-Hex) was readily soluble in many organic solvents including methylene chloride, chloroform, and tetrahydrofuran. The combination of both types of pendant chains within the same polymer resulted in no apparent advantage: the solubility of poly(CTTH) in organic solvents was not higher than that of poly(Dat-Tyr-Hex).

The thermal properties of tyrosine-derived poly(iminocarbonates) were also investigated. Based on analysis by DSC and thermogravimetric analysis, all poly(iminocarbonates) decompose between 140 and 220°C. The thermal decomposition is due to the inherent instability of the iminocarbonate bond above 150°C and is not related to the presence of tyrosine derivatives in the polymer backbone. The molecular structure of the monomer has no significant influence on the degradation temperature as indicated by the fact that poly(BPA-iminocarbonate) also decomposed at about 170°C, while the structurally analogous poly(BPA-carbonate) is thermally stable up to 350°C.

The low thermal stability of many poly(iminocarbonates) limits the use of melt fabrication techniques such as injection molding or extrusion. For example, among all six polymers tested, only poly-(Dat-Tyr-Hex) and poly(CTTH) had low enough softening points to be compression moldable without a significant degree of thermal decomposition.

The mechanical properties of tyrosine-derived poly(iminocarbonates) were investigated using the procedures described in ASTM standard D882-83 (Table 2). Solvent-cast, thin polymer films were prepared, cut into the required shape, and tested in an Instron stress strain tester. Since the films were unoriented, noncrystalline samples, the results are representative of the bulk properties of the polymers. In order to put these results into perspective, several commercial polymers were tested under identical conditions. In addition, some literature values were included in Table 2.

Poly(BPA-carbonate) is known for its high strength and toughness (24,28). Only recently, significantly stronger polymers have become available [e.g., poly(ether ether ketones) or polyaramides]. Within the limits of experimental error, poly(BPA-iminocarbonate) had the same tensile strength as commercial poly(BPA-carbonate), confirming an earlier report that the replacement of the carbonyl oxygen by an imino group has no noticeable effect on the mechanical properties of the resulting polymers (41). Interestingly, the tensile strength as determined in the ASTM standard test was quite insensitive to variations in polymer composition. For example, the replacement of 50% of all BPA units by either thiodiphenol (TDP) or CTTH had no significant effect on the tensile strength of the poly(iminocarbonate) copolymers. Likewise, poly(iminocarbonates) containing tyrosine-derived monomers exhibited similarly high values of tensile strength (about 500 kg/cm^2) and Young's modulus (about 20,000 kg/cm^2). For comparison, biodegradable polymers containing only aliphatic monomers were considerably weaker: for unoriented poly(lactic acid) films a <u>bulk</u> tensile strength of only 115−390 kg/cm^2 and a tensile modulus of only 6600 kg/cm^2 has been reported (49-51), while a film of poly(γ-methyl-D-glutamate) had only about 10% of the tensile strength of most poly(iminocarbonates) (52).

E. Biological Properties

Initially, the cytotoxicity against chick embryo fibroblasts of BPA, tyrosine, tyrosine dipeptide, and the dipeptide derivatives used in the synthesis of the polymers shown in Fig. 7 were evaluated in a comparative experiment (43). The surface of standard tissue culture wells was coated with 5 mg of each test substance. Then the adhesion and proliferation of the fibroblasts was followed over a 7-day period. Among all test substances, BPA was clearly the most cytotoxic material. Monomeric tyrosine derivatives containing the benzyloxycarbonyl group were also cytotoxic, while tyrosine itself, tyrosine dipeptide, and most of the protected dipeptide derivatives did not noticeably interfere with cell growth and adhesion and were therefore classified on a preliminary basis as possibly "nontoxic."

In a second series of experiments, 17 mg of each of the above monomers was implanted subcutaneously in the back of rabbits. After 7 days, the acute inflammatory response at the implantation site was evaluated. Bisphenol A resulted in a moderate level of irritation at the implantation site and was clearly the least biocompatible test substance. Tyrosine derivatives containing the benzyloxycarbonyl group caused a slight inflammatory response, while all other tyrosine derivatives produced no abnormal tissue response at all. These observations indicate that tyrosine dipeptide derivatives, even if fully protected, are more biocompatible than BPA, a synthetic diphenol.

In order to test the tissue compatibility of tyrosine-derived poly-(iminocarbonates), solvent cast films of poly(CTTH) were subcutaneously implanted into the back of outbread mice. In this study, conventional poly(L-tyrosine) served as a control (42). With only small variations, the experimental protocol described for the biocompatibility testing of poly(N-palmitoylhydroxyproline ester) (Sec. III. E) was followed. By week 16 the poly(CTTH) implants had been completely encapsulated by a thin layer of fibrous, connective tissue. In two of four implantation sites the presence of some multinucleated giant cells, lymphocytes, and polymorphonuclear leukocytes indicated a mild inflammatory response. Overall, however, a typical foreign body response was seen with little evidence for any pathological tissue damage at the implantation site. After a year, only a small amount of a white residue could be detected in two out of four implantation sites. Again, no gross pathological changes were evident from visual inspection of these sites. Histological examination revealed no major change in the tissue response since week 16: the small polymeric residues were surrounded by a thin (one to three cell layers) capsule of fibrous connective tissue. The majority of implants were localized in the subcutis but in some instances seemed to have gravitated into the dermis. Only very few lymphocytes and macrophages were present within or adjacent to the connective tissue layer.

Since poly(L-tyrosine) cannot be processed into shaped devices, compressed pellets rather than solvent cast films were used as control implants. Poly(L-tyrosine) formed strikingly yellow, moderately inflamed patches that remained at the implantation site throughout the 1-year study. Contrary to soluble proteins or peptides that ar rapidly degraded by enzymes, implants of conventional poly(L-tyro-sine) were evidently nondegradable over a 1-year period. At wee 56 all poly(L-tyrosine) implants were infiltrated by a moderate n ber of inflammatory cells.

Poly(CTTH) (Figs. 6 and 7) was also used as a model compound for the preliminary evaluation of the in vitro degradability of tyrosine-derived poly(iminocarbonates): Solvent cast films of poly(CTTH) became turbid and crumbled into small pieces within 10 months when exposed to phosphate buffer (pH 7.4) at 37°C. During the decomposition process, the release of ammonia was confirmed by quantitatively monitoring the ammonium ion concentration in the release buffer. This observation is a strong indication for the hydrolysis of the iminocarbonate backbone linkage during the degradation process (Scheme 3). There is also a good correlation between the degradation time in vitro and in vivo.

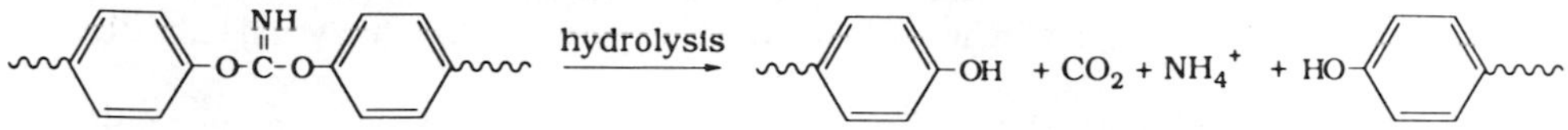

SCHEME 3

At this point, the major conclusions of our studies are that poly-(CTTH) degrades clearly more rapidly than conventional poly(L-tyrosine) and that poly(CTTH) elicits a typical foreign body response, while poly(L-tyrosine) causes a moderate inflammatory response over a 1-year implantation period. The preliminary evaluation of the tissue response toward poly(CTTH) is encouraging. Polymers not containing the benzyloxycarbonyl group are presently under investigation. In addition, a comprehensive study is now being conducted, designed to investigate the effect of the monomer structure and the presence of side chains on the rate of biodegradation and the tissue compatibility of several tyrosine-derived poly(iminocarbonates).

Finally, we selected poly(CTTH) as a model for the evaluation of the adjuvant properties of tyrosine-derived poly(iminocarbonates) (42). Solvent cast films of poly(CTTH) were loaded with 50 µg of bovine serum albumin (BSA) and implanted subcutaneously into outbread CD-1 mice. Blood samples were obtained at predetermined intervals and used to monitor the titers of circulating anti-BSA antibodies by a hemagglutination assay. Films of poly(BPA-iminocarbonate), loaded with an identical dose of 50 µg of BSA, were implanted as a control. As an additional control, a group of mice were subjected to a conventional immunization protocol consisting of a primary injection of 25 µg of BSA in physiological saline solution, followed by an identical booster injection of 25 µg of BSA at day 28. Throughout the entire course of the experiment the mean anti-BSA antibody titers in the animals treated with poly(CTTH) implants were significantly higher (P < 0.006) than the titers in the animals treated with poly(BPA-iminocarbonate) implants. Since the two types of implants were fabricated in an identical fashion, contained an identical dose of antigen, were shown to have comparable release profiles in vitro, and share the same iminocarbonate backbone structure, the observed higher antibody titers obtained for poly(CTTH) can conceivably be attributed to the intrinsic adjuvanticity of the monomeric repeat unit, CTTH.

These results open the exciting possibility of using degradable, tyrosine-derived polymers as "custom-designed" antigen delivery devices. On the other hand, our results indicate that the immunological properties of tyrosine-derived polymers will have to be carefully evaluated before such polymers can be considered for use as drug delivery systems or medical implants.

V. CONCLUDING REMARKS

Among all biomaterials described in this volume, the pseudopoly(amino acids) represent the latest and therefore least advanced entry into the growing field of promising biomaterials. Only a few pseudopoly-(amino acids) have been synthesized and characterized, and no clinical tests have so far been conducted. The presently available results, however, clearly indicate that the incorporation of nonamide linkages into the backbone of poly(amino acids) leads to a significant improvement of the processibility and the physicomechanical properties of the resulting polymers. A particularly promising finding is that none of the presently available pseudopoly(amino acids) exhibited gross toxicity or tissue incompatibility upon subcutaneous implantation in mice, rats, or rabbits.

Furthermore, our results on the characterization of the physicomechanical properties of tyrosine-derived poly(iminocarbonates) provide preliminary evidence for the soundness of the underlying experimental rationale: The incorporation of tyrosine into the backbone of poly(iminocarbonates) did indeed result in the formation of mechanically strong yet apparently tissue-compatible polymers.

Consequently, pseudopoly(amino acids) are now being actively investigated in several laboratories for applications ranging from biodegradable bone nails to implantable adjuvants. Considering that literally hundreds of structurally new pseudopoly(amino acids) can be designed from trifunctional amino acids or dipeptides as monomeric starting materials, one can predict with confidence that this synthetic approach will give rise to a variety of biomaterials that will find useful applications in veterinary or human medicine.

ACKNOWLEDGMENT

The editorial assistance of Mrs. Shirley Maimone in the preparation of this manuscript is gratefully acknowledged. The author also gratefully acknowledges the contributions of his students Satish Pulapura, Chun Li, Farzana Haque, and the contributions of his coworkers Israel Engelberg and Fred Silver. This work was supported by NIH grant GM 39455, by Zimmer Inc., and by NSF grant DMR-8902468.

REFERENCES

1. Hench, L. L., and Ethridge, E. C., Biomaterials: An Inter-
    facial Approach, Academic Press, New York, 1982, pp. 52-54.
2. Lyman, D. J., Polymers in medicine: An overview, in Poly-
    mers in Medicine (E. Chiellini and P. Giusti, eds.), Plenum
    Press, New York, 1983, pp. 215-218.
3. Yolles, S., and Sartori, M. F., Degradable polymers for sus-
    tained drug release, in Drug Delivery Systems (R. L. Juliano,
    ed.), Oxford University Press, New York, 1980, pp. 84-111.
4. Sanders, H. J., Improved Drug Delivery, Special Report,
    C&EN, 1985, April 1, pp. 31-48.
5. Anderson, J. M., Spilizewski, K. L., and Hiltner, A., Poly-
    α-amino acids as biomedical polymers, in Biocompatibility of
    Tissue Analogs, Vol. 1 (D. F. Williams, ed.), CRC Press,
    Boca Raton, 1985, pp. 67-88.
6. See Ref. 5, page 82: "polymers containing 3 or more different
    types of amino acids are not generally considered as being can-
    didates for biomedical materials."
7. A thorough and still valid discussion of the basic physical prop-
    erties of synthetic poly(amino acids) was written by C. H.
    Bamford, A. Elliott, and W. E. Hanby, Synthetic Polypeptides,
    Academic Press, New York, 1956.
8. Block, H., Poly(γ-benzyl-L-glutamate) and Other Glutamic Acid
    Containing Polymers, Gordon and Breach, New York, 1983,
    p. 10.
9. Sidman, K. R., Schwope, A. D., Steber, W. D., Rudolph, S.
    E., and Poulin, S. B., Biodegradable, implantable sustained
    release systems based on glutamic acid copolymers, J. Membr.
    Sci., 7, 277-291, 1980.
10. Arnold, L. J., Dagan, A., and Kaplan, N. O., Poly(L-lysine)
    as an antineoplastic agent and a tumor-specific drug carrier,
    in Targeted Drugs (E. P. Goldberg, ed.), John Wiley and sons,
    New York, 1983, pp. 89-112.
11. See Ref. 5, tables 3 and 4.
12. Kohn, J., and Langer, R., Polymerization reactions involving
    the side chains of α-L-amino acids, J. Am. Chem. Soc., 109,
    817-820, 1987.
13. Spatola, A. F., Peptide backbone modifications, in Chemistry
    and Biochemistry of Amino Acids, Peptides, and Proteins (B.
    Weinstein, ed.), Marcel Dekker, New York, 1983, pp. 268-357.
14. Greenstein, J. P., Studies on multivalent amino acids and pep-
    tides, J. Biol. Chem., 118, 321-329, 1937.
15. Fasman, G. D., Acyl N$\rightarrow$O shift in poly-DL-serine, Science,
    131, 420-421, 1960.
16. Jarm, V., and Fles, D., Polymerization and properties of opti-

cally active α-(p-substituted benzenesulfonamido)- β-lactones, J. Polym. Sci., 15, 1061-1071, 1977.

17. Kohn, J., and Langer, R., A new approach to the development of bioerodible polymers for controlled release applications employing naturally occurring amino acids, in Proceeding of the ACS Division of Polymeric Materials, Science and Engineering, American Chemical Society, 1984, Vol. 51, pp. 119-121.

18. Kohn, J., and Langer, R., Non-peptide poly(amino acids) for biodegradable drug delivery systems, in Proceedings of the 12th International Symposium on Controlled Release of Bioactive Materials (N. A. Peppas and R. J. Haluska, eds.), Controlled Release Society, Lincolnshire, IL, 1985, pp. 51-52.

19. Zhou, Q. X., and Kohn, J., Preparation of poly(L-serine ester): A structural analog of conventional poly-L-serine, Macromolecules, In Press.

20. Saegusa, T., Hirao, T., and Ito, Y., Polymerization of (4S,5R)-4-carbomethoxy-5-methyl-2-oxazoline, Macromolecules, 8, 87, 1975.

21. Yu, H., Lin, J., and Langer, R., Preparation of hydroxyproline polyesters, Proc. Int. Symp. Control. Rel. Bioact. Mater., 14, 109-110, 1987.

22. Yu, H., Pseudopoly(amino acids): A study of the synthesis and characterization of polyesters made from α-L-amino acids, Ph.D. Thesis, Massachusetts Institute of Technology, Cambridge, MA, 1988.

23. Yu, H., and Langer, R., Pseudopoly(amino acids): A study of the synthesis and characterization of poly(acyl-hydroxyproline esters), Macromolecules, 22, 3250-3255, 1989.

24. Goodman, I., and Rhys, J. A., Polyesters, American Elsevier, 1965, Vol. 1, and also: Vieweg, R., and Goerden, L., Polyester, in Kunststoff-Handbuch, Vol. 8, Carl Hanser Verlag, Munich, 1973.

25. Peppas, N. A., A model of dissolution-controlled solute release from porous drug delivery polymeric system, J. Biomed. Mater. Res., 17, 1079, 1983.

26. Saltzman, W. M., Pasternak, S. H., and Langer, R., Microstructural models for diffusive transport in porous polymers, in Controlled-Release Technology, ACS Symposium Series 348 (P. I. Lee and W. R. Good, eds.), American Chemical Society, 1987, pp. 16-33.

27. Langer, R., Brem, H., and Tapper, D., Biocompatibility of polymeric delivery systems for macromolecules, J. Biomed. Mater. Res., 15, 267-277, 1981.

28. Billmeyer, F. W., Jr., Textbook of Polymer Science, Wiley-Interscience, New York, 1971.

29. Black, W. B., and Preston, J., High-Modulus Wholly Aromatic Fibers, Marcel Dekker, New York, 1973.

30. Attwood, T. E., Dawson, P. C., Freeman, J. L., Hoy, L. R.
    J., Rose, J. B., and Staniland, P. A., Synthesis and proper-
    ties of polyaryletherketones, Polymer, 22, 1096-1103, 1981.
31. Rosen, H. B., Chang, J., Wnek, G. E., Linhardt, R. J., and
    Langer, R., Bioerodible polyanhydrides for controlled drug de-
    livery, Biomaterials, 4, 131-133, 1983.
32. Hay, D. L., von Fraunhofer, J. A., Chegini, N., and Master-
    son, B. J., Locking mechanism strength of absorbable ligating
    devices, J. Biomed. Mater. Res., 22, 179-190, 1988.
33. Kreamer, J. W., Intralumenal Graft, U.S. Patent 4,740,207,
    April 26, 1988.
34. King, S. B., Vascular stents and atherosclerosis, Circulation,
    79(2), 1-3, 1989.
35. Sela, M., Studies with synthetic polypeptides, Adv. Immunol.,
    5, 29, 1966.
36. Sela, M., Synthetic antigens and recent progress in immunology,
    in Peptides, Polypeptides and Proteins, Proceedings of the
    Rehovot Symposium on Poly(Amino Acids), Polypeptides and
    Proteins, John Wiley and sons, New York, 1974, pp. 495-509.
37. Miller, A. C. M. L., and Tees, E. C., A metabolizable adju-
    vant: Clinical trial of grass pollen-tyrosine adsorbate, Clin.
    Allergy, 4, 49-55, 1974.
38. Sela, M., and Katchalski, E., Biological properties of poly-α-
    amino acids, Adv. Prot. Chem., 14, 391-479, 1959.
39. See additional references cited in Ref. 38, page 460.
40. Schminke, H. D., Gobel, W., Grigat, E., and Pütter, R.,
    Polyiminocarbonic Esters and Their Preparation, U.S. Patent
    3,491,060, January 20, 1970.
41. Kohn, J., and Langer, R., Poly(iminocarbonates) as potential
    biomaterials, Biomaterials, 7, 176-181, 1986.
42. Kohn, J., Niemi, S. M., Albert, E. C., Murphy, J. C., Langer,
    R., and Fox, J. G., Single step immunization using a controlled
    release, biodegradable polymer with sustained adjuvant activity,
    J. Immunol. Meth., 95, 31-38, 1986.
43. Silver, F., and Kohn, J., Unpublished Results.
44. Hedayatullah, M., Cyanates et iminocarbonates d'aryle, Bull.
    Soc. Chem. France, 2, 416-421, 1967.
45. Grigat, E., and Pütter, R., Umsetzung von Cyansäureestern
    mit hydroxylgruppen haltigen Verbindungen, Chem. Ber., 97,
    3018-3021, 1964.
46. Li, C., and Kohn, J., Synthesis of poly(iminocarbonates):
    Degradable polymers with potential applications as disposable
    plastics and as biomaterials, Macromolecules, 22, 2029-2036,
    1989.
47. Kohn, J., and Langer, R., Backbone modification of synthetic
    poly-α-L-amino acids, in Peptides (G. R. Marshall, ed)., Escom,
    Leiden, Netherlands, 1988, pp. 658-660.

48. Kohn, J., Unpublished Results.
49. Feng, X. D., Song, C. X., and Chen, W. Y., Synthesis and evaluation of biodegradable block copolymers of ε-caprolactone and D,L-lactide, J. Polym. Sci., 21, 593-600, 1983.
50. Cohen, D., and Younes, H., Biodegradable PEO/PLA block copolymers, J. Biomed. Mater. Res., 22, 993-1009, 1988.
51. Zhu, K. J., Xiangzhou, L., and Shilin, Y., Preparation and properties of D,L-lactide and ethylene oxide copolymer: A modifying biodegradable polymeric material, J. Polym. Sci. Part C: Polym. Lett., 24, 331-337, 1986.
52. Mohadger, Y., and Wilkes, G. L., The effect of casting solvent on the material properties of poly(γ-methyl-D-glutamate), J. Polym. Sci. Polym. Phys. Ed., 14, 963-980, 1976.

7

Natural Polymers as Drug Delivery Systems

SIMON BOGDANSKY   Nova Pharmaceutical Corporation, Baltimore, Maryland

I. INTRODUCTION

New technological advances have brought many innovative drug delivery systems to the market and others to the brink of commercial-
ization. A variety of approaches have been investigated for the
controlled release of drugs and their targeting to selective sites:
polymeric prodrugs, drug conjugates, liposomes, monoclonal antibodies, and microcapsules (1-3). Synthetic and naturally occurring
absorbable polymers in the form of matrix (monolith) devices, hydrogels, microspheres, nanoparticles, films, and sponges are finding increasing use in drug delivery systems (4). They hold the promise
of providing better drug efficacy, reducing toxicity, and improving
patient compliance. Manufacturers have also recognized the potential
benefit of reformulating existing products in new delivery systems
as an effective tactic to extend their proprietary position for drugs
coming off patent.

The use of natural biodegradable polymers to deliver drugs continues to be an area of active research despite the advent of synthetic biodegradable polymers (5-12). The desirable characteristics
of polymer systems used for drug delivery, whether natural or synthetic, are minimal effect on biological systems after introduction
into the body; in vivo degradation at a well-defined rate to nontoxic
and readily excreted degradation products; absence of toxic endogenous impurities or residual chemicals used in their preparation,
e.g., crosslinking agents. Natural polymers remain attractive primarily because they are natural products of living organisms, readily

available, relatively inexpensive, and capable of a multitude of chemical modifications. The pharmaceutical and biopharmaceutical considerations in developing delivery systems based on natural polymers
have been described by Tomlinson and Burger (83) using albumin
as an example, and are as follows:

1. Purity of albumin
2. Route of preparation with respect to 3–7
3. Size, related to 13
4. Drug incorporation
5. Payload
6. Drug release (in vitro and in vivo)
7. Drug stability during 2 and 9
8. Particle stability (in vitro and in vivo)
9. Effect of storage on 6 and 7
10. Surface properties as they relate to 4,6 and 11
11. Presentation (e.g., free-flowing, freeze-dried powder or emul-
     sified suspension)
12. Antigenicity
13. Adsorption of plasma proteins
14. Carrier fate and toxicity
15. Drug and carrier biokinetics

A majority of investigations of natural polymers as matrices in
drug delivery systems have centered on proteins (e.g., collagen,
gelatin, and albumin) and polysaccharides (e.g., starch, dextran,
inulin, cellulose, and hyaluronic acid) (13). Most protein-based delivery systems have been formulated as solid crosslinked microspheres in which the drug is dispersed throughout the polymer matrix, although one recent report describes the preparation of an
enzyme-digestible disc made from an albumin-crosslinked hydrogel
(14). The formulation of proteins into microspheres has been dic-
tated to a great extent by considerations related to their mechanical
strength, dimensional and conformational stability in biological fluids,
and conditions under which processing is possible.

Collagen, because of its unique structural properties, has been
fabricated into a wide variety of forms including crosslinked films,
meshes, fibers, and sponges. Solid ocular inserts have also been
prepared from purified animal tissues.

Polysaccharides for drug delivery systems have been prepared
by a variety of routes. They will be discussed only briefly in this
chapter and the reader is directed to publications on the subject,
which appear as references.

Starch is usually derivatized by the introduction of acrylic
groups, prior to polymerization and manufacture into microspheres.
Poly(acryl) starch microspheres, as they are referred to, are an
example of a semisynthetic polymer system. Their extensive use as

drug carriers has been the subject of a recent review (15) and continues to be an area of active research (16-18). In addition to starch, dextran, inulin, and cellulose have frequently been used as drug carriers by covalently bonding the drugs, antibiotics, and enzymes to reactive derivatives of available functional groups (19-22). Questions still remain about the immunological properties of these polysaccharide derivatives and their fate in the body.

Hyaluronic acid is a linear polysaccharide found in the highest concentrations in soft connective tissues where it fills an important structural role in the organization of the extracellular matrix (23,24). It has been used in ophthalmic preparations to enhance ocular absorption of timolol, a beta blocker used for the treatment of glaucoma (25), and in a viscoelastic tear formulation for conjunctivitis (26). The covalent binding of adriamycin and daunomycin to sodium hyaluronate to produce water-soluble conjugates was recently reported (27).

Partially deacetylated chitin, a cellulose-like biopolymer consisting predominantly of $\underline{N}$-acetyl-D-glucosamine chains, in the form of films or crosslinked hydrogels has been used for the delivery of drugs (28,29). The suitability of chitin as a vehicle for the sustained release of drugs was examined using indomethacin and papaverine hydrochloride as model drugs (30). In vitro studies showed that over 80% of the indomethacin was released within 7 hr, whereas papaverine hydrochloride dissolved almost immediately.

Since the purpose of this book is to describe applications of biodegradable polymers to drug delivery systems, particularly from the perspective of the materials employed, the approach taken in this chapter has been to focus on the natural biodegradable polymers which have been used most extensively as matrices for the delivery of drugs. Consideration was also given to the fact that collagen has not been the subject of any recent reviews.

II. COLLAGEN

Collagen is a major structural protein found in animal tissues. It is usually present as aligned fibers in tissues such as skin and tendons, and serves to limit tissue deformation and to prevent mechanical failure. Because of its unique structural properties, collagen has been used in a wide variety of biomedical applications as homostatic fleece, absorbable sutures, sponge wound dressings, composite tissue tendon allografts, injectables for facial reconstructive surgery, and as drug delivery vehicles (31). Its characteristics as a biomaterial offer several advantages: it is biocompatible and nontoxic in most tissues (32); it is readily isolated and purified in large quantities; it has well-documented structural, physical, chemical, and immunological properties (33,34); and it can be processed into a variety of forms.

However, certain properties of collagen have adversely influenced its use as a drug delivery vehicle and its development has been somewhat overshadowed by advances made in both synthetic absorbable polymers (e.g., polylactide and polyglycolide) and nonabsorbable polymers such as silicone rubber and hydrogels. These properties include poor dimensional stability due to swelling in vivo; poor in vivo mechanical strength and low elasticity; possible occurrence of an antigenic response; tissue irritation due to residual aldehyde crosslinking agents; poor patient tolerance of ocular inserts; and variability in drug release kinetics.

A. Films

In 1973, Rubin et al. first described the ocular delivery of pilocarpine from a collagen film (35). Pilocarpine is a topically applied drug used for controlling elevated intraocular pressure associated with glaucoma. Bloomfield et al. demonstrated that the antibiotic gentamicin could be delivered from a soluble succinylated collagen ocular insert (36). Slater et al. used an insoluble collagen film crosslinked by glutaraldehyde to release gentamicin for the treatment of infectious bovine keratoconjunctivitis (37). A good correlation was found between in vitro and in vivo release in bovine eyes. However, in vivo retention of the inserts was difficult and could not be tolerated for more than several hours. These studies indicated that collagen film could have use as a biodegradable matrix for the delivery of gentamicin, provided the retention problems can be overcome.

The effect of various chemical modifications on the mechanical properties of reconstituted collagen and the diffusion rates of the steroid medroxyprogesterone was investigated (38). Formaldehyde-treated films, which are heavily crosslinked, have high moduli and low rates of drug release. Films treated with chrome quickly become hydrated in solution and have low moduli and very rapid drug release characteristics.

In vitro studies (39) further elucidated the parameters controlling drug release from collagen films. In those studies, collagen gel prepared from beef hide was rendered soluble by succinylation, and air-dried to produce a film. Films with different dissolution rates were produced by crosslinking with buffered formaldehyde of varying concentrations. Insoluble films were prepared by exposing samples to ammonia and glutaraldehyde. In vitro release rates were determined, after incorporation of tritium-labeled gentamicin, by exposing the film to buffered phosphate and counting radioactivity in solution.

Insoluble films did not change weight after 120 hr of immersion. For those samples, the initial rate of release of gentamicin was very rapid but decreased within 4 hr. The succinylated films exhibited

a correlation between the amounts of succinic anhydride added and
the time to disintegrate (collagen solubility). Disintegration times
also increased with increasing formaldehyde concentrations and could
be varied from 1 hr to 18 days. The initial rate of gentamicin release from a film with a 2-day disintegration time was significantly
lower than from insoluble films, but was equivalent at 2 hr. From
8 to 24 hr, however, the rate was again higher. While these studies
demonstrated that an antibiotic could be released from a collagen
film, drug release was difficult to control, was very dependent on
preparation methods, and could not be extended beyond 24 hr.

Since these earlier studies, there have been numerous investigations of collagen films, primarily for the delivery of ophthalmic drugs
(40,41,45). Shell (42) in his review on the ocular delivery of drugs
contends that continuous delivery will have major beneficial advantages over current therapy employing drops. Besides avoiding pulse
delivery with its side effects, sustained release will improve patient
compliance and allow for around-the-clock therapy, considered important for the treatment of glaucoma. He points out, however, that
there remain significant unanswered questions regarding the pharma-
codynamic aspects of continuous delivery. For example, 12 days
after the continuous administration of pilocarpine via an insert was
stopped, the hypotensive effect of the drug could be measured. The
introduction of new, potent ophthalmic drugs, which may have more
serious systemic side effects, will be a factor in the design and development of controlled delivery devices.

Succinylated soluble collagen (SSC) and insoluble collagen (IC)
films with procaine penicillin, erythromycin, or erythromycin estolate
were prepared and their in vitro release rates measured (40). Results showed that procaine penicillin was not released from IC, probably due to inactivation by the ammonia vapor or glutaraldehyde
used during the film preparation. Differences in release of drugs
from the two different collagen preparations were thought to be related to ionic repulsion and solubility factors. The preparation and
evaluation in cattle eyes of ring-shaped insoluble collagen inserts by
three different methods was detailed. The rings lacked elasticity
and folded or tore as insertion into the conjunctival sac was attempted. Rings prepared by a lyophilization procedure absorbed
water, swelled, and collapsed, and could not be inserted. While
the study demonstrated that sustained release of insoluble antibiotics for up to 24 hr is possible, it confirmed the design and fabrication problems encountered previously (37,43).

Drug delivery problems associated with pilocarpine, most notably
low ocular bioavailability and short duration of action, continue to
be significant (44). In an effort to prolong delivery and to avoid
undesirable effects, an investigation was carried out on the incorporation and in vitro release of pilocarpine from a soluble film

prepared from calf skin (45). For plain and crosslinked films,
"burst" release occurred within the first 4 hr, after which release
was zero order for 5 days (plain) and 7 days (crosslinked). Covalent bonding of the drug to collagen resulted in the prolongation
of the zero-order release up to 15 days.

B. Inserts

To overcome the problems associated with reconstituted collagen,
particularly its poor mechanical properties and nonreproducible delivery rates, investigators recently turned to another form of collagen. Collagen shields, made from intact porcine scleral tissue, were
designed for relieving the discomfort of epithelial trauma and pro-
moting epithelial healing in patients with chronic defects, recurrent
erosions, corneal abrasions, and following corneal surgery. Since
it had also been demonstrated that soft contact lenses could be used
to enhance ocular penetration of topical medications, including antibiotics (46-48), the concept of using a collagen corneal shield as a
drug delivery device seemed feasible.

The collagen shield, fabricated from procine scleral tissue, is a
spherical contact lens-shaped film whose thickness can be made to
vary from 0.027 to 0.071 mm. It has a diameter of 14.5 mm and a
base curve of 9 mm. Once the shield is hydrated by tear fluid and
begins to dissolve, it softens and conforms to the corneal surface.
Dissolution rates can be varied from 2 to as long as 72 hr by exposing the shields to ultraviolet radiation in order to achieve varying
degrees of crosslinking.

Treatment with topical tobramycin was compared with and without
the presence of a corneal shield fabricated from porcine scleral tissue in a rabbit eye model of Pseudomonas keratitis (41). A significant, 30-fold increase was found in the penetration of the antibiotic
into the anterior chamber in the eyes with the shield in place. After
12 hr of topical dosing, eyes with the shield had a statistically significant decrease in colony-forming units compared to treated eyes
without the shield and control eyes. In these studies the drug was
not incorporated into the shield. Instead, the shield acted either
by creating a tear reservoir, thereby increasing the time and maintaining drug concentration, or by absorbing the drug as it is dosed
and then gradually releasing it into the tear film as the shield dis-
solves.

An alternative approach was reported (49) in which tobramycin
was directly incorporated into a porcine scleral shield with a 72-hr
dissolution time by immersion in a solution of the drug. After implantation in rabbit eyes for up to 8 hr, the shields were removed
and the concentration of antibiotic in the corneas and aqueous humor

was determined. The controls consisted of animals which received
subconjunctival injections of tobramycin.

At the highest drug-loading level, corneal concentrations were
significantly higher than controls for up to 4 hr but equivalent at
8 hr. However, at 8 hr, eyes in which the shields were present
exhibited epithelial defects and mucus production. At a lower drug-
loading level the shields were well tolearted, with no signs of toxicity, and drug concentrations in the cornea were significantly higher
than controls for up to 4 hr. Concentrations of drug in the aqueous
humor of animals with shields were significantly higher than controls
for up to 1 hr (low dose) and 4 hr (high dose). Tobramycin concentrations in all corneal and aqueous humor samples at all intervals
assayed were above the mean inhibitory concentration for most sensitive strains of Pseudomonas. The authors speculated that the corneal distribution of antibiotic may be more uniform with the shields,
although no direct evidence was presented.

A similar study reports on the release of gentamicin and vancomycin from a collagen shield derived from porcine scleral tissue (50).
The majority of the gentamicin was released in vitro within 1 hr,
while vancomycin release took about 6 hr. In the experiments carried out in a rabbit eye model, a full-thickness epithelial defect was
created by application of n-heptanol. After soaking in the appropriate drug solution, the shields were applied to one eye, while
drops containing the drug were applied to the other eye as controls.
Drug levels in the cornea, aqueous humor, and tears were determined
at appropriate time periods. For the gentamicin-treated eyes, the
corneal antibiotic levels were significantly higher than controls at
30 min and 1 hr, but at no other period up to 6 hr. Furthermore,
the levels were judged to be considerably above the therapeutic dose.
By contrast, there was no significant difference in vancomycin levels
between shield- or drop-treated eyes at any site. Combination of
the two drugs generally resulted in higher gentamicin levels. The
differences in release characteristics of the two drugs were thought
to be due to reversible collagen binding by vancomycin. While the
findings are encouraging, extrapolation to human therapy is somewhat
complicated by the differences between human and rabbit eyes.

The clinical use of the collagen shield for the delivery of a number of drugs has been reported (51). The drugs studied—tobramycin, gentamicin, pilocarpine, dexamethasone, and flurbiprofen—
were incorporated by soaking the shields in a solution of the drug.
The devices were administered postsurgically to patients following
keratoplasty or cataract extraction. The patients tolerated the
shields without significant adverse effects and the reservoir effect
could be supplemented by application of additional drops of the drug
to reload the shield.

C. Sponges

Since the theory and medical applications of collagen sponge were
described by Chvapil et al. (52-54), they have been used to deliver
steroid hormones (55), the anticancer drugs trans-retinoic acid
(TRA) (56,57), and 5-fluorouracil (5-FU) (58), and antibiotics (59).
The effects of progesterone and several other steroids delivered
from a collagen matrix as a means for preventing allograft skin re-
jection in mice and rats was reported (55). Results showed that the
half-life of both progesterone and estradiol preparations was about
7 days, and about 90% of all steroid present was absorbed in 15-30
days.

All-TRA, delivered via a collagen sponge fitted into a cervical
cap, has been used in the treatment of intraepithelial cervical dysplasia, a precancerous lesion (57). The sponge, with an average
pore size of 400 Å, was made from pure collagen which was isolated
from bovine skin. Glutaraldehyde was used as a crosslinking agent
to provide high resilience and fluid-building capacity. Because of
the highly crosslinked nature of the collagen, it did not degrade in
the 24 hr in which it was in place. The sponges were cut into thin,
round wafers, approximately 3-4 mm thick and 7 mm in diameter.
The uptake of TRA into cervical tissue was determined by measuring
tissue radioactivity following insertion of the collagen sponge cervical cap containing tritium-labeled TRA. The TRA concentrations
peaked at 4 hr and then diminished rapidly by 24 hr. Since measurements of blood samples revealed that no systemic absorption had
occurred, high local concentrations over an extended period of time
may be possible without systemic side effects.

Clinical trials (60,61) have shown that TRA can be safely delivered topically using the cervical cap and collagen device. In
these studies the TPA was applied as a cream to the sponge, which
was then inserted into the vaginal vault. A new device was used
daily for up to 4 days. Systemic effects were minimal, and local
toxicity was dose-related and acceptable.

The release kinetics of tetracycline, incorporated into a collagen
sponge device prepared from acid-swollen collagen paste, have been
studied (59). After adjusting the pH to 5, "acidic" matrix, or 7.5,
"alkaline" matrix, aqueous formaldehyde was added and the slurry
was stirred to a foamlike consistency. Air-drying produced a sponge-
like material. Acidic collagen was a dense, feltlike material, while
alkaline collagen was highly porous and absorbant. Tetracycline
loading levels were dependent on slurry pH and the initial ratio of
collagen to drug in the slurry. Basic preparations released significantly higher amounts of the antibiotic than acidic preparations.
Potency was determined by the ability of sponge samples to inhibit
growth of E. coli. All sponges effectively inhibited grwoth, indicat-
ing that antibacterial activity was not lost during the preparation

process and that sufficient antibiotic diffused from the collagen matrix to provide a measurable antibiotic effect.

In vivo release was determined by implanting samples of collagen-
loaded tetracycline subcutaneously in rabbits for up to 7 days and
monitoring plasma levels. Initial levels were about 3 ng/ml on day
1, decreasing to 1.3 ng/ml on day 5 and then rising to 4.3 ng/ml
on day 7, leading to the speculation that nonuniform release resulted
from disintegration of the sponge, although additional studies would
be needed to verify this. In any case, since pharmacologically effective levels of tetracycline are between 0.4 and 1 μg/ml of plasma,
these devices could provide such levels for at least 7 days.

5-Fluorouracil (5-FU) and bleomycin have been shown to be effective in preventing fibroblast proliferation following ophthalmic
surgery. Currently, the drugs are administered by daily subconjunctival injections. The drawbacks to this approach are the necessity for frequent injections, pain associated with injections, corneal
epithelial defects, and needle tract leaks. Both drugs were incorporated into a crosslinked sponge made from purified bovine skin
collagen (58). The sponges were implanted in rabbit eyes for up
to 14 days (5-FU) and 50 days (bleomycin). Treated eyes were
judged to be significantly better than controls, which received the
sponge without drug. However, the absence of controls in which
the drug was directly injected makes the assessment of the efficacy
of this treatment difficult. A chronic inflammatory reaction was
elicited by the sponge, even in the absence of drug, but had no
adverse effect on treatment outcome.

D. Peptide and Protein Delivery

The development of peptides and proteins as therapeutic agents has
been greatly accelerated by advances in the fields of biotechnology.
A class of peptides receiving much attention are the various growth
factors, many of which have been synthesized by genetic engineering
or isolated from natural sources. Currently, the clinical administration of these drugs requires repeated intramuscular or subcutaneous
injections because oral administration is not suitable due to degradation caused by proteolytic enzymes in the gastrointestinal tract
(62). Clearly, there is a need for novel drug delivery systems
capable of prolonged release in order to minimize dosing frequency
and improve patient compliance (63-66).

Recently, water-soluble protein fractions, isolated from extracts
of bone matrix, were incorporated into a collagen matrix and shown
to induce bone (67,68) and cartilage formation both in vitro and in
vivo (69,70). In the latter studies, in the absence of the collagen
delivery system, the proteins were incapable of inducing cartilage
formation in vivo when implanted intramuscularly into mice. The
success of this approach appears to depend on delivering the active
agents at an effective dose over an extended time period.

III. NONCOLLAGENOUS PROTEINS

As discussed earlier, noncollagenous proteins, particularly albumin
and to a lesser extent gelatin, in the form of microspheres and nanoparticles continue to be exploited as drug delivery systems. Oppenheim (71) and Speiser (72) reviewed the technology developed to
produce ultrafine particles, often referred to as nanoparticles.

In a recent review, Oppenheim (73) proposed a functional defi-
nition based on particle size for classifying solid particulate systems
used to achieve controlled release of drugs. Nanoparticles were defined as solid colloidal particles ranging in size from 10 to 1000 nm
where the active ingredient is dissolved, entrapped, encapsulated,
adsorbed, or attached. Nanoparticles therefore have direct interaction with individual target cells, either with the external membrane
or inside the cell at the site of action. Microspheres, typically
greater than 4 μm, can be targeted to specific organ sites depending on their size and route of administration. A concise summary
of these relationships can be found in a review by Tomlinson (74).
In practice the terms microspheres, nanoparticles, and microbeads
are often used synonymously.

A. Albumin Microspheres

Albumin is a major plasma protein constituent, accounting for about
55% of the total protein in human plasma. Since they were first described by Kramer (75), albumin microspheres have been extensively
investigated in controlled release systems and as vehicles for the
delivery of therapeutic agents to local sites. Albumin microspheres
have been used extensively in diagnostic nuclear medicine for the
evaluation of organ function and circulatory studies following administration by a variety of routes (76). The exploitable features of
albumin include its reported biodegradation into natural products
(77-79), its lack of toxicity and nonantigenicity (80,81), and its
ready availability.

The delivery of drugs from albumin microspheres was reviewed
extensively by Davis et al. (82) and Tomlinson and Burger (83).
Morimoto (104), describing studies in which the mechanism of biodegradation was elucidated, showed that exposure of microspheres
to proteolytic enzymes such as trypsin, papain, and protease resulted in significant weight loss and surface topographic changes.

The preparation of microspheres can be accomplished by either
of two methods: thermal denaturation, in which the microspheres
are heated to between 95 and 170°C, and chemical crosslinking with
glutaraldehyde in a water-in-oil emulsion. Well-defined microspheres
can be easily prepared using these methods in large batches which
are usually physically and chemically stable. Newer preparation
methods for the preparation of albumin microspheres have been described by several authors (84-88).

Tomlinson (89) reported that over 90 drugs have been incorporated into various microsphere systems, while human serum albumin
(HSA) microspheres have been used to deliver more than 40 drug
and drug types (90). Burger et al. (91), Sokolski et al. (92,93),
and Gallo et al. (94) studied the manufacturing variables affecting
the properties of microspheres such as size, size distribution, surface and colloidal characteristics, and storage distribution, surface
and colloidal characteristics, and storage stability. Tomlinson and
Burger et al. (87,95) investigated the relationship between these
variables and drug incorporation and release. They concluded that
the incorporation of drug into microspheres influences the routes
available for microsphere manufacture; the appearance, size, and
surface charge of the microspheres; and the extent of swelling in
vivo.

As development activities reach the clinical trial stage, a key
product requirement is good shelf life stability with reproducible
properties. Freeze-drying has been investigated as a technique for
storing drug-containing microspheres (96). After 11 days of storage at 4°C, the particle size, state of aggregation, drug content,
and release characteristics were shown to be unaffected when compared to freshly prepared product. Freeze-drying was considered
suitable for the storage of clinical supplies.

The factors reported to influence drug release from microspheres
have been discussed (89) and are as follows (95):

Drug

Position in the microsphere
Molecular weight
Physicochemical properties
Concentration
Interaction between drug and matrix

Microsphere

Type and amount of matrix material
Size and density of the sphere
Extent and nature of crosslinking, denaturation, and polymerization
Presence of adjuvants

Environment

pH
Polarity
Presence of enzymes

Both processes used to manufacture microspheres produce relatively porous structures. Consequently, release of drugs is generally biphasic, with an initial fast release phase followed by a slower

first-order release. For highly water-soluble drugs, burst release
can lead to 95% of the incorporated drug being released within 5
min. Furthermore, the relatively high temperatures used in the
thermal degradation process preclude the use of albumin for thermally
sensitive drugs, proteins, and hormones. Since albumin is susceptible
to cleavage by proteases, in vivo release performance will be highly
tissue-specific and difficult to predict from in vitro tests.

While the major application of albumin microspheres is in the
area of chemotherapy, there have been studies reporting the release
of such varied compounds as 1-norgestrel (97), insulin (98), and
hematoporphyrins (99) from bovine serum albumin, and the antibacterial sulfadiazine from ovalbumin (100). In general, burst phenomena are found for all systems studied. However, the results from
the insulin study are worthy of note in that blood glucose levels
were depressed for more than 14 days following the administration
of insulin-containing BSA microspheres to diabetic rats. The smaller
microspheres were absorbed by day 28 and the larger particles by
day 56.

Corticosteroid drugs show a high incidence of adverse side effects. To minimize these effects, intraarterial therapy was investigated (101). Rabbit serum albumin (RSA) microspheres containing
corticosteroids injected into rabbit knee joints were evaluated for
the treatment of rheumatoid arthritis (102,103). The in vitro and
in vivo release characteristics of microspheres, prepared using both
chemical crosslinking and heat denaturation, were determined. With
respect to prednisolone, high glutaraldehyde content resulted in
rapid drug release (within 10 min), while low content exhibited
slower release. Diffusion of drug through the swelled microsphere
was considered to be the rate-controlling step. Chemical crosslinking was not effective in reducing diffusion of a low molecular weight
drug, while heat stabilization did result in the slow release of drug,
thereby avoiding the burst effect. Drug release decreased significantly as heating time increased, presumably due to an increase in
disulfide bond reformation which yielded a tighter structure.

The in vivo portion of the study employed relatively large microspheres (23 μm diameter) to prevent their being phagocytosed by
macrophages in the joint synovium. Heat-stabilized microspheres
containing either tritium-labeled prednisolone or triamcinolone were
injected either into the intraarticular space of the knee or into the
thigh muscle of rabbits. Radioactivity was measured in plasma as
a function of time. All preparations exhibited biphasic first-order
kinetics, with the microspheres having significantly slower release
for both drugs than controls in which animals were injected with suspensions of the drugs. Plasma levels were sustained for 15 days
with triamcinolone and for 8 days with prednisolone.

The delivery of drugs using albumin microspheres has been concentrated in the area of chemotherapy primarily because it offers the
capability of producing high local drug levels, while minimizing systemic toxicity, making it possible to maintain higher drug concentrations over a specific area for a longer period of time.   Morimoto et al.
(104) reviewed research on the use of albumin microspheres as drug
carriers for cancer chemotherapeutic agents such as 5-fluorouracil,
doxorubicin (adriamycin), and mitomycin C. Kim et al. (105), reporting on their use with cytarabine in rabbits, found burst release
during the first 10 min after intravenous administration.

Willmott et al. (106-108) studied bovine serum albumin (BSA)
microsphere residence time in the rat lung and the release characteristics of microspheres containing adriamycin following intravenous injection.   Drug release exhibited a burst phenomenon within the first
few hours, typical of these systems, and was attributed to rapid
efflux of drug by desorption from the surface.   A slower release
over a 24-hr period followed, which appeared to be related to biodegradation of the albumin matrix.   The data also suggested that
biodegradation could be controlled by varying the conditions of microsphere preparation, particularly the degree of crosslinking.

In a series of papers, Gupta et al. (109-112) studied the in
vitro release properties of heat-stabilized BSA microspheres containing adriamycin.   The biphasic release of drug was attributed to its
location in the microsphere.   The initial release results from surface
desorption and diffusion through pores, while the later release arises
from drug within the microsphere, which becomes available as the
microsphere hydrates.

Adriamycin release from BSA chemically crosslinked with tere-
phthaloyl chloride was studied by Sawaya et al. (113).   The unusual
aspect of these studies is that incorporation was accomplished by
immersing the crosslinked microspheres in a solution of the drug.
The structure formed in weakly crosslinked microcapsules is highly
porous, allowing for significant drug penetration and binding.   Increasing the extent of crosslinkage of the microcapsule walls by increasing either the concentration of the crosslinking agent or the
reaction time significantly reduced the ability of the microcapsules
to absorb drug.   In vitro release was highly dependent on the ionic
strength of the medium.   In deionized water less than 5% of the drug
was released in 24 hr, whereas in isotonic saline (0.9%) about 70%
was released within 1 hr.   These observations are consistent with
the fact that adriamycin is bound to the microcapsules and that an
ion exchange process is involved in the release (114).   It was concluded that at low and moderate sodium ion concentrations, release
kinetics are governed by a film diffusion process and by a particle
diffusion process at high ion concentrations.

Epirubicin, a successor to adriamycin, was studied in ovalbumin microspheres (115). The microspheres, prepared by a heat denaturation process, were 20 μm in diameter and contained 12.5% drug. In vitro dissolution was virtually complete after 6–8 hr. Since the plot of cumulative drug release versus time is hyperbolic, the authors attempted to fit the data to the Higuchi matrix dissolution model (116,117), which predicts a linear correlation between cumulative drug release and the square root of time. Linearity occurred only between 20 and 70% release.

The efficacy of the drug-containing microspheres was studied in mice with Ehrlich ascites carcinoma and in rats with a Walker 256 carcinoma. The difference between the mean survival time of mice treated with the microsphere preparation on day 0 (tumor inoculation day) was significantly greater than that of controls. However, when the drug was introduced at day 7 following inoculation, survival times were not significantly different. This raises the question of treatment efficacy, since in a clinical setting there is always a delay between the onset of tumor growth and therapy. In the Walker 256 tumor-bearing rats, the final metastatic incidence showed a significant difference between untreated controls and animals receiving either the drug-microsphere preparation or the free drug. However, no benefit to using microspheres was apparent, since the latter two groups had equivalent incidences of metastases.

Despite the intensive research efforts employing albumin microspheres for the delivery of drugs, little information has been available on the stability of drugs in heat-denatured human serum albumin microspheres. Adriamycin-containing albumin microspheres were examined for chemical integrity of the incorporated drug by Cummings and Willmott (118). No degradation was found to occur when microspheres were made at 120–160°C in the presence of glutaraldehyde. By contrast, in a recent study (119), several degradation products of mitomycin C (MMC) were identified after microspheres were heated to 120°C for 20 min, conditions generally employed in their preparation. The three products identified arose via the opening of the aziridine ring, which occurred during microsphere preparation. It was concluded that MMC is extremely unstable to the processes used in the preparation of HSA microspheres and that an interaction with either the protein or the crosslinking agent at high temperatures leads to activation of the drug and the formation of inactive but potentially toxic products. These findings have significant implications for research employing MMC for cancer chemotherapy.

There have been few reports on the clinical use of albumin microspheres for the delivery of drugs. The limited clinical trials, using chemotherapeutic agents, employ a technique referred to as arterial chemoembolism in which the drug-containing microspheres

are infused through a catheter. Preliminary studies using heat
denatured or crosslinked microspheres (45 μm in diameter) containing
5% MMC, showed that in vitro release was slow, with only 10% of the
MMC release after 24 hr and about 25% after 5 days. Microspheres
were intraarterially infused into rabbits and rats, as a preclinical
model of infusion treatment for patients with inoperable hepatic tumor.
In vivo MMC blood levels following intraarterial infusion of microspheres in either rabbit femoral arteries or rat hepatic arteries were
significantly higher than controls in which the drug was administered
intravenously (120,121).

This approach has been used primarily in Japan to treat patients
with malignant, inoperable hepatic cancer. Mitomycin C contained in
albumin microspheres was administered to patients (122) by percutaneous intraarterial catheterization. Tumor reduction was seen in
over 68% of the cases. By contrast, the control group, receiving
infusion therapy, had a poorer response. Survival times were also
greater for patients receiving the microspheres.

Intraarterial infusion of microspheres containing adriamycin was
used for the local treatment of breast cancer and recurrent breast
cancer with liver metastases (123). A reduction in tumor size was
noted when the microspheres were injected into the internal and
lateral thoracic arteries for treatment of the primary tumor. However, hepatic artery injection for liver metastases resulted in improvement in only one of three patients treated.

In reviewing intraarterial infusion of microencapsulated anticancer agents and its clinical applications, Nemoto (124) concluded
that this mode of treatment can be used for a wide variety of tumors,
providing remarkable therapeutic effects with minimal systemic toxicity. However, 25% of the treated tumors failed to respond, which
was thought to be attributable to either inadequate catheterization,
inadequate dose relative to tumor size, insufficient tumor vascularity,
low drug sensitivity, or a combination of these factors. More carefully designed studies will be necessary before this technique is
likely to meet with widespread acceptance.

In a novel application, Illum et al. (125) explored the possibility
of achieving slow nasal clearance by employing several bioadhesive
systems including rabbit serum albumin microspheres. They reported
on the clearance, in human subjects, of intranasally administered
microspheres containing the antiallergic agent sodium cromoglycate
and labeled with technetium-99m. Three hours after administration
50% of the initial activity was still at the site of deposition, while
the half-time of clearance in the controls, who were given the drug
either as a powder or a solution, was 15 min. The slower clearance
was assumed to result from two factors: mucoadhesion of the swelled
microspheres following water uptake and deposition of the microspheres in the anterior portion of the nose where there are fewer

cilia. Preliminary unpublished work on sheep has shown that gentamicin applied intranasally in starch microspheres gave a 30-fold
higher blood level than controls.

B. Magnetic Albumin Microspheres

An alternative use of albumin microspheres for delivering chemotherapeutic agents involves the incorporation of colloidal, magnetic
particles. The technique entails the incorporation of ultrafine particles of magnetite, Fe_3O_4, into the microspheres, making them responsive to an externally applied magnetic field. The application of
a magnetic field on the target site, such as a tumor, followed by the
intraarterial injection of the magnetically responsive albumin microspheres (MRAMs) containing drug results in the microspheres concentrating in the target site and releasing the drug. This sytem of site-
specific delivery eliminates the adverse side effects associated with
systemic drug delivery and is particularly useful for cytotoxic drugs,
which have a high systemic toxicity. Since Widder et al. (126,127)
first reported on the use of magnetic albumin microspheres and Morimoto (104) and Widder and Senyei (128) extensively reviewed their
preparation and drug release properties, numerous papers have appeared on the subject (129-132).

Lee et al. (133) studied the size distribution and morphology of
BSA microspheres containing magnetite after their introduction and
residence in the lungs and liver of mice. Cavities found on the
surface of the microspheres retrieved from the liver were attributed
to biodegradation, not to dissolution of entrapped drug as postulated
by Gupta et al. (109). This study offers clear evidence of the degradation process in vivo, probably initiated by tissue enzymes.

Papisov et al. (134-136) studied the in vivo kinetics of radio-
labeled, magnetically targeted drug carriers including albumin microspheres and developed a mathematical model for carrier capture in
tissue. He stressed that the efficacy of magnetic targeting should
be very low for drugs with long half-lives in the systemic circulation. By contrast, the maximal effect of targeting can be expected
for drugs with short half-lives, which are rapidly inactivated upon
release into the blood.

The release of adriamycin from BSA microspheres both with and
without magnetic particles present as magnetite was investigated
(137). While the presence of drug and/or magnetite had no effect
on the size of the hydrated or unhydrated microspheres, the stabilization temperature affected the size of hydrated microspheres.
Drug incorporation was also influenced by stabilization temperature
for different reasons. Low incorporation was noted at high tempera-

tures due to drug decomposition and at low temperatures because of
a high surface-adsorbed fraction which was subsequently washed
off. The presence of magnetite reduced the amount of drug
incorporated, probably by occupying space normally available for
drug. Consistent with previous studies, biphasic drug release was
observed, which was dependent on stabilization temperature as well
as the presence of Fe_3O_4.

The tumoricidal activity of MRAMs containing either doxorubicin
or protein A, a constituent of the cell wall of <u>Staphylococcus aureus</u>,
was tested in rats with induced mammary adenocarcinoma (138).
The microsphers, 1 µm in diameter, were injected into the tail artery
and localized to the implanted tumor using a magnet. Survival was
significantly greater than that in control animals that received intravenously or intraarterially administered drug-containing microspheres.
The authors point out that the major hurdle to human studies is
the development of a suitable magnetic source capable of directing
the MRAMs in the microcirculation to the site of deep-seated tumors.

The disposition of adriamycin following intraarterial administration
in rats via MRAMs was investigated (139). As in the previous study,
microspheres (0.75 µm diameter) were injected into the tail artery
of rats and localized in one segment designated as the target site.
Control animals received intraarterial injections of the drug in solution. The concentration of drug was measured in the heart, lung,
liver, spleen, kidney, small intestine, and serum as a function of
time for up to 48 hr. Increased tissue exposure to the drug and
higher targeting efficiency, as determined by AUC measurements,
were found for the animals receiving the localized magnetic microspheres. Increased exposure to the target site with a concomitant
reduction in drug concentration in systemic tissues should reduce
the dose-related toxicity of adriamycin.

In a related study, the ultrastructural disposition of adriamycin-
loaded MRAMs at the target site was followed over a 72-hr period
using transmission electron microscopy (TEM) to ascertain if this
carrier could traverse the vascular lining of healthy tissue. As the
authors point out, if the drug delivery device crosses the endothelial
barrier in the target tissue, then most of the drug is likely to be
released at the cellular or subcellular level, thereby increasing the
efficacy of the therapy and reducing the systemic drug toxicity.

Target tissue sections from animals sacrificed at 8 hr and later
after dosing showed the presence of microspheres in the extravas-
cular interstitial tissue. Changes in red blood cells and damage to
other cellular components suggest that the cytotoxic properties of
adriamycin have been retained. The microspheres appeared to still
be intact for up to 72 hr.

C. Casein Microspheres

Prior to the study by Chen et al. (140), only one publication discussed the use of the protein casein as a drug carrier (141). Chen
et al. systematically compared the many features of albumin and
casein microspheres—morphology, drug (adriamycin) incorporation,
and release—in an effort to identify important factors in the antitumor effect of this delivery system.

Casein microspheres containing adriamycin were produced by
dispersing casein in heated phosphate buffer and then adding the
drug which was in a 2% w/v solution with lactose as excipient. This
dispersed phase was added to a cottonseed, light petroleum mixture,
stirred at high speed to which was then added a solution of glutaraldehyde. Human serum albumin microspheres, used for comparison,
were also prepared by chemical crosslinking using glutaraldehyde,
by a process described previously (106,107).

From this comparison of the two proteins it was concluded that
there is a different distribution of incorporated drug, leading to a
different surface charge and drug release rate. In addition, there
is evidence within the microspheres of a complexed form of the drug.
The antitumor effects of the two protein systems were similar, although less drug was present in casein and the release rate was
slower. In this instance, the slower rate of release is apparently
sufficient to compensate for the lower total dose of drug administered.

D. Gelatin Microspheres

Gelatin is obtained by the partial hydrolysis of collagenous animal
tissue, which converts the tough fibrous collagen into an unoriented
water-soluble protein. This process has been described in detail
(142). Because it has poor mechanical properties when wet, it is
used primarily in surgery as a hemostatic agent in the form of a
foam or film. Gelatin as the basis of drug delivery systems has
been the subject of several review articles by Oppenheim (71,73).
A comprehensive review of their preparation and use as carriers for
antineoplastic agents has been reported by Sezaki et al. (143).

As Oppenheim points out in his extensive review, although the
literature contains many examples of the use of albumin microspheres,
there are far fewer reports describing gelatin systems (144). Despite this, gelatin offers the advantages of ready availability, relatively low antigenicity, and a good history in parenteral formulations.
By comparison with albumin, gelatin offers several advantages in
drug delivery systems: weaker binding to drugs, less potential for
drug degradation because of low temperature preparation techniques,
and lower antigenicity.

Gelatin has been used as a coating material to microencapsulate
drugs and as a matrix, usually in the form of microspheres. The

microspheres, used to deliver chemotherapeutic agents including mitomycin C, adriamycin, 5-fluorouracil, and bleomycin, have been manufactured using emulsification (145-150) and desolvation (151-153) techniques. There is also a report on microspheres being produced by reverse micellization (154). The influence of pH, gelatin concentration, stirring rate, and surfactant concentration on the size distribution and drug content of microspheres prepared by emulsification has been described (155). Despite claims for an improved process, the microspheres were inhomogeneous and agglomerated. Drug content was low, with the majority of free drug entrapped near the surface where it presumably would burst-release, although no release data were presented.

The use of gelatin micropellets in a controlled oral release dosage form has been reported (156). Micropellets, 0.4 to 1.5 mm in diameter, are produced by a spray congealing method. The drug is dispersed in a gelatin sol which is then poured into a hydrophobic medium such as warm liquid paraffin. This results in coalescence of the gel to form spherical droplets. After chilling and washing, the micropellets are crosslinked with a mixture of glutaraldehyde and formaldehyde (157,158). The factors responsible for producing good micropellets are described: paraffin and drug-gelatin viscosities, stirring speed, temperature, and crosslinking conditions. However, the particles produced are quite porous, as evidenced from SEM, and very hydrophilic, resulting in typical first-order release. Initial drug release occurs from dissolution of drug adsorbed on the surface (burst), followed by diffusion of solvated drug from the interior of the micropellet.

Oppenheim (144), in reviewing his studies on drug release from gelatin microspheres produced by emulsion, observed that particle size and porosity play a major role in determining release kinetics. Submicrometer size particles released drug more rapidly than 15-μm particles of the same porosity. By contrast, desolvation techniques are capable of producing nonporous submicrometers which do not exhibit burst release. This is consistent with the findings of El-Samaligy and Röhdewald (153), who concluded that release appears to be due primarily to degradation of the microsphere with some contribution from surface desorption.

Data on the biocompatibility of gelatin microspheres is extremely limited. Drug-free microspheres elicited no untoward effects when injected intravenously into mice over a 12-week period (159). When albumin microspheres were injected repeatedly into the knee joints of rabbits, pronounced rapid joint swelling occurred after the second and subsequent injections. By comparison, no swelling occurred when gelatin microspheres were administered (160).

Magnetically responsive gelatin microspheres, prepared using methods adopted from albumin preparations (161), were evaluated

as a delivery system for the antineoplastic drug aclarubicin (162).
Burst release, reaching 45%, was noted in vitro. Microspheres injected intravenously into the mouse tail vein were observed by fluorescence microscopy to be localized in the lung. Biodegradation in
the pulmonary capillaries began by 36 hr and was complete 10 days
after administration. When the microspheres were injected into mouse
femoral muscle, mild inflammation was observed.

As was the case with albumin, reports on clinical studies employing gelatin microspheres are limited and confined to Japan. Bleomycin, contained in a microsphere-in-oil formulation, was targeted to
the lymph nodes for the treatment of infantile cystic hygroma, a
benign tumor (143). Successful treatment was noted in all cases.
Chen et al. (163) targeted magnetic microspheres containing 5-
fluorouracil to the esophagus by application of an external magnet.
The concentration of the drug in the esophageal mucosa was over
30 times the systemic levels after intravenous administration. Here
again, more carefully controlled studies using gelatin will be required.

E. Fibrinogen Microspheres

A number of studies have been reported on the use of microspheres
made from fibrinogen for the delivery of the anticancer agents doxorubicin and 5-FU (164-166). Fibrinogen, a soluble plasma protein
of molecular weight 340,000, is commonly used as a coagulant. Microspheres were prepared from bovine blood fibrinogen by an emulsification technique followed by thermal denaturation at either 90 or
160°C. Diameter and drug loading were dependent on temperature
and the microspheres were shown to be susceptible to degradation
by protease enzyme. In vitro studies of 5-FU showed that after an
initial burst release there was a slow continuous release over 5 days.

In a similar study, adriamycin was incorporated into fibrinogen
microspheres. In contrast to 5-FU, adriamycin released slowly from
the matrix for up to 7 days, with little evidence of burst. This
difference is probably attributable to the lack of surface release of
adriamycin as evidenced by the unchanged nature of the microsphere
surface.

In vivo studies involving intraperitoneal injections of drug-
containing microspheres into mice with Ehrlich ascites carcinoma gave
varying results. Tumor-bearing mice given the microspheres with
adriamycin lived longer than control animals given the free drug.
However, animals receiving microspheres containing 5-FU did not
have significantly better survival times than controls receiving the
drug alone. It would appear that, as was the case in vitro, the in
vivo release of adriamycin is slower than that of 5-FU, resulting in
a greater antitumor effect.

IV. CONCLUSIONS

Natural polymers, particularly in the form of microspheres, have an important role in the controlled release of drugs and their targeting to selective sites. Yet there remain many issues to be addressed before they will have widespread use in clinical situations. Among these issues are better understanding of the kinetics of drug release; more effective ways to control burst phenomena; greater understanding of drug-polymer interactions and their effect on shelf life stability; additional animal studies to determine local tissue response, biodegradation rates, and metabolic fate; and, most importantly, as it relates to cancer chemotherapy, well-designed clinical studies to assess efficacy in relation to current therapies.

In the area of drug targeting, there needs to be continuing emphasis on understanding the interaction between polymeric particles and biological systems such as blood components, cell types (e.g., phagocytes), and cell receptors.

For ocular drug delivery, much progress has been made in developing and commercializing novel drug delivery systems. However, as Shell points out (42), these systems have not gained widespread acceptance because of patient compliance and cost-benefit issues. Ocular delivery will gain greater acceptance as progress is made in educating physicians and patients to the benefits of the approach, particularly the reduction in systemic side effects and better patient compliance. Treatment costs should also become more competitive with current therapy through improvements in the processes used to manufacture these new delivery systems.

REFERENCES

1. Juliano, R. L. (ed.), Biological Approaches to the Controlled Delivery of Drugs, Ann. N.Y. Acad. Sci., 507, New York, 1987.
2. Gardner, C. R., in Drug Delivery Systems: Fundamentals and Techniques (P. Johnson and J. G. Lloyd-Jones, eds.), Horwood, Chichester, 1987.
3. Poznansky, M., and Juliano, R. L., Pharmacol. Rev., 36, 277, 1984.
4. Linhardt, R. J., in Controlled Release of Drugs: Polymers and Aggregate Systems (M. Rosoff, ed.), VCH, New York, 1989.
5. Rosen, H. B., Kohn, J., Leong, K., and Langer, R., in Controlled Release Systems: Fabrication Technology, Vol. 2 (D. Hsieh, ed.), CRC Press, Boca Raton, 1988.

6. Illum, L., and Davis, S. S. (eds.), Polymers in Controlled
    Drug Delivery, Wright, Bristol, 1987.
7. Langer, R., and Peppas, N., J. Macromol. Sci., 23, 61, 1983.
8. Juni, K., and Nakano, M., Crit. Rev. Therapeut. Drug Car-
    rier Syst., 3, 209, 1986.
9. Holland, S. J., Tighe, B. J., and Gould, P. L., J. Control.
    Rel., 4, 155, 1986.
10. Mills, S. N., and Davis, S. S., in Polymers in Controlled Drug
    Delivery (L. Illum and S. S. Davis, eds.), Wright, Bristol,
    1987.
11. Juni, K., and Nakano, M., in Polymers in Controlled Drug
    Delivery (L. Illum and S. S. Davis, eds.), Wright, Bristol,
    1987.
12. Maulding, H. V., J. Control. Rel., 6, 167, 1987.
13. Heller, J., in Hydrogels in Medicine and Pharmacy, Vol. 3,
    Properties and Applications (N. A. Peppas, ed.), CRC Press,
    Boca Raton, 1987.
14. Park, K., Biomaterials, 9, 435, 1988.
15. Edman, P., Artursson, P., Laakso, T., and Sjöholm, I., in
    Polymers in Controlled Drug Delivery (L. Illum and S. S.
    Davis, eds.), Wright, Bristol, 1987.
16. Laakso, T., Edman, P., and Brunk, U., J. Pharm. Sci., 77,
    138, 1988.
17. Laakso, T., Stjarnkvist, P., and Sjöholm, I., J. Pharm. Sci.,
    76, 134, 1987.
18. Baille, A. J., Coombs, G. H., Dolan, T. F., and Hunter,
    C. A., J. Pharm. Pharmacol., 39, 832, 1987.
19. Schacht, E., in Polymers in Controlled Drug Delivery (L.
    Illum and S. S. Davis, eds.), Wright, Bristol, 1987.
20. Schacht, E., Vermeersch, J., Vandoorne, F., and Vercauteren,
    R., J. Control. Rel., 2, 245, 1985.
21. Molteni, L., Optimization of Drug Delivery (H. Bundgaard,
    A. B. Hansen, and H. Kofod, eds.), Munksgaard, Copen-
    hagen, 1982.
22. Sjöholm, I., and Edman, P., in Microspheres and Drug Ther-
    apy: Pharmaceutical, Immunological and Medical Aspects (S. S.
    Davis, L. Illum, J. G. McVie, and E. Tomlinson, eds.), Else-
    vier, Amsterdam, 1984.
23. Laurent, T. C., Acta Otolaryngol. Suppl. (Stockholm), 442,
    7, 1987.
24. Mason, R. M., Prog. Clin. Biol. Res., 54, 87, 1981.
25. Chang, S. C., Chien, D. S., Bundgaard, H., and Lee, V. H.,
    Exp. Eye Res., 46, 59, 1989.
26. Limberg, M. B., McCaa, C., Kissling, G. E., and Kaufman,
    H. E., Am. J. Ophthalmol., 103, 194, 1987.

27. Cera, C., Terbojevich, M., Cosani, A., and Palumbo, M., Int. J. Biol. Macromol., 10, 66, 1988.
28. Pangburn, S. H., Trescony, P. V., and Heller, J., Seminar on Chitin, Chitosan, and Related Enzymes (J. P. Zikakis, ed.), Academic Press, Orlando, 1984.
29. Yamaguchi, R., Hirano, S., Arai, Y., and Ito, T., Agric. Biol. Chem., 42, 1981, 1977.
30. Miyazaki, S., Ishii, K., and Nadai, T., Chem. Pharm. Bull., 29, 3067, 1981.
31. Arem, A., Clin. Plast. Surg., 12, 209, 1985.
32. DeLustro, F., Cendell, R. A., Nguyen, M. A., and McPherson, J. M., J. Biomed. Mat. Res., 20, 109, 1986.
33. Kühn, K., and Timpl, R., Prog. Clin. Biol. Res., 154, 45, 1984.
34. Timpl, R., in Biochemistry of Collagen (G. N. Ramachandran and A. H. Reddi, eds.), Plenum Press, New York, 1976.
35. Rubin, A. L., Stenzel, K. H., Miyata, T., White, M. J., and Dunn, M. J., Clin. Pharmacol., 13, 309, 1973.
36. Bloomfield, S. E., Miyata, T., Dunn, M. W., Bueser, N., Stenzel, K. H., and Rubin, A. L., Arch. Ophthalmol., 96, 885, 1978.
37. Slater, D. H., Costa, N. D., and Edwards, M. E., Aust. Vet. J., 59, 4, 1982.
38. Bradley, W. G., and Wilkes, G. L., Biomater. Med. Devices Artif. Organs, 5, 159, 1977.
39. Punch, P. I., Slater, D. H., Costa, N. D., and Edwards, M. E., Aust. Vet. J., 62, 79, 1985.
40. Punch, P. I., Costa, N. D., Edwards, M. E., and Wilcox, G. E., J. Vet. Pharmacol. Therap., 10, 37, 1987.
41. Sawusch, M. R., O'Brien, T. P., Dick, J. D., and Gottsch, J. D., Am. J. Ophthalmol., 106, 279, 1988.
42. Shell, J. W., in Drug Delivery Systems: Fundamentals and Techniques (P. Johnson and J. G. Lloyd-Jones, eds.), Horwood, Chichester, 1987.
43. Punch, P. I., Doctoral Thesis, Murdoch University, 1984.
44. Saettone, M. F., Giannaccini, B., Marchesini, G., Galli, G., and Chiellini, E., in Polymers in Medicine, Vol. 2, Biomedical and Pharmaceutical Applications (E. Chiellini, P. Giusti, C. Migliaresi, and L. Nicolais, eds.), Plenum Press, New York, 1986.
45. Vasantha, R., Sehgal, P. K., and Rao, K. P., Int. J. Pharm., 47, 95, 1988.
46. Matoba, A. Y., and McCuley, J. P., Ophthalmology, 92, 97, 1985.
47. Busin, M., Goebbels, M., and Spitnas, M., Ophthalmology, 94 (suppl.), 124, 1987.

48. Kaufman, H. E., Utoila, M. H., Gassett, A. R., Wood, T. O.,
     and Ellison, E. D., Trans. Am. Acad. Ophthalmol. Otolaryngol.,
     75, 361, 1971.
49. Unterman, S. R., Rootman, D. S., Hill, J. M., Parelman, J.
     J., Thompson, H. W., and H. E. Kaufman, J. Cataract Refract.
     Surg., 14, 500, 1988.
50. Phinney, R. B., Schwartz, S. D., Lee, D. A., and Mondino,
     B. J., Arch. Ophthalmol., 106, 1599, 1988.
51. Aquavella, J. V., Ruffini, J. J., and LoCascio, J. A., J.
     Cataract Refract. Surg., 14, 492, 1988.
52. Horakova, Z., Krajicek, M., Chvapil, M., and Boissier, J. R.,
     Therapie, 22, 1455, 1967.
53. Chvapil, M., and Holusa, J., J. Biomed. Mater. Res., 2, 245,
     1968.
54. Chvapil, M., J. Biomed. Mater. Res., 11, 721, 1977.
55. Kincl, F. A., and Ciaccio, L. A., Endocrinol. Exp., 14, 27,
     1980.
56. Meyskens, F. L., Graham, V., Chvapil, M., Dorr, R. T.,
     Alberts, D. S., and Surwit, E. A., J. Natl. Cancer Inst.,
     71, 921, 1983.
57. Peng, Y. M., Alberts, D. S., Graham, V., Surwit, E. A.,
     Weiner, S., Meyskens, F. L., Invest. New Drugs, 4, 245,
     1986.
58. Kay, J. S., Litin, B. S., Jones, M. A., Fryczkowski, A. W.,
     Chvapil, M., and Herschler, J., Ophthalmic Surgery, 17, 796,
     1986.
59. Kincl, F. A., Ciaccio, L. A., and Henderson, S. B., Arch.
     Pharmazie, 317, 657, 1984.
60. Graham, V., Surwit, E. S., Weiner, S., and Meyskens, F. L.,
     West. J. Med., 145, 192, 1986.
61. Weiner, S. A., Surwit, E. A., Graham, V. E., and Meyskens,
     F. L., Invest. New Drugs, 4, 241, 1986.
62. Lee, V. H. L., Crit. Rev. Therapeut. Drug Carrier Syst.,
     5, 69, 1988.
63. Banga, A. K., and Chien, Y. W., Int. J. Pharm., 48, 15,
     1988.
64. Eppstein, D. A., Longenecker, J. P., Crit. Rev. Therapeut.
     Drug Carrier Syst., 5, 99, 1988.
65. Siddiqui, O., and Chien, Y. W., Crit. Rev. Therapeut. Drug
     Carrier Syst., 3, 195, 1986.
66. Lee, V. H. L., Pharmaceut. Technol., 11, 26, 1987.
67. Deatherage, J. R., Matukas, V. J., Miller, E. J., Int. J.
     Oral Maxillofac. Surg., 17, 395, 1988.
68. Deatherage, J. R., and Miller, E. J., Collagen Related Res.,
     7, 225, 1987.

69. Lucas, P. A., Syftestad, G. T., Goldberg, V. M., Caplan, A. I., J. Biomed. Mater. Res., 23, 23, 1989.
70. Seyedin, S., Thomas, T., Bentz H., Ellingsworth, L., and Armstrong, R., U.S. Patent 4,810,691, 1989.
71. Oppenheim, R. C., in Drug Delivery Systems: Characteristics and Biomedical Applications (R. L. Juliano, ed.), Oxford University Press, New York, 1980.
72. Speiser, P., in Optimization of Drug Delivery (H. Bundgaard, A. B. Hansen, and H. Kofod, eds.), Munksgaard, Copenhagen, 1982.
73. Oppenheim, R. C., in Controlled Release Systems: Fabrication Technology, Vol. 2 (D. Hsieh, ed.), CRC Press, Boca Raton, 1988.
74. Tomlinson, E., in Drug Delivery Systems: Fundamentals and Techniques (P. Johnson and J. G. Lloyd-Jones, eds.), Horwood, Chichester, 1987.
75. Kramer, P. A., J. Pharm. Sci., 63, 1646, 1974.
76. Davis, S. S., Mills, S. N., and Tomlinson, E., J. Control. Rel., 4, 293, 1987.
77. Zolle, I., Rhodes, B. A., and Wagner, Jr., H. N., Int. J. Appl. Radiat. Isot., 21, 155, 1970.
78. Bernard, N. G., Shaw, W. V., Kessler, Landolt, R. R., Peck, G. E., Dockerty, G. H., Can. J. Pharm. Sci., 15, 30, 1980.
79. Raju, A., Mani, R., Divekar, K. D., Narasimhan, D. V. S., Rahalkar, R. L., Sundararajan, R. L., Kotrappa, P., Isotopenpraxis, 2, 57, 1978.
80. Rhodes, B. A., Zolle, I., Buchanan, J. W., and Wagner, Jr., H. N., Radiology, 92, 1453, 1969.
81. Taplin, G. V., Johnson, D. E., Dore, E. K., and Kaplan, H. S., J. Nucl. Med., 5, 259, 1964.
82. Davis, S. S., Illum, L., McVie, J. G., and Tomlinson, E. (eds.), Microspheres and Drug Therapy: Pharmaceutical, Immunological and Medical Aspects, Elsevier, Amsterdam, 1984.
83. Tomlinson, E., and Burger, J. J., in Polymers in Controlled Drug Delivery (L. Illum and S. S. Davis, eds.), Wright, Bristol, 1987.
84. Ishizaka, T., Endo, K., and Koishi, M., J. Pharm. Sci., 70, 358, 1981.
85. Longo, W. E., Iwata, H., Cindheimer, T. A., and Goldberg, E. P., J. Pharm. Sci., 71, 1323, 1982.
86. Longo, W. E., and Goldberg, E. P., in Methods in Enzymology, Drug and Enzyme Targeting, Part A, Vol. 112 (K. J. Widder and R. Green, eds.), Academic Press, New York, 1985.
87. Burger, J. J., Tomlinson, E., and McVie, J. G., in Drug Targeting (P. Buri and A Gumma, eds.), Elsevier, Amsterdam, 1985.

88. Sokoloski, T. D., and Royer, G. P., in Microspheres and Drug Therapy: Pharmaceutical, Immunological and Medical Aspects (S. S. Davis, L. Illum, J. G. McVie, and E. Tomlinson, eds.), Elsevier, New York, 1984.

89. Tomlinson, E., Int. J. Pharm. Technol. Prod. Manuf., 4, 49, 1983.

90. Yapel, A. F., U.S. Patent 4,147,767, 1979.

91. Burger, J. J., Tomlinson, E., Mulder, E. M. A., and McVie, J. G., Int. J. Pharm., 23, 333, 1985.

92. Lee, T. K., Sokoloski, T. D., and Royer, G. P., Science, 213, 233, 1981.

93. Sheu, M. T., Moustafa, M. A., and Sokoloski, T. D., J. Parent. Sci. Technol., 40, 253, 1986.

94. Gallo, J. M., Hung, C. T., and Perrier, D. G., Int. J. Pharm., 22, 63, 1984.

95. Tomlinson, E., Burger, J. J., Schoonderwoerd, E. M., and McVie, J. G., in Microspheres and Drug Therapy: Pharmaceutical, Immunological and Medical Aspects, Elsevier, Amsterdam, 1984.

96. Willmott, N., and Harrison, P. J., Int. J. Pharm., 43, 161, 1988.

97. Sheu, M. T., and Sokoloski, T. D., J. Parent. Sci. Technol., 40, 259, 1986.

98. Goosen, M. F. A., Leung, Y. F., O'Shea, G. M., Chou, S., Sun, A. M., Diabetes, 32, 478, 1983.

99. Margalit, R., and Silbiger, E., J. Microencapsul., 2, 183, 1986.

100. Shah, M. V., DeGennaro, M. D., and Suryakasuma, H., J. Microencapsul., 4, 223, 1987.

101. Ratcliffe, J. H., Hunneyball, I. M., Wilson, C. G., Smith, A., and Davis, S. S., J. Pharm. Pharmacol., 39, 290, 1987.

102. Burgess, D. J., Davis, S. S., and Tomlinson, E., Int. J. Pharm., 39, 129, 1987.

103. Burgess, D. J., and Davis, S. S., Int. J. Pharm., 46, 69, 1988.

104. Morimoto, Y., and Fujimoto, S., Crit. Rev. Therapeut. Drug Carrier Syst., 2, 19, 1985.

105. Kim, C.-K., Jeong, E. J., Yang, J. S., Kim, S. H., and Kim, Y. B., Arch. Pharmaceut. Res. (Seoul), 8, 159, 1985.

106. Willmott, N., Cummings, J., Stuart, and Florence, A. T., Biopharmaceutics Drug Dispos., 6, 91, 1985.

107. Willmott, N., Cummings, J., and Florence, A. T., J. Microencapsul., 2, 293, 1985.

108. Willmott, N., Kamel, H. M. H., Cummings, J., Stuart, J. F. B., and Florence, A. T., in Microspheres and Drug Therapy: Pharmaceutical, Immunological and Medical Aspects, Elsevier, Amsterdam, 1984.

109. Gupta, P. K., Hung, C. T., and Perrier, D. G., Int. J. Pharm., 33, 137, 1986.

110. Gupta, P. K., Hung, C. T., and Perrier, D. G., Int. J. Pharm., 33, 147, 1986.

111. Gupta, P. K., Hung, C. T., and Perrier, D. G., J. Pharm. Sci., 76, 141, 1987.

112. Gupta, P. K., Gallo, J. M., Hung, C. T., and Perrier, D. G., Drug Dev. Indust. Pharmacy, 13, 1471, 1987.

113. Sawaya, A., Benoit, J.-P., and Benita, S., J. Pharm. Sci., 76, 475, 1987.

114. Sawaya, A., Fickat, R., Benoit, J.-P., Puisieux, F., and Benita, S., J. Microencapsul., 5, 255, 1988.

115. Leucuta, S. E., Risca, R., Daicoviciu, D., and Porutiu, D., Int. J. Pharm., 41, 213, 1988.

116. Higuchi, T., J. Pharm. Sci., 50, 874, 1961.

117. Higuchi, T., J. Pharm. Sci., 52, 1145, 1963.

118. Cummings, J., and Willmott, N., J. Chromatogr., 343, 208, 1985.

119. Mehta, R. C., Hogan, T. F., Mardmomen, S., and Ma, J. K. H., J. Chromatogr., 430, 341, 1988.

120. Fujimoto, S., Miyazaki, M., Endoh, F., Takahashi, O., Shrestha, R. D., Morimoto, Y., and Terao, K., Cancer, 55, 522, 1985.

121. Fujimoto, S., Miyazaki, M., Endoh, F., Takahashi, O., Okui, K., Sugibayashi, K., and Morimoto, Y., Cancer Drug Deliv., 2, 173, 1985.

122. Fujimoto, S., Miyazaki, M., Endoh, F., Takahashi, O., Okui, K., and Morimoto, Y., Cancer, 56, 2404, 1985.

123. Asaishi, K., Toda, K., Watanabe, Y., and Mikami, T., Gan To Kagaku Ryoho, 15, 2484, 1988.

124. Nemoto, R., and Kato, T., in Microspheres and Drug Therapy: Pharmaceutical, Immunological and Medical Aspects (S. S. Davis, L. Illum, J. G. McVie, and E. Tomlinson, eds.), Elsevier, New York, 1984.

125. Illum, L., Jørgensen, H., Bisgaard, H., Krogsgaard, O., and Rossing, N., Int. J. Pharm., 39, 189, 1987.

126. Widder, K. J., Senyei, A. J., and Scarpelli, D. G., Proc. Soc. Exp. Biol. Med., 158, 141, 1978.

127. Widder, K. J., Senyei, A. E., and Ranney, D. F., Adv. Pharmacol. Chemo., 16, 213, 1979.

128. Widder, K. J., and Senyei, A. E., in Microspheres and Drug Therapy: Pharmaceutical, Immunological and Medical Aspects (S. S. Davis, L. Illum, J. G. McVie, and E. Tomlinson, eds.), Elsevier, New York, 1984.

129. Driscoll, C. F., Morris, R. M., Senyei, A. E., Widder, K. J., and Heller, G. S., Microvasc. Res., 27, 353, 1984.

130. Ishii, F., Takamura, A., and Noro, S., Chem. Pharm. Bull., 32, 678, 1984.
131. Zimmerman, U., in Targeted Drugs (E. P. Goldberg, ed.), John Wiley and Sons, New York, 1983.
132. Ranney, D. F., Biochem. Pharmacol., 35, 1063, 1986.
133. Lee, K. C., Koh, I. B., Kim, W. B., and Lee, Y. J., Int. J. Pharm., 44, 49, 1988.
134. Papisov, M. I., Savelyev, V. Y., Sergienko, V. B., and Torchilin, V. P., Int. J. Pharm., 40, 201, 1987.
135. Papisov, M. I., Torchilin, V. P., Int. J. Pharm., 40, 207, 1987.
136. Papisov, M. I., Savelyev, V. Y., Sergienko, V. B., and Torchilin, V. P., Antibiotiki I Khimioterapiya, 33, 744, 1988.
137. Gupta, P. K., Hung, C. T., Lam, F. C., and Perrier, D. G., Int. J. Pharm., 43, 167, 1988.
138. Rettenmaier, M. A., Stratton, J. A., Berman, M. L., Senyei, A., Widder, K., White, D. B., DiSaia, P. J., Gyn. Oncol., 27, 34, 1987.
139. Gallo, J. M., Gupta, P. K., Hung, C. T., and Perrier, D. G., J. Pharm. Sci., 78, 190, 1989.
140. Chen, Y., Willmott, N., Anderson, J., and Florence, A. T., J. Pharm. Pharmacol., 39, 978, 1987.
141. Desoize, B., Jardillier, J. C., Kanoun, K., Guerin, D., and Levy, M. C., J. Pharm. Pharmacol., 38, 8, 1986.
142. Veis, A. (ed.), The Macromolecular Chemistry of Gelatin, Academic Press, New York, 1964.
143. Sezaki, H., Hashida, M., and Muranishi, S., in Optimization of Drug Delivery (H. Bundgaard, A. B. Hansen, and H. Kofod, eds.), Munksgaard, Copenhagen, 1982.
144. Oppenheim, R. C., in Polymers in Controlled Drug Delivery (L. Illum and S. S. Davis, eds.), Wright, Bristol, 1987.
145. Tanake, N., Takimo, S., and Utsumi, I., J. Pharm. Sci., 52, 664, 1963.
146. Sjögren, H. O., Nilsson, K., Malmström, P., and Axelsson, B., J. Immunol. Meth., 56, 285, 1983.
147. Nakamoto, Y., Hasida, M., Muranishi, S., and Sezaki, H., Chem. Pharm. Bull., 23, 3125, 1975.
148. Hashida, M., Muranishi, S., Sezaki, H., Tanigawa, N., Satomura, K., and Hikasa, Y., Int. J. Pharm., 2, 245, 1979.
149. Yoshioka, T., Hashida, M., Muranishi, S., and Sezaki, H., Int. J. Pharm., 8, 131, 1981.
150. Ratcliffe, J. H., Hunneyball, I. M., Smith, A., Wilson, C. G., and Davis, S. S., J. Pharm. Pharmacol., 36, 431, 1984.
151. Oppenheim, R. C., Int. J. Pharm., 8, 217, 1981.
152. El-Samaligy, M. S., and Röhdewald, P., J. Pharm. Pharmacol., 35, 537, 1983.

153. El-Samaligy, M. S., and Röhdewald, P., _Pharm. Acta Helv._, 57, 201, 1982.
154. Luisi, P. L., Imre, V. E., Jaeckle, H., and Pande, A., in _Topics in Pharmaceutical Sciences_ (D. D. Breimer and P. Speiser, eds.), Elsevier, Amsterdam, 1983.
155. Chemtob, C., Assimacopoulos, T., and Chaumeil, J. C., _Drug Dev. Indust. Pharmacy_, 14, 1359, 1988.
156. Das, S. K., and Gupta, B. K., _Drug Dev. Indust. Pharmacy_, 14, 1673, 1988.
157. Das, S. K., and Gupta, B. K., _Drug Dev. Indust. Pharmacy_, 11, 1621, 1985.
158. Sa, B., Roy, S., Das, S. K., _Drug Dev. Indust. Pharmacy_, 13, 1267, 1987.
159. Marty, J. J., and Oppenheim, R. C., _Aust. J. Pharm. Sci._, 6, 65, 1977.
160. Kennedy, K. T., M. Pharm. Thesis, Victorian College of Pharmacy Ltd. (Aust), 1983.
161. Lee, K. C., Koh, I. B., Oh, I. J., _Arch. Pharmaceut. Res._, (Seoul), 9, 145, 1986.
162. Lee, K. C., and Koh, I. B., _Arch. Pharmaceut. Res._, 10, 42, 1987.
163. Chen, Q., Sun, S-Y., Gu, X-Q., Li, Z-W., and Li, Z-X., _Acta Pharmacol. Sin._, 4, 273, 1983.
164. Miyazaki, S., Hashiguchi, N., Sugiyama, M., Takada, M., and Morimoto, Y., _Chem. Pharm. Bull. (Tokyo)_, 34, 1370, 1986.
165. Miyazaki, S., Hashiguchi, N., Hou, W. M., Yokouchi, C., and Takada, M., _Chem. Pharm. Bull. (Tokyo)_, 34, 3384, 1986.
166. Miyazaki, S., Hashiguchi, N., Yokouchi, C., Takada, M., and Hou, W. M., _J. Pharm. Pharmacol._, 38, 618, 1986.

8

Liposomes

ULLA K. NÄSSANDER, GERT STORM*, PIERRE A. M. PEETERS[†] and
DAAN J. A. CROMMELIN Department of Pharmaceutics, Faculty of
Pharmacy, University of Utrecht, Utrecht, The Netherlands

I. INTRODUCTION

Liposomes have been widely used as model membranes and their
physicochemical properties have therefore been studied extensively.
More recently, they have become important tools for the study of
membrane-mediated processes (e.g., membrane fusion), catalysis of
reactions occurring at interfaces, and energy conversion. Besides,
liposomes are currently under investigation as carrier systems for
drugs and as antigen-presenting systems to be used as vaccines.
The last two topics will be the subject of this chapter.

Liposomes are members of a family of vesicular structures which
can vary widely in their physicochemical properties. Basically, a
liposome is built of one or more lipid bilayers surrounding an aqueous
core. The backbone of the bilayer consists of phospholipids; the
major phospholipid is usually phosphatidylcholine (PC), a neutral
lipid. Size, number of bilayers, bilayer charge, and bilayer rigidity
are critical parameters controlling the fate of liposomes in vitro and
in vivo. Dependent on the preparation procedure unilamellar or
multilamellar vesicles can be produced. The diameter of these vesicles can range from 25 nm up to 50 μm—a 2000-fold size difference.

Current affiliation
*Visiting scientist at Liposome Technology, Inc., Menlo Park,
California
†Institute for Pharmaceutical and Biomedical Consultancy,
Pharma Bio-Research International B. V., Assen, The Netherlands

Charged phospholipids [like phosphatidylglycerol (PG) or phosphatidylserine (PS)] or other charged amphipatic molecules (like choles-
terylhemisuccinate or stearylamine) are often included in the bilayer.
The physical stability of liposomes tends to be improved by the presence of these charge-inducing agents in the bilayer because electro-
static repulsion forces hamper close contact between neighboring vesicles.   Bilayer rigidity depends strongly on the selected bilayer components.   In general, long (chain length > C14) and saturated acyl
chains in bilayers of phosphatidylcholine give so-called gel-like structures at body temperature; the bilayer is relatively rigid.   Above the
transition temperature (Tc) a lower rigidity of the bilayer is found:
the "fluid" state.   Phosphatidylcholine isolated from natural sources
contains unsaturated fatty acids.   Then Tc values lie below 37°C,
giving rather loosely packed fluid state bilayers at body temperature.
Gel-like liposomes tend to be less prone to leakage of liposome-
encapsulated drugs than liposomes with fluid-like structures.   Bilayers
with a low rigidity (fluid bilayers) can be "rigidified" by inclusion
of cholesterol.   If specific interactions with target cells are required
(e.g., with immunoliposomes) anchor molecules which are needed to
connect a homing device to the vesicles are included in the bilayer
structure [e.g., phosphatidylethanolamine (PE)].

Liposomes are currently under investigation as carrier systems
for drugs and as antigen-presenting systems to be used as vaccines.
In general, hydrophilic drugs are located in the aqueous phase inside
the liposomes, either in between the subsequent bilayers or in the
core.   Lipophilic drugs are associated with the bilayers.   The aim
of studies with liposome-drug combinations is to assess whether the
therapeutic index of existing and newly developed drugs can be improved by manipulating the disposition of the drug in the body via
liposome encapsulation.   As alternatives for liposomes other particulate systems are being studied.   These particulate systems are based
on alkylcyanoacrylate (nanoparticles), albumin spheres, mixed micelles, or oil/water emulsions.   Liposomes offer advantages over these
"alternative" systems because of their relatively low toxicity and their
high versatility in terms of physicochemical properties.   The limited
stability during storage of certain types of liposomes may, on the
other hand, pose problems.

In vivo reproducible results can only be achieved if the liposome-
drug or -antigen combinations are thoroughly characterized upon
preparation in terms of their physical and chemical properties and,
besides, if the stability during storage is ensured.   In this chapter
both the pharmaceutical (preparation, characterization, and stability)
aspects and the therapeutic potentials and limitations of drug and
antigen delivery with liposomes will be discussed.

II. PREPARATION OF LIPOSOMES

The methodology for liposome preparation has evolved rapidly during
the last few years as a response to the need to prepare well-defined
liposomes for specific applications. The methodological aspects of
liposome preparation have been thoroughly reviewed by Szoka and
Papahadjopoulos (1980), Gregoriadis (1984), Machy and Leserman
(1987), and Lichtenberg and Barenholz (1988). Obviously, the
choice of a specific method (or combination of methods) depends on
the intent of the investigator and the intended application. In this
section, we will give references to the more prominent liposome preparation methods and discuss them briefly. The use of multilamellar
vesicles (MLV) is especially beneficial when lipophilic compounds are
to be entrapped or if the efficiency of aqueous encapsulation is not
a critical issue. For some applications, small unilamellar vesicles
(SUV) are good candidates because of their small size and high
surface-to-lipid ratio. However, there must be no need for efficient
encapsulation of water-soluble compounds, which is very low in the
case of SUV preparations. In contrast, large unilamellar vesicles
(LUV) are particularly useful for the efficient encapsulation of water-
soluble substances because of their favorable volume-to-surface ratio.

Below, we define small liposomes as those smaller than 0.1 μm
and large liposomes as t. se larger than 0.1 μm.

A. Large Liposomes

1. Multilamellar Liposomes

Multilamellar vesicles were first described by Bangham and coworkers
in 1965 (Bangham et al., 1965). (They reported on the spontaneous
formation of liposomes by lipid hydration.) This simple method is
still widely used. Lipids are dissolved in an organic solvent in a
round-bottomed flask. A thin lipid layer is formed on the inside
wall of the flask by removal of the organic solvent by rotary evaporation at reduced pressure. Finally, an aqueous buffer is added
and the dry film is hydrated above the transition temperature (Tc)
of the lipids. Shaking (by hand or by vortex mixer) yields a dispersion of MLV. (Duration and the intensity of shaking, the presence of charge-inducing agents in the bilayer, ionic strength of the
aqueous medium, and lipid concentration are important parameters
influencing the size and the encapsulating efficiency of MLV.) The
liposomes formed are quite heterogeneous both in size and in number
of lamellae. Extrusion through a polycarbonate filter with well-
defined pore sizes or exposing the dispersions to high shear stress
as occurs in a microfluidizer or equipment designed for the processing of dairy products (decreases the size and number of lamellae and
narrows down the particle size distribution of the dispersion) (Olson
et al., 1979; Mayhew et al., 1987a; Jousma et al., 1987; Talsma et

particle size of liposomes after extrusion

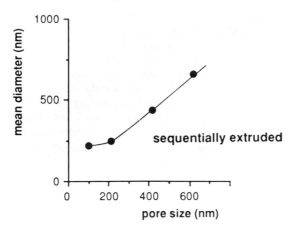

FIGURE 1   Effect of (sequential) extrusion of MLV dispersions
through polycarbonate membrane filters (Unipore) with pore sizes
of 1.0, 0.6, 0.4, 0.2, and 0.1 μm on the mean liposome diameter.
DXR-containing MLV (phosphatidylcholine/phosphatidylserine/
cholesterol 10:1:4); mean diameter of nonextruded dispersion: about
2 μm; pH 4. Mean particle size determined by dynamic light scattering (Nanosizer, Coulter Electronics). (From Crommelin and Storm,
1987.)

al., 1989). The average particle size of liposomes in dispersions
consecutively extruded through polycarbonate filters with decreasing
pore diameter is shown in Fig. 1.

Some techniques to produce small, mainly unilamellar vesicles
from MLV (sonication, French pressure cell) are discussed below in
separate paragraphs.

Liposomes can be prepared from pure lipids or mixtures of lipids.
(Cholesterol is known to serve as a "fluidity buffer": it enhances
the fluidity of the gel state bilayer, while it decreases the fluidity
of the fluid state bilayer. Increasing concentrations of cholesterol
in bilayers cause a broadening and gradual disappearance of the
phase transition)(Demel and De Kruyff, 1976).

When liposomes are prepared from a molecular mixture of lipid
components it is important that all lipids be homogeneously dissolved
in an organic solvent in order to obtain bilayers with evenly distributed lipids after hydration. For example, the solubilities of phosphatidylcholine and cholesterol in chloroform are similar; their solubility in benzene differs. Upon removal of benzene from the lipid
solution an inhomogeneous lipid film is formed on the glass wall and

upon hydration liposome dispersions containing liposomes with different compositions will be formed (Estep et al., 1978, 1979).

Repeated freezing and thawing is a convenient method for increasing the trapped volume of MLV preparations (Cullis et al., 1987).
The vesicle structure is dramatically altered as determined by freeze-
fracture electron microscopy and NMR studies (Mayer et al., 1985a,
1986a); multivesicular structures were observed. This might be explained by a combination of both head group dehydration and physical disruption of the multilamellar structure by ice crystals formed
during the freezing process.

A method resulting in improved encapsulation of aqueous phase
by MLV is the so-called dehydration-rehydration procedure (Kirby
and Gregoriadis, 1984; Shew and Deamer, 1985). The lipid (usually
preformed liposomes) is dried (by either lyophilization or evaporation)
in the presence of the aqueous solute to be entrapped, thus forming
a mixed film with solute trapped between layers. Subsequent gradual rehydration with a minimum of aqueous phase leads to the formation of MLV with a high entrapment of the aqueous solutes added.

2. Large Unilamellar Vesicles

Reverse Phase Evaporation   Szoka and Papahadjopoulos (1978)
developed the so-called reverse phase evaporation method. Vesicles
prepared with this technique (REV) show higher encapsulation efficiencies of hydrophilic compounds than unextruded MLV.

The lipids are dissolved in an organic solvent (diethyl ether,
diisopropyl ether, or a mixture of one of the two with chloroform
depending of the solubility properties of the lipids used). The
aqueous phase is added to the organic phase at a ratio of 1:3 when
diethyl ether is used and at a ratio of 1:6 when a mixture of diisopropyl ether and chloroform is used. The mixture is sonicated in
order to form an emulsion, followed by slow removal of the organic
phase via rotary evaporation under reduced pressure.

The actual characteristics of REV produced depend on a number
of factors such as choice of lipids (% cholesterol and charged lipids),
lipid concentration used in the organic solvent, rate of evaporation,
and ionic strength of the aqueous phase (Szoka and Papahadjopoulos,
1980). Modifications of this REV technique were proposed by several
groups. The SPLV (stable plurilamellar vesicles) method consists of
bath-sonicating an emulsion of the aqueous phase in an ether solution of lipid while evaporating the ether (Grüner et al., 1985).

Solvent Injection   The solvent injection technique involves the
injection of solutions of lipid in solvents with high vapor pressure
(ether, fluorocarbons, ethanol) into excess aqueous phase under
reduced pressure. In general, the aqueous phase is maintained
above the phase transition of the lipids (Tc) and a reduced pressure

is used to vaporize the lipid solvent during the injection. This method can produce LUV with a relatively high entrapped aqueous volume per mole of lipid (Deamer and Bangham, 1976; Schieren et al., 1978). A slow rate of injection combined with rapid solvent removal and low lipid concentration in the solvent favors LUV formation. Compared to the ethanol vaporization method (see below), the size of the liposomes prepared by the ether vaporization method is large (150–250 nm) and not dependent on the lipid concentration in the ether.

Fusion   Addition of bivalent cations as Ca^{2+} and Mn^{2+} to small unilamellar vesicles (SUV) containing acidic lipids (e.g., PS, PG, or PA) changes the structure of the vesicles to large, cylindrical, spiral multilamellar structures (Papahajopoulos et al., 1975). These structures precipitate from the solution and can subsequently be removed. After addition of EDTA the cylindrical structures will transform into a rather heterogeneous population of essentially unilamellar vesicles with a size ranging from 0.2 to 1.0 μm (Papahadjopoulos and Vail, 1978). The entrapped volume of these large vesicles can be up to 7 liters/mol phospholipid. One of the advantages of this method is that molecules to be encapsulated in liposomes are not exposed to organic solvents.

pH Adjustment Method   The pH adjustment method was introduced by Hauser and Gains (1982). The pH of a dispersed phosphatidyl-choline/phosphatic acid mixture in water was increased from pH 3 to pH 7–8 resulting in the formation of liposomes. The mechanism behind the formation of vesicles by this method involves the ionization of the phosphate group of PA. The maximal pH reached and the velocity of pH titration are important parameters in determining the particle size (Gaines and Hauser, 1983). By this method large (>100 nm) liposomes of the LUV and MLV type are formed along with smaller SUV structures (30 nm). The distribution over these two populations can be varied depending on the experimental conditions used.

A modification of this method was described by Aurora et al. (1985). The LUV vesicles (300–600 nm) were prepared by dispersing dioleoylphosphatidic acid (DOPA) and cholesterol at a high salt concentration. The preparation formed contained only about 3% SUV vesicles.

The pH adjustment method has the advantage of being fast and simple, but its use is rather limited as currently this method has been demonstrated to work with only a few acidic lipids (Hauser, 1987).

Formation from Template Surfaces   Recently, a new method for the preparation of LUV was reported by Lasic et al. (1988). The method is based on a simple procedure which leads to the formation of homogeneous populations of LUV with a diameter of around 1 μm. Upon addition of solvent to a dry phospholipid film deposited on a template surface, vesicles are formed instantly without any chemical or physical treatment. The formation of multilamellar structures is prevented by inducing a surface charge on the bilayers. The size of the vesicles is controlled by the topography of the template surface on which the phospholipid film was deposited (Lasic, 1988).

3.   Cell Size Liposomes

Multivesicular Liposomes   Kim and his colleages described a method for the preparation of cell size liposomes with high encapsulation efficiency: the so-called multivesicular liposomes (Kim et al., 1983). The lipid phase consists of a combination of amphiphatic lipids and a small amount of triglycerides (triolein or trioctanoin) dissolved in chloroform-diethyl ether (1:1). The aqueous phase is slowly added to the organic phase and after vigorous shaking a water-in-lipid emulsion is formed (Fig. 2A-B). Via a narrow Pasteur pipet the emulsion is subsequently added to a sucrose solution. This mixture is shaken intensively to form spherules (Fig. 2C-D). Finally, multivesicular liposomes gradually form when a stream of nitrogen is led over the dispersion removing chloroform and diethyl ether (Fig. 2E-F). With this method, liposomes have been made with average diameters ranging from 5.6 to 29 μm. Depending on the lipid composition and on the nature of the entrapped material, percentages of encapsulation above 80% have been reached. It is not completely clear as to the function of the "triglycerides" in these vesicles. The authors suggest that triglycerides might stabilize the corners or edges formed in the multivesicular liposomes.

Hydration of Phospholipids with Solutions of Very Low Ionic Strength   Very large unilamellar and oligolamellar vesicles can be prepared when a thin lipid film is dispersed in a solution of very low ionic strength (Reeves and Dowben, 1969). The formation of vesicles with diameters up to 300 μm enclosing latex beads with a diameter of 20 μm have been reported (Antanavage et al., 1978). The enormous size of these liposomes enables measurements to be made on a single lipid bilayer. However, the conditions required for their preparation, their fragility, and their size preclude their use in therapy.

Freeze-thawing   If SUV are freeze-thawed in the presence of a high concentration of electrolytes and subsequently dialyzed against a low electrolyte concentration, unilamellar liposomes with a diameter of more than 10 μm are formed. The formation of these giant

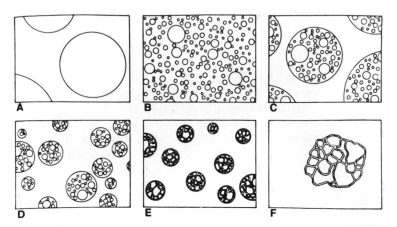

FIGURE 2 Schematic representation of multivesicular liposome formation. Panel A depicts the initial water-in-lipid emulsion. B shows
the water-in-lipid emulsion after 9 min of mechanical shaking. C
shows the initial formation of chloroform-ether spherules by addition
of the water-in-lipid emulsion to a 0.2 M sucrose solution. D shows
the establishment of spherule size by the second mechanical shaking.
E represents liposome formation via evaporation of the chloroform
and ether, resulting in thinning of the lipid phase to a bilayer
membrane as the solvents are removed. F is an enlarged representation of the final multivesicular liposome product, obtained by trac-
ing an actual electron micrograph. Each line represents a monolayer
of amphiphatic lipids; those in panels A–D represent lipid monolayers
at the interphase separating water and lipid phases, and those in
panels E and F represent monolayers of bilayer unit membrane.
(From Kim et al., 1983.)

liposomes (GUV) is caused by the influx of water due to differences
in osmotic pressure (Oku and MacDonald, 1983a,b).

B. Small Unilamellar Liposomes

1. Sonication

Sonication of MLV dispersions above the Tc of the lipids results in
the formation of SUV (Saunders, et al., 1962). Sonication can be
performed with a bath sonicator (Papahadjopoulos and Watkins, 1967)
or a probe sonicator (Huang, 1969). During sonication the MLV
structure is broken down and small unilamellar vesicles with a high
radius of curvature are formed. In case of SUV with diameters of
about 20 nm (maximum radius of curvature), the outer monolayer
can contain over 50% of the phospholipids and in the case of lipid

mixtures an asymmetrical distribution over the inner and outer monolayer can occur (Szoka and Papahadjopoulos, 1980). The probe sonicator gives fast and reproducible results and is therefore often used.
However, probe sonication has a number of drawbacks. Oxidation
of unsaturated acyl chains can occur if oxygen is not properly removed from the system (Klein, 1970; Kemps and Crommelin, 1988a,b)
and metal particles originating from the probe often contaminate the
dispersion (Hauser, 1971). Both problems can be avoided by using
a bath sonicator, which also has the advantage of being a closed
system in which the temperature of the preparation can be carefully
monitored. However, there are also several drawbacks associated
with the use of bath sonicators: (it is much more time consuming;
besides, sample size, container, and location in the bath are critical,
so that reproducibility can be problematic;) and, finally, the residual
concentration of large particles is greater (Szoka and Papahadjopoulos, 1980). The size of SUV obtained by sonication depends on the
phospholipid composition, sonication time, and cholesterol concentration (Huang, 1969; Johnson, 1973). Minimum diameters of 22 nm
were reported. The encapsulated volume per mole phospholipid is
limited and ranges from 0.2 to 1.5 liters/mol lipid. The introduction
of cholesterol and charged lipids increases the encapsulated volume
(Johnson, 1973).

2. Detergent Removal

Removal of detergent from mixed micelles, formed by solubilization
of dried lipid mixtures or preformed liposomes with a detergent-
containing aqueous phase, results in the formation of unilamellar
vesicles. This is a gentle method where no strong mechanical forces
and no high temperatures are applied. Consequently, this method
is popular for the encapsulation of labile molecules such as peptides
and proteins. Kagawa and Racker (1971) first introduced this method,
removing cholate or deoxycholate by dialysis from lipid-protein mixtures, to incorporate proteins in liposomes. Liposome diameters
strongly depend on the experimental conditions. This is illustrated
in Fig. 3 where the effect of cholesterol content and dilution rate on
the particle size of vesicles is shown (Jiskoot et al., 1986a). The
preparation procedure should include a step to minimize residual
detergent levels after liposome formation. The techniques reported
for the removal of detergents include dialysis (Rhoden and Goldin,
1979; Schwendener et al., 1981), gel filtration chromatography
(Brunner et al., 1976; Allen et al., 1980), adsorption to hydrophobic resin beads (Ueno et al., 1984), and dilution (Schurtenberger
et al., 1984; Fischer and Lasic, 1984; Jiskoot et al., 1986a; Van
Dalen et al., 1988).

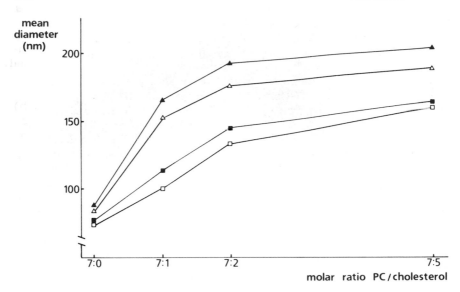

FIGURE 3   Effect of the amount of cholesterol on the particle size.
Phosphatidylcholine/cholesterol liposomes were prepared by the octyl
glucoside dilution technique.   The begin concentration of the mixed
micelles was 150 mM octyl glucoside and 10 mM phosphatidylcholine
in 10 mM tris(hydroxymethyl)aminomethane and 0.9% NaCl, pH 7.4.
Dilution was performed with an automatic titration unit at a dilution
rate (= dilution factor, relative to the initial volume, per unit of
time) of 0.026 sec^{-1} (▲ and △) or 0.69 sec^{-1} (■ and □).   Mean
diameters after dilution (▲ and ■) and after filtration (△ and □)
are represented.   (Adapted from Jiskoot et al, 1986a.)

3.   Ethanol Injection

Injection of phospholipid dissolved in ethanol into excess water heated
above the Tc of the lipids results in the formation of mainly unilamellar vesicles (Batzri and Korn, 1973).   The remaining ethanol can
be removed by dialysis or by gel filtration (Nordlund et al., 1981).
A disadvantage of the ethanol injection method to produce SUV is
the need to use a low lipid concentration, resulting in a low encapsulation efficiency of the aqueous phase.   The dispersions can be
concentrated by ultracentrifugation, ultrafiltration, or removal of
water by evaporation.

Depending on the lipid concentration, vesicles with different
particle sizes can be prepared (Kremer et al., 1977).   At a low
lipid concentration (3 mM), vesicles with a diameter of 30 nm are
formed, whereas 110-nm vesicles are formed at higher concentrations

(36 mM). The vesicle size is also dependent on the rate of ethanol injection. Kremer et al. (1977) could produce rather large vesicles by slow injection, whereas rapid injection resulted in small vesicles (about 30 nm) (Batzri and Korn, 1973). The type of phospholipids used is also affecting the particle size. Saturated lipids and mixtures of lipids produce larger liposomes than unsaturated lipids under the same conditions (Nordlund et al., 1981). If concentrated solutions of lipid in ethanol are injected into aqueous media, MLV are formed (Lichtenberg and Barenholz, 1988). This method could be useful for the large-scale production of liposomes.

4. French Press Extrusion

A French pressure cell can be used to reduce the size of MLV by extrusion under high pressure. Four extrusions of egg PC-MLV at 4°C resulted in the formation of small unilamellar vesicles; 94% of the lipid was found in 31- to 52-nm vesicles (Barenholz et al., 1979). It was shown that these vesicles were more stable and had a larger osmotic activity than SUV of minimal size produced by sonication. Advantages of this technique are the fact that high lipid concentrations can be used, resulting in a relatively high encapsulation efficiency of the aqueous phase; the method is mild and nondestructive; and it does not introduce impurities such as organic solvents and organic solvents and/or detergents into the liposome dispersion (Lichtenberg and Barenholz, 1988).

5. Homogenization

Homogenization of MLV or other lipid dispersions by commercially available high-shear homogenizers, e.g., the microfluidizer (Mayhew et al., 1987a; Talsma et al., 1989), produces unilamellar vesicles. The SUV formed are larger than the minimal size formed by sonication and significant amounts of larger particles are also present. In fact, the size of the vesicles produced by the microfluidizer depends on the pressure used, on the number of passes of the preparation through this device, and on the liposomal lipid composition.

C. Encapsulation of Drugs and Other Compounds in Liposomes

Upon the spontaneous rearrangement of anhydrous phospholipids in the presence of water into a hydrated bilayer structure, a portion of the aqueous phase is entrapped within a continuous, closed bilayer structure. By this process water-soluble compounds are passively entrapped in liposomes. The efficiency of encapsulation varies and depends, for example, on the method of preparation of liposomes and the phospholipid concentration during preparation. Different parameters can be used to describe the encapsulation efficiency:

1. Efficiency of encapsulation; amount of the encapsulated compound
    divided by the total amount added to the preparation.
2. Encapsulation ratio; amount of compound (mole) encapsulated per
    mole of (phospho)lipid.
3. Entrapped volume; the volume (liters) encapsulated per mole of
    (phospho)lipid.

In addition to encapsulation within the aqueous phase of vesicles,
amphiphilic and lipophilic compounds can interact with the bilayer.
The interaction of such compounds with the bilayer can result in al-
teration in vesicle properties such as permeability and stability of
the bilayer structure. Amphiphatic compounds such as detergents
(e.g., Triton and lysophospholipids) can intercalate in the bilayer
below their critical micelle concentration (CMC) (Kitagawa et al.,
1976), thereby increasing the permeability to entrapped compounds,
whereas concentrations above the CMC can lead to disruption of the
bilayer.
Apart from the passive encapsulation methods, different active
entrapment techniques are described in the literature. Nichols and
Deamer (1976) prepared liposomes with a pH gradient across the
membrane (acidic interior with respect to the external buffer). These
liposomes efficiently incorporated several catecholamines added to the
external buffer. The same technique has been used to concentrate
doxorubicin (DXR) in pH gradient liposomes (Mayer et al., 1986b).
The same group also demonstrated that doxorubicin and vinblastine
can be rapidly accumulated into egg-PC LUV in response to a valino-
mycin-dependent K^+ diffusion potential (Mayer et al., 1985b).

D. Removal of Free, Nonencapsulated Compounds

The final step in liposome preparation deals with the removal of free,
nonencapsulated drug from the liposome dispersion. Dialysis and gel
chromatography are the classical techniques. Dialysis is rather time
consuming and gel filtration brings about a considerable dilution of
the dispersion. This dilution is in many cases undesirable and ne-
cessitates the introduction of an additional step to concentrate the
dispersion afterward. For certain drug-containing dispersions the
use of an ion exchange resin offers a preferable alternative (Moro
et al., 1980; Van Bommel and Crommelin, 1984). The proposed
mechanism for the removal of a positively charged drug (like DXR)
from a dispersion of negatively charged liposomes is presented in
Fig. 4. For this particular drug (DXR) the fraction (in percentage)
of DXR which is liposome-associated after treatment with Dowex 50
W-X4 (in sodium form) was minimally 95% and dilution of the dispersion was negligible. This procedure took much less time than dialysis or gel filtration (Storm et al., 1985).

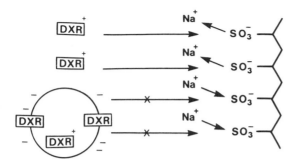

FIGURE 4   Proposed mechanism for the separation of "free" DXR from DXR in negative liposomes with a cation exchange resin.   (From Crommelin and Storm, 1987.)

Free drug can also be removed by ultracentrifugal sedimentation or by ultrafiltration where apart from removing the free drug the liposome dispersion can be concentrated.   A possible drawback of the ultracentrifugal sedimentation method is the risk of inducing aggregation or fusion due to mechanical forces on the vesicles during sedimentation.

E.   Conclusion

As illustrated above there exist a large variety of techniques for preparing liposomes.   From a pharmaceutical point of view, optimum liposome preparation techniques would avoid the use of organic solvent and detergents (which are difficult to remove), would exhibit a high trapping efficiency, would yield well-defined vesicles which can be produced in a reproducible way, and would be rapid and amenable to scale-up procedures (see Sec. VIII).

III.   PHYSICAL AND CHEMICAL CHARACTERIZATION
       OF LIPOSOMES

In the introduction the need for proper evaluation of liposomes to be administered has been stressed.   Physical parameters affecting the behavior of liposomes in vivo are type (multilamellar, oligolamellar, unilamellar, multivesicular), size, charge, bilayer rigidity, and "surface hydrophobicity" (Senior, 1987).   In the following paragraphs the pros and cons of the different techniques used to characterize liposomes will be dealt with briefly.   More detailed information can be found, for example, in the books of Knight (1981) and Gregoriadis (1984) and in a review article by Lichtenberg and Barenholz (1988).

A. Size

Liposome size can range from around 20 nm to around 50 μm. To a certain extent, the mean diameter and distribution of the diameters can be controlled by sizing procedures after the formation of the initial liposome dispersion or by a careful selection of the preparation conditions (cf. Sec. II). Several techniques can be used to determine mean particle size and particle size distribution (Groves, 1984).

Large liposomes, those with diameters over 1 μm, can be adequately measured by light microscopy and the Coulter counter. Light microscopy offers the possibility of collecting information on particle shape. This technique cannot be used for particles with diameters smaller than 1 μm because of lack of resolution. With a Coulter counter the volume distribution of liposomes (>1 μm) in dispersions can be determined. For particles in the submicrometer range, particle size can be assessed by electron microscopic analysis. Both negative staining and freeze-fracture techniques have been described. Sample preparation procedures should be properly validated. In particular with the negative staining technique, artifacts occur easily (Szoka and Papahadjopoulos, 1981). With the freeze-fracture technique, artifacts occur easily (Szoka and Papahadjopoulos, 1981). With the freeze-fracture technique, the fracture plane passes through liposomes which are randomly positioned in the frozen sample. Some liposomes will be cut far from their midplane sections, others through their midplane section. Therefore, the analysis of freeze-fracture pictures requires corrections for nonequatorial fracture. Besides, corrections have to be made for the size-dependent probability of a vesicle being in the fracture plane (Jousma et al., 1987; Guiot et al., 1980). Recently, results with a new technique based on electron microscopy was discussed; this technique allows analysis not only of liposome size, but also of the number of bilayers (Lautenschläger et al., 1988).

Light-scattering techniques and turbidity measurements have been used extensively for liposome size analysis (Groves, 1984; Lichtenberg and Barenholz, 1988). Dynamic light scattering is rapidly growing in popularity for size analysis of liposome dispersions. This relatively new technique has proven to be an excellent tool to determine the mean diameter of dispersions with a relatively narrow particle size distribution. It can be used routinely, readings can be taken in minutes, 1 ml of material (in dilute form) is needed, and it is not destructive. The distribution analyses obtained for heterodisperse colloidal systems (e.g., bimodal or skewed) need critical evaluation as the technique is biased toward the large particles in a dispersion.

Size exclusion chromatography has been used to analyse the size distribution of liposomes. For example, SUV can be separated from MLV, which elute in the void volume, by using a Sepharose 4B gel.

Vesicles with diameters over 1 μm tend to stick to the top of the column. This method has been used to monitor aggregation processes of liposomes.

Techniques which seem less suitable for routine size analysis are (1) analytical ultracentrifugation combined with a Schlieren optical system (Mason and Huang, 1978; Weder and Zumbuehl, 1984); (2) the sedimentation field flow fractionation (SFFF) technique to separate heterogeneous dispersions (e.g., Kirkland et al., 1982).

B. Charge

The surface potential can play an important role in the behavior of liposomes in vivo and in vitro (e.g., Senior, 1987). In general, charged liposomes are more stable against aggregation and fusion than uncharged vesicles. However, physically stable neutral liposomes have been described (e.g., Van Dalen et al., 1988). They are sufficiently stabilized by repulsive hydration forces, which counteract the attractive van der Waals forces.

In order to obtain an estimate of the surface potential, the ζ potential of individual liposomes can be measured (>0.2 μm) by microelectrophoresis (e.g., Crommelin, 1984). This technique also offers the opportunity to detect the presence of structures with deviating electrophoretic mobility and, therefore, deviating composition.

An alternative approach is the use of pH-sensitive fluorophores (Lichtenberg and Barenholz, 1988). These probes are located at the lipid-water interface and their fluorescence behavior reflects the local surface pH, which is a function of the surface potential at the interface. This indirect approach allows the use of vesicles independent of their particle size. Recently, techniques to measure the ζ potential of liposome dispersions on the basis of dynamic light scattering became commercially available (Müller et al., 1986).

The presence of impurities like free fatty acids in egg or soybean phosphatidylcholine, or in the (semi)synthetic phosphatidylcholines (e.g., DMPC, DPPC, DSPC) can be detected by monitoring the electrophoretic behavior of liposome dispersions of these phospholipids in aqueous media with low ionic strength: a negative charge will be found on these liposomes when free fatty acids are present in the bilayers.

C. Bilayer Rigidity and Homogeneity of Liposome Dispersions

Bilayer rigidity is a parameter which influences biodistribution and biodegradation of liposomes. In vitro a hydrophilic marker molecule (carboxyfluorescein) leaked much faster from the vesicles with bilayers in a fluid state than from bilayers in a gel state (Crommelin and Van Bommel, 1984). An indication of the bilayer rigidity can

be obtained by using fluorescence techniques. Fluorescence polari-
zation of a probe which interacts with the liposome bilayer is determined; 1,6-diphenyl-1,3,5,-hexatriene (DPH) is often used as a
probe. Data obtained with these fluorescing probes have to be in-
terpreted with care as the probe may locally disturb the bilayer or
accumulate in certain domains of a nonhomogeneous dispersion.

Inhomogeneity can occur both within liposome structures themselves (asymmetrical distribution of bilayer components) and between
liposomes in one dispersion (different liposomes), even when the
liposomes are prepared via the same preparation procedure. For
unilamellar vesicles the distribution of phospholipids over the inner
and outer "monolayer" of the bilayer can be studied by chemical
labeling. Aminolipids like PE react with trinitrobenzenesulfonate
(TNBS) or fluorescamine; in general, these two compounds are incapable of penetrating lipid bilayers (Lichtenberg and Barenholz,
1988). Alternatively, enzymatic approaches (for membrane proteins)
or ^{31}P-NMR techniques are available (Van Dalen et al., 1988). Phase
separation of lipid components in the bilayers can be detected by
differential scanning calorimetry (DSC) (Mabrey-Gaud, 1981). Different liposome structures in one dispersion can be identified on the
basis of density differences (gradient centrifugation, SFFF) or size
differences (gel chromatography, dynamic light scattering).

D. Number of Lamellae

The number of lamellae can be determined by ^{31}P-NMR. The ^{31}P-
NMR spectrum of egg phosphatidylcholine in liposomal form consists
of an asymmetrical peak. Upon addition of Mn^{2+} (impermeable to the
membrane) to the external phase of the liposome dispersion the peak
surface is quenched, and only the contribution of the phospholipid
head groups inside the vesicles is left (Browning, 1981; Jousma et
al., 1987). From the change in spectrum the average number of
bilayers can be calculated. The assays with TNBS or fluorescamine,
mentioned in Section III.C, also offer the opportunity to determine
the mean number of bilayers. Then aminolipids have to be present
in the bilayers to couple the reagents; it is assumed that these
aminolipids are evenly distributed over the bilayers. In addition,
small-angle X-ray analysis of liposomes can give an indication of the
number of bilayers (Jousma et al., 1987). Finally, electron microscopy can shed light on the question of the number of lamellae.
Electron microscopy can be particularly helpful in discriminating between multilamellar and multivesicular liposomes (Cullis et al., 1987;
Lautenschläger et al., 1988).

IV. PHYSICAL AND CHEMICAL STABILITY OF LIPOSOMES

Long-term stability of liposomes is a major issue in the development of liposomes as drug carrier systems. Drug-containing liposomes can be physically unstable: they aggregate or fuse, or they may lose the associated drug by leakage. They may be chemically unstable because of (1) hydrolysis reactions and (2) peroxidation reactions (Lichtenberg and Barenholz, 1988; Kemps and Crommelin, 1988a,b).

A. Physical Stability

The physical forces which stabilize liposomes have been discussed extensively (e.g., Parsegian, 1973; Parsegian et al., 1979; Rand, 1981; Nir et al., 1983; Boden and Sixtl, 1986). Depending on the bilayer structure (charge, rigidity), the ionic strength of the aqueous phase, and the presence of additives in the aqueous phase which can interact with bilayer components and temperature, liposomes may aggregate and/or fuse. Ca^{2+} ions tend to interact with negatively charged inducing agents (e.g., PS) in the bilayer, causing aggregation and even fusion. A freezing-thawing cycle also can induce aggregation and fusion.

Lipophilic or amphiphatic molecules are fully solubilized by the bilayers and tend not to leak (e.g., Weiner, 1987; Van Bloois et al., 1987). Leakage of encapsulated hydrophilic material from liposomes depends on bilayer rigidity and structure of the liposomes (unilamellar, multilamellar, or multivesicular). An example demonstrating the effect of bilayer composition on the release of carboxyfluorescein (CF) from liposomes is given in Fig. 5 (Crommelin and Van Bommel, 1984). Obviously, the presence of cholesterol had a condensing effect on the egg phosphatidylcholine bilayer and reduced release kinetics. The liposomes with the highly rigid DSPC/DPPG bilayers, with and without cholesterol, did not lose CF over a period of 9 months in the refrigerator (results not shown). Neither aggregation nor fusion could be detected.

Many attempts have been made in recent years to stabilize liposome dispersions by (freeze)-drying (Özer et al., 1988). The results strongly depend on the conditions chosen for the lyophilization process [e.g., freezing rate, freezing time, freezing temperature (Strauss, 1984; Fransen et al., 1986)], the composition of the aqueous phase, and the physicochemical characteristics of the drug. The selection of proper cryoprotectants like trehalose, lactose, glucose, and glycerol can prevent aggregation and fusion. Several suggestions have been made for the mechanism by which cryoprotectants stabilize liposomes during freeze-drying (reviewed by Özer et al., 1988). Successful attempts to freeze-dry liposome dispersions containing lipophilic compounds were reported. However, after a freeze-

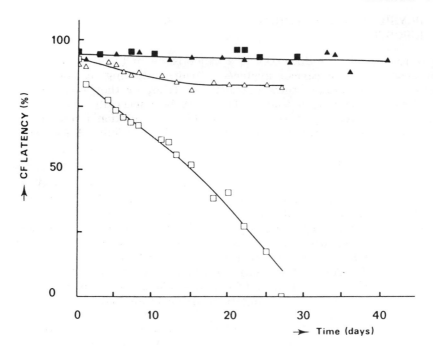

FIGURE 5 CF latency in REV stored at 4–6°C. The aqueous phase consisted of iso-osmotic sodium chloride/0.01 M Tris solution, pH 7.4. □ PC/PS 9:1; △ PC/PS/chol 10:1:4; ■ DSPC/DPPG 10:1; ▲ DSPC/ DPPG/chol 10:1:5. (From Crommelin and Van Bommel, 1984.)

drying/rehydration cycle the retention of hydrophilic compounds like CF is low (Crommelin and Van Bommel, 1984). The issue of stabilization of liposomes by freeze-drying resembles the problems encountered in cryopreservation of membranes (Crowe et al., 1983; Rudolph and Crowe, 1985); therefore, one can take advantage of the insights already gained in the past in the field of cryobiology. Cryomicroscopy, DSC, NMR, and electrical conductivity measuring at low temperatures ("freezing analyzer") are techniques which have proven valuable in elucidating the fundamentals of freezing/thawing and freeze-drying/rehydration of liposomes. Attempts have been made to avoid the freezing step in the drying process of liposome dispersions and thus to remove the water phase at temperatures above 0°C by, for example, (spray)-drying (Crommelin et al., 1986; Hauser and Strauss, 1987).

An alternative approach to avoid drug leakage from the liposomes on storage is to minimize the volume of the external water phase by concentrating the liposomes in a wet pellet or paste; this can be done by ultracentrifugation or by ultrafiltration.

B. Chemical Stability

The stability of liposomes depends on the chemical stability of their lipid components. The main bilayer constituents of liposomes designed for carrying a drug are phospholipids, like phosphatidylcholine and phosphatidylglycerol, often in combination with cholesterol or cholesterol esters like cholesteryl hemisuccinate. Commonly, hydrolysis and peroxidation are the two degradation processes which occur with phospholipids. The first step in the hydrolysis process of phosphatidylcholine is the formation of the 2-lysophosphatidylcholine and free fatty acid. This process can be either acid- or base-catalyzed. The lyso compounds can be further degraded by acyl or phosphoryl migration and hydrolysis to glycerophospho compounds (Kemps and Crommelin, 1988a,b). The degradation process can be monitored by TLC, HPLC, and NMR. Hardly any information is available on the effect of hydrolytic processes on the physical properties of the liposomes, e.g., the permeability of the bilayer.

Apart from the pH, other experimental conditions like temperature, ionic strength, buffer species, and ultrasonication were reported to influence hydrolysis reactions (e.g., Hauser, 1971; Frøkjaer et al., 1984; Grit et al., 1989). An example of the pH dependency on the hydrolysis rate of liposomes consisting of soybean phosphatidylcholine is presented in Fig. 6. Hydrolysis kinetics changed rather abruptly around the phase transition temperature. For liposomes with bilayers in either the gel or fluid state, hydrolysis kinetics could be adequately described by the Arrhenius equation (Frøkjaer et al., 1984; Grit et al., 1989). This finding opens the opportunity to perform accelerated stability tests to predict liposome stability at ambient temperatures or in the refrigerator provided that no fluid-to-gel transition of the bilayer occurs in the temperature range under investigation.

Peroxidation reactions occur primarily at unsaturated bonds in the acyl chains of the phospholipids. Peroxidation can be induced via either autooxidation (via a free radical mechanism) or photooxidation (Konings, 1984; Lichtenberg and Barenholz, 1988; Kemps and Crommelin, 1988b). Peroxidation of phospholipids with unsaturated acyl chains produces a set of substances varying widely in their chemical nature, like hydroperoxides, hydroperoxy cyclic peroxides, epoxyhydrodienes, and malon dialdehyde. The following measures can be taken to avoid peroxidation: (1) Only phospholipids with saturated acyl chains can be selected for liposome formation. (2) Oxygen can be removed from the dispersions by applying a vacuum or by substituting oxygen for argon or nitrogen. (3) The dispersions should be stored in the dark to prevent photooxidation.

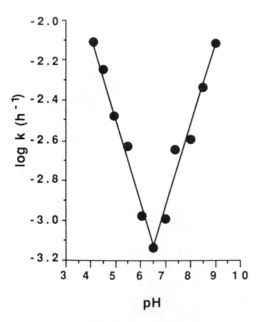

FIGURE 6 Effect of pH on the degradation of PC at 72°C (buffer concentration = 0). The lines were calculated with linear regression analysis. (From Grit et al., 1989.)

(4) As metal ions catalyze peroxidation reactions, glass-distilled water should be used and chelating agents can be added to the medium. (5) The dispersions should not be exposed to γ irradiation. (6) Antioxidant can be added to the system. α-Tocopherol, butylated hydroxytoluene, butyl hydroxyanisole, and ascorbic acid have been proposed as antioxidants.

Cholesterol autooxidation can occur. Not much information is available on autooxidation of cholesterol in liposomes (Lichtenberg and Barenholz, 1988).

V. IN VIVO BEHAVIOR OF LIPOSOMES: GENERAL ASPECTS

This section will briefly discuss the clearance kinetics and tissue distribution of liposomes in vivo (reviewed by Senior, 1987; Hwang, 1987; Gregoriadis, 1988a; Hwang and Beaumier, 1988; Juliano, 1988). Liposomes, when administered in vivo by a variety of routes, rapidly accumulate in the mononuclear phagocyte system (MPS), also referred to as the reticuloendothelial system (RES). The major sites of accumulation are liver and spleen (Senior, 1987; Gregoriadis, 1988a), in terms of both total uptake and uptake per gram of tissue. Abundance of phagocytic macrophages and a rich blood supply are the primary reasons for the preponderant uptake of liposomes by liver and spleen.

Thus far, intravenous injection is the most common route used to administer liposomes. Studies on the in vivo fate of intravenously administered particles also provide information on other routes of injection, since liposomes, which are injected via other routes, usually reach the circulation to some extent (see Sec. VI.E; Senior, 1987). Generally, all routes of administration other than intravenous injection reduce the rate of uptake of liposomes by liver and spleen and reduce the total amount of liposomes in the MPS cells in liver and spleen. In addition to the MPS cells in the liver and spleen, bone marrow (macrophages, sinusoidal lining cells), lymph nodes (free and fixed macrophages and sinusoidal lining cells), lung (alveolar macrophages), serous cavity (peritoneal macrophages), bone tissue (osteoclast), and connective tissue (histiocytes) all possess highly phagocytic macrophages (Bradfield, 1984), which may play a role in the tissue distribution of liposomes administered by routes other than the intravenous one. The following paragraphs of this section will only deal with the intravenous route of administration.

The behavior of liposomes in vivo can be influenced to a considerable extent by varying chemical composition and physical properties. Parameters affecting rate of clearance from the blood and tissue distribution include size, composition, dose, and surface characteristics (e.g., charge, hydrophobicity, presence of homing devices such as antibodies).

It has been early recognized that size and surface charge are critical factors in the clearance kinetics of liposomes (Juliano and Stamp, 1975). Small liposomes are cleared more slowly than larger ones. Positively charged and neutral liposomes circulate longer than negatively charged ones with similar size. Very large liposomes (above 12 μm) are trapped in the first capillary bed (usually that in the lungs) encountered upon injection into the bloodstream (Poste and Kirsh, 1983). Several studies demonstrated that very small liposomes (SUV; <0.1 μm) are able to extravasate through the fenestrations of the liver sinusoids and interact with liver parenchymal cells (Poste et al., 1982; Roerdink et al., 1981; Scherphof et al., 1983). Therefore, SUV may partially avoid the efficient degradative machinery of specialized phagocytic cells (e.g., Kupffer cells in the liver). This might be the main reason why large liposomes (which are taken up by Kupffer cells) are degraded about three times faster than SUV in the liver in vivo (Hwang and Beaumier, 1988).

In addition to size and charge, the chemical composition of liposomes affects their in vivo fate. For example, it has been shown that incorporation of cholesterol in the liposomal bilayer, inclusion of sphingomyelin, and/or the use of phospholipids with long and saturated acyl chains (gel-like bilayers) prolong the vascular circulation time of liposomes (Senior, 1987). Recent studies showed that ganglioside-containing liposomes have a long circulation time (see Sec. VI.A.2) (Allen, 1989).

The total amount of liposomal lipid administered is also an important determinant of clearance rate and, to a certain degree, of tissue distribution (Abra and Hunt, 1981; Kao and Juliano, 1981). Thus injection of larger doses slows down the clearance rate, probably as a result of saturation or "blockade" of the MPS (especially the Kupffer cells in the liver). Such MPS blockade also shifts the distribution of liposomes; spleen uptake increases at the expense of hepatic uptake. Splenic retention thus represents "spillover" of liposomes from the liver. Saturation of the ability of splenic macrophages to remove liposomes results in turn in spillover of liposomes to bone marrow macrophages. Multiple injections of liposomes can eventually exhaust the capabilities of macrophages in liver, spleen, and bone marrow macrophages to clear liposomes and other particles, which may result in an unusually high sensitivity to, for example, pathogens due to impaired MPS function (see Section VII) (Senior, 1987; Allen, 1988).

Immediately upon injection into the blood compartment liposomes encounter high concentrations of plasma proteins (Senior, 1987). Depending on the chemical composition and rigidity of the bilayer, a scala of proteins (e.g., albumin, immunoglobulins, complement, fibronectin, clotting factors, and lipoproteins) can adsorb. This may lead to opsonization which promotes liposome uptake by macrophages or destabilization. These interactions between liposomes and plasma proteins were recently reviewed by Senior (1987) and Juliano (1988). As a consequence of plasma protein binding, liposomes may bind to circulating cells. For example, liposome surfaces rich in adsorbed fibrinogen and/or Von Willebrand factor tend to promote the adhesion of platelets. Adsorption of IgG and complement components to the liposomal surface may promote binding to and uptake by phagocytic neutrophils and monocytes (Hsu and Juliano, 1982; Derksen, 1987).

In addition to the reports of uptake of intact small liposomes (SUV) by hepatocytes (Scherphof et al., 1983), there is some evidence of uptake of intravenously administered liposomes as intact structures by cells other than mononuclear phagocytes of the MPS. Recently, the integrity of the capillary endothelial barrier in several pathophysiological conditions was discussed (Bodor and Brewster, 1986). Several studies already indicated an increased capillary permeability during inflammation both in animals (Lopez-Berestein et al., 1984a) and in man (Morgan et al., 1985; Williams et al., 1987).

Increased accumulation of liposomes, especially small liposomes, has been reported to occur at sites of inflammation (Williams et al., 1986) and in tumors (Turner et al., 1988; Gabizon and Papahadjopoulos, 1988) (cf. Sec. VI.B). However, it is well established that endocytosis of liposomes by MPS cells, primarily those located in liver and spleen, accounts for most of the uptake of liposomes—and, in general, uptake of particulate matter—from the blood circulation. From the limited data on the kinetics and distribution of liposomes in man (Zonneveld and Crommelin, 1988) it can be concluded that the overall behavior of liposomes in man is similar to that observed in animals. Studies in cancer patients showed predominant liver and spleen uptake of technetium-labeled liposomes (Richardson et al., 1979; Lopez-Berestein et al., 1984b; Perez-Soler et al., 1985).

The functions of the human MPS were excellently reviewed by Becker (1988). Removing particles from the bloodstream is just one of the important biological functions of mononuclear phagocytes. Uptake usually takes place by nonspecific endocytosis; it can be increased by linking ligands to the liposomal surface which are recognized by receptors expressed on macrophages (receptor-mediated endocytosis). In Fig. 7 several other possible mechanisms of liposome-cell interaction are presented (including endocytosis). It is generally accepted that endocytosis and adsorption are the major mechanisms of interaction between liposomes and cells in vivo (Poznansky and Juliano, 1984).

VI. SELECTED EXAMPLES OF POTENTIAL MEDICAL APPLICATIONS OF LIPOSOMES

Because of their ability to carry a wide variety of pharmaceuticals, liposomes have been studied for many different therapeutic situations. Therefore, the literature on this topic is abundant. Excellent reviews are available (e.g., Poznansky and Juliano, 1984; Gregoriadis, 1984, 1988b). We will restrict ourselves to the description of certain applications which illustrate the potential benefits of the use of liposomes in the field of drug delivery.

A. Intravenous Injection

1. MPS-Directed Drug Delivery

One approach where the characteristics of the liposomal carrier system are well matched to the intended therapeutic application is the delivery of drugs to the MPS. Because of their particulate nature, the major route of clearance of liposomes, when administered in vivo by a variety of routes, is phagocytosis by MPS cells, especially macrophages in liver and spleen. Obviously, this "natural" fate of liposomes in vivo is an advantage if one attempts to treat diseases

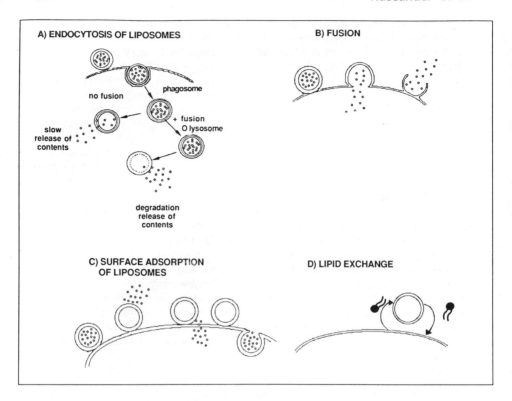

FIGURE 7   Various mechanisms of liposome-cell interaction.   (Adapted from Pagano and Weinstein, 1987.)

involving the MPS or in cases where drug delivery to macrophages is required.   For instance, when macrophages themselves become infected by foreign organisms, liposomes can be efficient, targeted drug delivery systems.   Intracellular infections are caused by organisms which are capable of surviving inside the cells.   In some cases, they are capable of multiplying in the extracellular environment as well; these organisms are known as facultative intracellular parasites. Poor drug penetration into affected cells in combination with toxicity of and resistance to antiinfectious compounds often makes such intracellular infections difficult to treat.

The first reports on this approach were published independently by three different groups using liposomes as carriers of antimonial drugs like meglumine antimoniate in experimental leishmaniasis infection (Alving et al., 1978; Black et al., 1977; New et al., 1978). The Leishmania parasites are lodged inside the lysosomes of phagocytic cells, precisely the intracellular site where liposomes end up after intravenous injection. Liposomal encapsulation of the antimonial drug resulted in a more than 700-fold increase in therapeutic efficacy of this drug (Alving et al., 1978). Although this parasitic disease is infecting an estimated 100 million individuals throughout the world (Ostro, 1987), unfortunately in the near future the high therapeutic potential of this formulation is not expected to have a major impact on the treatment due to lack of current commercial interest for a mostly "third world" disease.

A highly successful use of a liposomal carrier system in infectious diseases which seems to be more promising from a commercial point of view is the therapy of systemic fungal infections with amphotericin B (AMB) incorporated into liposomes. Systemic fungal infections occur frequently in patients suffering from cancer or an immunodeficiency disorder (e.g., AIDS). For example, fungi, in particular Candida albicans, are responsible for about 20% of the lethal infections in leukemia patients. The use of AMB, a polyene antibiotic, is hindered by acute (fever, chills) and chronic toxicity to kidneys, central nervous system, and hematopoietic system (Miller and Bates, 1969). Amphotericin B has a certain preference to interact with fungal membranes due to their ergosterol content. However, it also binds to the cholesterol-containing mammalian membranes resulting in toxicity. Despite the serious side effects, limited efficacy and the development of several new antifungals, AMB is still the cornerstone of therapy for systemic fungal infections. As MPS organs such as liver, spleen, and lung are a frequent target of systemic fungal infections, encapsulation of AMB into liposomes was proposed as an approach to increase its therapeutic index.

Liposomal AMB has been shown to be effective in experimental fungal and also parasitic diseases (reviewed by Emmen and Storm, 1987; Lopez-Berestein, 1988; and Wiebe and Degregorio, 1988). Liposomal AMB was shown to be effective in the treatment of experimental histoplasmosis (Taylor et al., 1982) and cryptococcosis (Graybill et al., 1982). Mehta and coworkers (1984) found that, while free AMB is extremely toxic to both fungal cells and mammalian cells in vitro, the liposomal drug remains toxic to fungal cells but has little effect on mammalian cells. The AMB entrapped in multilamellar or small unilamellar liposomes of different lipid compositions was shown to be active and less toxic than conventional AMB solubilized in deoxycholate (Fungizone) in the treatment of experimental candidiasis in mice. As a result much higher doses of AMB could be administered to the animals when given in liposomal form (Lopez-Berestein et al., 1983; Tremblay et al., 1984; Ahrens et al., 1984). Animals treated with these high doses survived for much longer periods. In mice made neutropenic by the administration of cyclophosphamide— conditions which relate to systemic fungal infections occurring in

cancer patients—AMB liposomes were more effective than AMB in the
treatment of advanced candidiasis (Lopez-Berestein et al., 1984c).
When injected intravenously, liposomal AMB induced higher antibiotic
concentrations in liver, spleen, lungs, and kidneys in normal and in
C. albicans-infected mice.   All these animal studies point to the su-
periority of AMB liposomes over AMB in the treatment of systemic
fungal infections.   However, it appears that a favorable outcome is
not necessarily dependent on the entrapment of the drug into liposome structures.   Recently, results were reported on an AMB colloidal dispersion (AMB-CD) being a size-controlled complex of AMB and
sodium cholesteryl sulfate at a stoichiometric molar ratio of 1:1.   The
AMB-CD was as effective as Fungizone against species of Candida
and Aspergillus.   As AMB-CD was significantly less toxic than the
free drug, a much higher maximum tolerated dose of AMB-CD could
be administered to rats and mice (Newman et al., 1989).

Striking results have been obtained in clinical trials by two
groups of investigators using AMB liposomes in cancer patients with
systemic fungal infections.   One group used multilamellar liposomes
made of dimiristoylphosphatidylcholine and dimiristoylphosphatidylglycerol (molar ratio 7:3) which were tested at the M.D. Anderson
Hospital and Tumor Institute at Houston, Texas (Lopez-Berestein et
al., 1985; Lopez-Berestein, 1988, 1989).   The other group used
sonicated liposomes made of egg phosphatidylcholine, cholesterol,
and stearylamine (molar ratio 4:3:1) which were tested at the Institut
J. Bordet, Brussels (Sculier et al., 1988).   Clinical data from these
studies indicate that two different types of liposomes, as far as
structure and lipid composition are concerned, were well tolerated
by cancer patients suffering from a systemic fungal infection.   The
liposomal product appears to be more effective and less toxic.   The
lack of serious toxicity at the dosage levels used is in agreement
with data obtained in the treatment of experimental infections in animals by liposome-entrapped AMB.   Hemolysis, kidney function deterioration, or central nervous system toxicity were not observed in
the patients.   Thus, upon intravenous administration, liposomes appear to be a safe carrier for AMB, a highly lipophilic compound
which is not soluble in water.   The true impact of liposomal AMB in
the treatment of systemic mycoses and parasitic diseases can be fully
evaluated once clinical trials are conducted in populations other than
cancer patients and immunocompromised hosts.   The mechanism behind the decreased toxicity and the enhanced activity is far from
clear.   Although natural targeting of liposomes to organs such as
liver, spleen, and lung, which are rich in macrophages and which
are a frequent target of systemic fungal infections, made the development of liposomal AMB formulations attractive, it must be noted that
fungal cells are by no means exclusively located in these macrophages.
Therefore, the improvement of the therapeutic index after encapsulation

is not solely due to passive targeting of AMB to macrophages but seems to be related to a preference of liposomal AMB to interact with fungal cells and a degree of localization in affected areas possibly resulting from passage of AMB liposomes through damaged capillaries or from transportation inside migrating peripheral phagocytic cells (Lopez-Berestein, 1989).

A variety of other clinically important infections, such as brucellosis, listeriosis, salmonellosis, and various Mycobacterium infections, are of interest as these are often localized in organs rich in MPS cells. Liposome encapsulation has been demonstrated to improve therapeutic indices of several drugs in a number of infectious models. The natural avidity of macrophages for liposomes can also be exploited in the application of the vesicles as carriers of immunomodulators to activate these cells to an microbicidal, antiviral, or tumoricidal state. These studies were recently reviewed by Emmen and Storm (1987), Popescu et al. (1987), and Alving (1988). In addition to the treatment of "old" infectious diseases, the concept of MPS-directed drug delivery is of considerable interest for the therapy AIDS, possibly enabling control of human immunodeficiency virus replication in human macrophages.

Besides infectious diseases, liposomes offer opportunities for the treatment of other MPS-associated diseases. In metal poisonings and metal storage diseases the intracellular accumulation of metals induces intoxication. The primary sites of metal deposition are MPS cells of the liver, bone marrow, and the lungs. Chelating agents, with strong electron donor groups, have difficulties in crossing cell membranes. Clearly, liposome encapsulation can be useful to overcome this problem by allowing passive targeting to the diseased sites (Rahman, 1988).

2. MPS Avoidance Drug Delivery

In addition to MPS-directed delivery, liposomes have the potential of providing a controlled "depot" release of an encapsulated drug in the blood compartment over an extended period of time, thereby reducing toxic side effects of the drug by avoiding toxic peak concentrations of free drug in the bloodstream as will be illustrated in the section on DXR delivery via liposomes. However, it was early recognized (Poste, 1983; Poznansky and Juliano, 1984) that the rapid accumulation in the MPS system can limit the fields of application of intravenously administered drug-containing liposomes. Therefore, several groups explored the idea of avoiding uptake by the MPS (Davis and Illium, 1986; Allen and Chonn, 1987; Gabizon and Papahadjopoulos, 1988). Liposomes which are capable of evading the MPS would provide two important advantages. First, the liposome circulation time in the blood is increased. This would prolong the time frame available for slow drug release in the bloodstream and

would also improve the prospect of directing liposomes to tissues
other than liver, spleen, and lung. The other advantage may
be decreased liposome loading of the MPS, thus avoiding blockade
and impairment of an important host defense system (Storm et al.,
1990a). One of the early approaches which have been followed
to increase liposome circulation time was "blocking" of the phagocytic uptake mechanisms of the MPS by predosing with high doses
of liposomes or other particles. Another approach was based on
improving liposome stability in plasma (Senior, 1987). During their
stay in the bloodstream liposomes can interact with serum proteins
(e.g., lipoproteins and opsonins) and sometimes become destabi-
lized, resulting in leakage of entrapped solute. In particular high-
density lipoproteins (HDL) appear to be responsible for the destabilization. Manipulations that reduce these interactions, e.g., by
rigidifying the bilayers by the inclusion of cholesterol and phospholipids yielding gel state bilayer structures, have been shown to decrease leakage of liposome contents in plasma and to prolong the
residence time of liposomes in the circulation (Senior and Gregoriadis,
1982; Allen and Everest, 1983; Senior, 1987). However, although
both approaches resulted in reduced clearance rates of circulating
liposomes, they were not very successful in achieving substantial
liposome uptake by tissues outside the MPS (Poste, 1983). Profitt
et al. (1983) reported a 50% increase in the uptake of labeled liposomes in tumors of mice with a suppressed MPS system by predosing
with "cold" liposomes compared to "nonsuppressed animals." However, MPS suppression seems to be of limited value in clinical practice since the MPS plays an important role in the body's defense
system. Thus, it is not feasible to close it down for prolonged
periods of time.

However, recent evidence from the laboratories of Allen (1989)
and Papahadjopoulos (Gabizon and Papahadjopoulos, 1988) suggests
that longer circulation of liposomes still may offer viable therapeutic
opportunities. It was found that inclusion of monosialoganglioside
(GM1), phosphatidylinositol, or sulfatides prolongs the circulation
times of liposomes, particularly in combination with liposomal formulations with rigid bilayers (e.g., containing sphingomyelin, distearoylphosphatidylcholine, cholesterol, or other rigidifying phospholipids).
Allen (1989) showed that as much as 50% of the injected dose remains
in the circulation for 8 hr with up to 20% of the injected dose still
circulating after 24 hr. It has been proposed that the phenomenon
of prolonged circulation and reduced MPS uptake might be related
to a decreased opsonization, probably obtained by mimicking important properties of the outer monolayer of red blood cells (Allen and
Chonn, 1987). Clearly, such liposomes have a far greater opportunity than conventional liposomes to be distributed to non-MPS sites.
A remarkable increase in tumor uptake (up to 25-fold as compared

to conventional liposomes) was observed when such liposome formulations, called "stealth" liposomes because of their ability to avoid detection and uptake by the MPS, were tested in mice bearing an implanted intramuscular tumor (Gabizon and Papahadjopoulos, 1988).

The "stealth" concept may offer two other opportunities for liposome application: (1) Conventional immunoliposomes (see Sec. VI.C)
have been shown to be removed rapidly from the circulation by the
MPS (Peeters et al., 1987). The combination of the stealth approach
for longer circulation with the attachment of antibodies or antibody
fragments may provide a means of delivery of drugs to their sites
of action with a high degree of specificity. This could be useful
for treating leukemia, graft-vs.-host diseases, and HIV disease.
(2) Extending the circulation time of relatively large liposomes (0.2–
0.4 μm) offers the potential for creating an encapsulated circulating
reservoir. With the benefit of the stealth coating such liposomes
would avoid MPS uptake for many hours and could be engineered to
release entrapped drugs as they circulate. The therapeutic availability of many drugs which are rapidly degraded or excreted in
their free form may be substantially increased using this approach.
Peptides often have short blood half-lives requiring large doses or
multiple daily injections to be effective. Such molecules might well
benefit from a long-circulating stealth microreservoir engineered to
gradually release drug. This might allow dosing frequency to be
reduced to only one or two injections daily. The circulating microreservoir concept may be preferred for the delivery of peptides over
the subcutaneous or intramuscular route of administration usually
used for these drugs, as the peptide would be released within the
blood compartment so that distances to most target sites are much
shorter thereby maximizing the likelihood of the peptide to reach
its site of action.

B. Diagnosis and Therapy of Cancer

As most antitumor drugs have a narrow therapeutic window, cancer
chemotherapy presents serious challenges for the design of drug
delivery systems. A large number of investigations using liposomes
as carriers for currently utilized antitumor drugs, such as anthra-
cyclines (Gabizon, 1989; Storm, 1987; Storm et al., 1989a,b), cisplatin (Yatvin et al., 1981; Steerenberg et al., 1988), methotrexate
(Kimelberg et al., 1976; Kosloski et al., 1978), and cytosine arabinoside (Kobayashi et al., 1977; Rustum et al., 1979; Mayhew et al.,
1982), have been published. Additionally, other compounds not
currently accepted as antitumor agents have been formulated in liposomes in order to enlarge the anticancer drug arsenal, e.g., lipophilic cisplatin analogs (Perez-Soler et al., 1986; Lautersztain et al.,
1986), nocodazole, a water-insoluble, experimental drug (Sculier et
al., 1986), so-called liposome-dependent drugs (referring to drugs

with poor ability to enter cells and whose toxicity can be significantly
enhanced by intracytoplasmic delivery) such as methotrexate-γ-aspar-
tate (Heath et al., 1985a) and 5-fluoroorotate (Heath et al., 1985b).
and valinomycin, an antibiotic with ionophoric activity (Daoud and
Juliano, 1986). Research on liposomal antitumor agents and the difficulties inherent to this approach have been reviewed in detail by
Weinstein and Leserman (1984), Poznansky and Juliano (1984), and
Gabizon (1989). In the initial literature on this subject, enthusiasm
was generated primarily by the idea that liposomes might be used
for specific delivery of the antitumor drug to tumor cells. However,
a better understanding of the biological barriers separating the drug-
liposome entity from the ultimate site of action led to the view that
uptake of liposomes by nonphagocytic cells, including most tumor
cells generally is questionable or, at best, of little significance
(Poznansky and Juliano, 1984). Nowadays the leading rationale for
liposome encapsulation of antitumor drugs is based on the expected
favorable effects on the pharmacokinetic and distribution characteristics of the encapsulated drug resulting in reduced toxicity without
loss of antitumor activity. To illustrate this valuable application of
liposomes in anticancer therapy, we will focus below on the description of results obtained in our group on the mechanism(s) underlying
the therapeutic effects obtained by encapsulation of doxorubicin
(DXR) and cisplatin (cDDP).

Different types of DXR- and cDDP-containing liposomes (extrusion MLV) were tested in rats bearing a solid IgM-immunocytoma on
the flank. The encapsulation of DXR in liposomes resulted in a
marked prolongation of the survival times (Table 1). In the animal
group treated with DXR-DPPC:DPPG:chol (10:1:10) liposomes no
deaths were scored 30 days after the onset of therapy; in the group
treated with PC:PS:chol (10:1:4) liposomes 20% of the animals died
within 30 days, while in the group treated with free DXR no long-
term survivors were scored. In Fig. 8 the antitumor activity of
DXR in free and in liposomal form is depicted. It can be derived
that the liposome-induced enhancement of survival time was not
caused by major changes in the antitumor activity as compared to
free DXR. Thus, for the dose regimen used, DXR encapsulation in
liposomes did not impair the antitumor activity of the drug, whereas
it apparently lowered the toxicity resulting in prolonged survival
and an overall increase in the therapeutic index.

Encapsulation of cDDP in liposomes did not show such favorable
effects. Liposome encapsulation of cDDP decreased the antitumor
effect (Fig. 9). It was demonstrated that administration of cDDP
liposomes resulted in a lower incidence as well as reduced severity
of focal alterations of the epithelium of the proximal tubuli compared
to administration of the free drug (Steerenberg et al., 1988). However, despite this reduction in renal toxicity the therapeutic index

TABLE 1   Effect of Treatment of Solid IgM Immunocytoma Bearing Lou/M Rats with Different Formulations of DXR

Type of treatment[a]	Long-term survivors[b] (>30 days)
Saline	0/10
Free DXR	0/10
DXR-DPPC/DPPG/chol	10/10
DXR-PC/PS/chol	8/10
DXR-DSPC/DPPG/chol	5/10

[a]Tumor-bearing rats were injected with 2 mg DXR/kg body weight i.v. daily for 5 days and on day 11.   Tumor diameter at the start of the experiment was 2–3 cm.
[b]Number of surviving versus the total number of rats.
Liposomes: DPPC/DPPG/chol: (molar ratio 10:1:10); size about 0.7 μm.   PC/PS/chol: (molar ratio 10:1:4); size about 0.3 μm.   DSPC/DPPG/chol: (molar ratio 10:1:10); size about 0.7 μm.
Source: Crommelin et al., 1990b.

was not improved due to the loss of antitumor activity.   Studies were performed in order to obtain a better understanding of the mechanism(s) underlying these therapeutic results.   For DXR liposomes, the results obtained pointed to sustained release as the primary mechanism behind the beneficial effects resulting from liposome encapsulation of DXR.   No direct DXR liposome accumulation in the tumor was found; DXR concentrations in the cardiac tissue were lower after administration of the free drug (Van Hoesel et al., 1984; Storm et al., 1989a).   Interestingly, two different pathways for the sustained release of DXR were identified: DXR is released directly from liposomes being present in the blood but also indirectly from the MPS following uptake and processing by macrophages (Storm et al., 1990b).   Peak concentrations of free DXR in organs which are particularly sensitive to the toxic action of the drug, like the heart, are avoided, while apparently the prolonged presence of free (i.e., nonliposomal) DXR levels in the blood can result in sufficient exposure levels for tumor cells.   Comparison of liposomal DXR delivery with DXR delivery via long-term infusions in the same tumor model confirmed the indicated concept of sustained release (Storm et al., 1989b).   Steerenberg et al. (1988) offered an explanation for the observed decrease in antitumor activity of cDDP after incorporation into liposomes (Fig. 9).   They reported that the reduction in antitumor activity might be related to conversion of liposomally delivered cDDP by macrophages of liver and spleen in a less cytotoxic form.

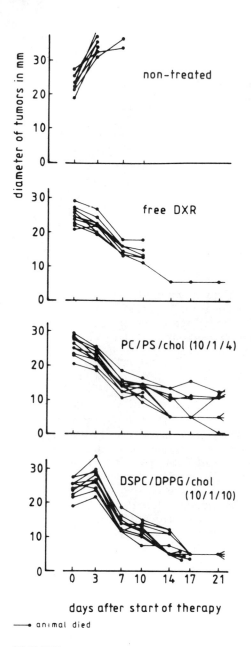

days after start of therapy

→ animal died

FIGURE 8 Antitumor activity of free DXR and DXR entrapped in different liposome types in solid IgM immunocytoma-bearing Lou/M Wsl rats. 2 mg DXR/kg body weight was injected i.v. daily for 5 days (0–4) followed by one additional injection at day 11 after start of therapy. Results obtained during the first 21 days after start of treatment are shown. Treatment groups consisted of 10 animals. For information on liposomes: see Table 1. (From Storm et al., 1987).

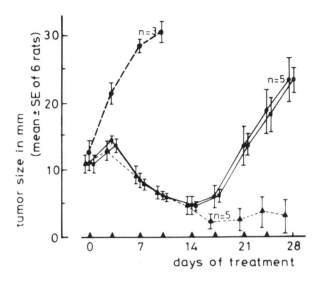

FIGURE 9 Antitumor activity of liposome-encapsulated cDDP and
free cDDP in solid IgM immunocytoma-bearing Lou/M rats. Tumor-
bearing rats were injected with 2 mg cDDP/kg body weight twice a
week (arrow). Tumor diameter at the start of the experiment was
2–3 cm. Tumor growth during treatment is presented (mean ± SD
of six animals unless otherwise indicated). Liposomes: DPPC/DPPG/
chol: (molar ratio 10:1:10); size about 0.7 μm. PC/PS/chol: (molar
ratio 10:1:4); size about 0.3 μm. ●---●:saline; ■___■:lip cis-Pt
(PC/PS/chol); ▲---▲:free cis-Pt; ●___●:lip cis-Pt (DPPC/DPPG/chol).
(Adapted from Steerenberg et al., 1988.)

Therefore, the difference between liposomal delivery of cDDP and
DXR seems to be that DXR is released in vivo in active form while
cDDP is (partly) inactivated.

Because of its clinical importance and the expected benefits of
the drug in liposomal form for cancer treatment, all three American
"liposome enterprises" (i.e., Liposome Technology Inc., Erbamont,
LyphoMed/Vestar joint ventures, and the Liposome Company, Inc.)
are developing a formulation of liposomal doxorubicin. Clinical studies already show promising results as far as the acute toxicity is
concerned (less vomiting, nausea, and hair loss) (Gabizon et al.,
1989; Treat et al., 1989).

For the successful treatment of human malignancy accurate stag-
ing and detection of primary and metastatic diseases is crucial.
Liposomes have been shown to be useful for oncological radionuclide
imaging. Profitt et al. (1983) demonstrated that stable, small,

unilamellar liposomes prepared from synthetic phospholipids and labeled with ^{111}In were specifically delivered to murine EMT-6 tumors.
This targeting was five- to eight-fold greater than the uptake of un-
entrapped radionuclide into tumors in mice. Subsequent experiments
of this group have confirmed that liposomes can be preferentially
taken up by many different murine tumors including EMT-6 mammary
carcinosarcoma, Lewis lung carcinoma, colon carcinoma 51, B-16 melanoma, mammary adenocarcinoma 16/c, and sarcoma 180 (Williams et
al., 1984). Recently, interesting clinical results on the use of [111]In
liposomes were reported (Presant et al., 1988). Patients with proven
primary and/or metastatic cancer received single intravenous injections of ^{111}In liposomes. Gamma camera scintigraphy 1–72 hr later
visualized tumors in 92% of the patients studied, including carcinomas
of breast, lung, colon, prostate, kidney, cervix, thyroid, and soft-
tissue sarcoma, lymphoma, and melanoma. Tumor sites were found
in soft tissues, bone, lung, liver, lymph node, and spinal cord.
There were only two false-positive and four false-negative images of
metastatic sites. Unsuspected areas of malignancy were seen in the
lumbar subdural space, pleura, liver, thyroid, and lung. Besides
tumor accumulations, uptake was observed in normal liver and spleen.
These results suggest broad clinical application of these vesicles by
permitting a wide variety of human tumors in primary and metastatic
sites to be imaged without serious toxicity and with radiation doses
comparable to that of other radionuclide scanning techniques. The
mechanism for tumor accumulation of the vesicles is unknown. As
illustrated above for DXR liposomes, attempts to target liposomes to
human tumors in vivo have generally been unsuccessful (Poznansky
and Juliano, 1984). According to the authors, altered vasculature
in the tumor as well as the chosen size (mean about 80 nm) and
chemical composition (neutral synthetic phospholipids) might be important factors involved (see Secs. V and VI.A.2).

C. Immunoliposomes

One of the approaches presently under investigation to achieve site-
specific drug delivery (Tomlinson, 1987) is the combination of liposomes and antibodies. These immunoliposomes consist of liposomes
to which antibodies or antibody fragments are attached; for reasons
of increased reproducibility, nowadays mostly preparations are used
in which the liposomes and antibody or antibody fragments are covalently linked (Toonen and Crommelin, 1983; Crommelin and Storm,
1990a). Maximum selectivity of drug delivery can be expected when
immunoliposomes are injected into the same compartment in which the
target tissue or cells are localized. Thus readily accessible target
cells are located in the blood compartment, the peritoneal cavity,
bladder, uterus, lungs, eye, and the lymphatic system (Poste, 1983;
Peeters et al., 1987; Crommelin and Storm, 1990a).

For a rational design of immunoliposomes a clear insight into the factors which control the disposition of liposomes in vivo is crucial (see Sec. V). In addition, there are other critical factors relevant to the use of immunoliposomes:

1. Inadequate performance may result from lack of antigen specificity, e.g., tumor-specific antigens are often differentiation antigens which also occur, although in a lower density on nontarget cells (Poste, 1984).
2. Often the target antigen is not expressed on all target cells. In that case a cocktail of antibodies directed to antigens located at surfaces of different cells is necessary (Poste and Kirsh, 1983; Papahadjopoulos et al., 1985).
3. Antigen structures might be shed from the parent cell and circulate in the blood. These circulating antigens form complexes with the immunoliposomes, thereby inactivating the homing device and preventing the specific interaction of immunoliposome and target tissue (Poste and Kirsh, 1983).
4. Immunogenicity of immunoliposomes is a point of concern (Alving, 1987). Multiple injections of immunoliposomes may raise antibodies against the homing device on the liposomal surface. These antibodies might then interact with the immunoliposomes thus compromising specific targeting. Connor et al. (1985) suggested that in the clinical setting human syngeneic monoclonal antibodies would be preferable to minimize the immunogenicity of immunoliposomes.
5. Several studies (Aragnol and Leserman, 1986; Konno et al., 1987) discussed the hypothesis that Fc-free antibodies as homing device might prolong the circulation time (less Fc-mediated uptake by macrophages), thereby increasing the therapeutic effect. However, thus far no hard data are available demonstrating significant therapeutic advantages of Fc-free fragments over whole IgG as a homing device.

To induce a therapeutic effect, successful targeting of drug-containing immunoliposomes is only the first step. To exert a certain pharmacological action the drug has to leave the immunoliposome and enter the target cell (assuming that the target cell cannot phagocytose immunoliposomes). There are several possibilities for how this can be achieved (Fig. 10). An interesting option opens up for circulating blood cells (e.g., erythrocytes and lymphocytes) (Peeters et al., 1988, 1989). After formation of the immunoliposome-target cell complex, the complex is taken up by macrophages. Within the macrophage the liposome is degraded and the drug liberated in the close proximity of the cophagocytosed target cell.

Recently we reviewed the literature (Peeters et al., 1987) dealing with the potential and limitations of immunoliposomes in the treatment

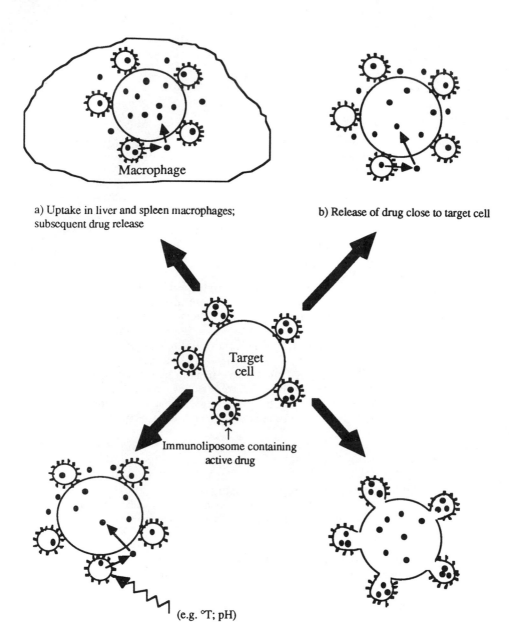

a) Uptake in liver and spleen macrophages;
subsequent drug release

b) Release of drug close to target cell

c) Release of drug close to target cell;
external triggering of release

d) Fusion with target cell;
subsequent drug release

FIGURE 10 Several pathways of drug internalisation after immuno-
specific binding of the immunoliposome to the appropriate cell.
(From Peeters et al., 1987.)

and diagnosis of life-threatening diseases such as cancer and malaria.
From this review it could be derived that immunoliposomes can be
helpful in improving the evaluation of gamma scintigraphy scans
(Begent et al., 1982; Keep et al., 1983; Barrat et al., 1983;
Udayachander et al., 1987). Promising results can also be expected
in the treatment of diseases with target cells located inside the blood
circulation as in the case of AIDS, lymphomas, leukemias, thrombosis,
sickle cell disease (Kumpati, 1987), and malaria (Agrawal et al., 1987;
Peeters et al., 1989). In the latter study (Peeters et al., 1989), it
was demonstrated that intravenously administered chloroquine-
containing immunoliposomes (Fab' liposomes) yielded enhanced therapeutic effects of the drug compared with free drug or drug entrapped
in conventional liposomes in rats infected intravenously with Plasmo-
dium berghei parasitized mouse red blood cells. Positive results
were also obtained in studies where the target cells and the injected
immunoliposomes were injected in the same compartment. Illustrative
examples of this approach were published by Hashimoto et al. (1983)
(in vivo local and systemic antitumor effect with actinomycin D-
containing immunoliposomes), Wang and Huang (1987) (gene transfer
with pH-sensitive immunoliposomes), and Straubinger et al. (1988)
(treatment of ovarian carcinoma); in all three cases the immunoliposomes were injected intraperitoneally, except in the study of Hashimoto et al., in which part of the antitumor experiments were performed to treat subcutaneously located tumors with intravenously
injected drug-containing immunoliposomes.

   Thus far no specific data on immunoliposome toxicity have been
published; study of the chronic toxicity has never been given high
priority because of the severe nature of the diseases and the relatively low toxicity of liposomes (see Sec. VII). However, data on
the possible induction of an immune response in the course of a
chronic multiple injection program should be collected. If an immune
response occurs, its effect on the general condition of the patient,
and on the fate and therapeutic effect of these immunoliposomes,
should be investigated. This information should be available before
immunoliposomes can be tested in long-term treatment regimens or
multiple-injection diagnostic programs.

D. Intrapulmonary Delivery

About a decade ago it was shown that a substantial fraction of intravenously administered liposomes larger than 1 μm in diameter is localized in the lungs, probably due to physical trapping in the lung
capillary bed (Sharma et al., 1977; Hunt et al., 1979; Abra et al.,
1984). Utilizing this observation, Fidler et al. (1980) as well as
Mayhew and Rustum (1983) demonstrated the potential usefulness of
such lung-accumulating liposomes for the treatment of lung tumors.
In addition to the intravenous route, inhalation is another possible

route for liposome-mediated drug delivery to the lung (Mihalko et al., 1988). Several anatomical and physiological properties of the lungs and airways render the lungs an attractive portal of entry for systemic drug delivery: large surface area (about 70 m^2), a very thin (less than 0.5 μm) diffusional barrier between the alveolar air space and its surrounding capillary bed, high blood flow, and relatively (as compared to the liver) low activity of enzymes (and consequently a low first-pass metabolism). It has been shown that liposomes consisting of DPPC and DPPG, the major phospholipid components of endogenous lung surfactant, administered to the lungs via the airways associate rapidly with the alveolar surfactant with no apparent physiological or pathological disturbance of lung function or architecture (Mihalko et al., 1988).

Several groups investigated the use of liposomes for the intrapulmonary delivery. Farr et al. (1985) showed that the deposition of aerosolized liposomes in the human lung depends on the aerosol particle size. Short-term retention profiles for MLVs and SUVs deposited in the lung were indicative of clearance via the mucociliary transport mechanism.

For the treatment of lung surfactant deficiency in premature human infants suffering from respiratory distress syndrome, limited clinical trials were performed showing that liposomes in the lung—instilled intratracheally either as an aerosolized mist (Ivey et al., 1977) or as a suspension via an endotracheal tube (Fujiwara et al., 1980)—rapidly improved lung function. No adverse effects were observed as a result of the supplementation with surfactant-like material. It appears, therefore, that liposomes are a suitable system for the delivery of major phospholipid components of endogenous lung surfactant.

In the case of drug delivery by inhalation, the direct administration of free drugs to the respiratory system usually suffers from the drawback of a short retention time in the lung. Juliano and McCullough (1980) demonstrated in rats that it is possible to prolong the retention of the antitumor drug cytarabine within the lung, and alter its pharmacokinetics, by instillation of the drug encapsulated in liposomes. The drug was confined to the lung with minimal distribution to other tissues. Woolfrey et al. (1988) showed that, after intracheal instillation, liposomal carboxyfluorescein appeared in the systemic circulation of rats at a slower rate than the free substance. Pulmonary absorption was dependent on dose and composition of the liposomes and did not involve the transfer of intact liposomes into the blood. Shek et al. (1988) demonstrated prolonged localization at the pulmonary site after encapsulation of atropine and glutathione. Prolonged absorption from the lungs was observed following encapsulation of benzyl penicillin and the peptide oxytocin (Mihalko et al., 1988). Since peptides generally have short biological half-lives, the latter observation suggests that liposomes might act as sustained release delivery systems.

Localized delivery to the lungs and airways via liposomes is of obvious interest for the treatment of asthma. The major current drugs used in inhalation therapy of bronchial asthma are bronchodilators such as albuterol and metaproterenol, antiinflammatories such as corticosteroids, and sodium cromoglycate. One problem with current bronchodilator therapy is that the rapid absorption of these drugs from the lung produces high initial plasma drug levels with accompanying side effects. Another problem is that drug half-life is too short to provide adequate duration of the therapeutic effect. Since in most patients acute asthma symptoms tend to worsen during the night, it is highly desirable to extend the duration up to 10–12 hr. Kellaway et al. (1988) administered sodium cromoglycate encapsulated in liposomes to human volunteers by nebulization. They found plasma concentration profiles consistent with first-order input of drug: lower peak plasma levels and a longer apparent half-life with significant plasma concentrations remaining more than 24 hr after administration. No loss in availability was observed for liposome-encapsulated cromoglycate relative to inhaled free drug. Of particular interest are recent studies on liposome-encapsulated β_2-agonist bronchodilators. These studies show that liposomal formulations of metaproterenol sulfate inhibited histamine-induced bronchospasm in anesthesized dogs and guinea pigs for longer periods than in saline (Mihalko et al., 1988; McCalden et al., 1989). Similar results were obtained for terbutaline (Fig. 11) (Fielding, 1989). Additionally, while liposomal formulations were as effective or even more effective as the free drug, they reduced cardiovascular side effects. Interestingly, it was also found that the liposomal lipid composition was critically important in determining the release rate of terbutaline in vivo (Fig. 12). The half-life of terbutaline absorption could be varied from 1.4 hr to 18 hr with 10% of the dose still in the lungs 2 days after administration by changing lipid composition. These studies indicate that liposomes can release their content at a controllable rate for extended periods of time suggesting that formulations can be tailored to each specific therapeutic application.

In light of the information presented, it appears that liposomes have much to offer as an improved delivery system for inhaled bronchodilators as they can provide sustained release, solubilization, stability, and safety in an inhalable formulation.

E. Intraperitoneal Administration of Liposomes

In this section the intraperitoneal route of liposome administration will be discussed. For a number of diseases this route of administration may be preferred over the intravenous route of administration of liposomes. For example, intraperitoneal injection of drug-

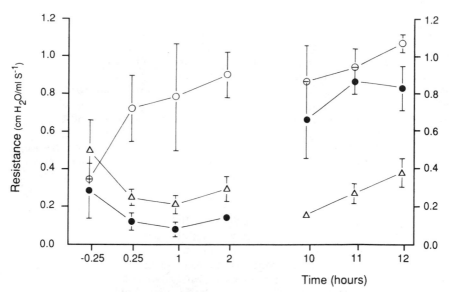

FIGURE 11 Duration of bronchodilator activity in an animal model. Anesthesized guinea pigs received 1 μmol/kg of terbutaline, free or in liposomes, via intracheal instillation at time zero. Bronchodilator activity was measured as the reduction in histamine-induced airway resistance. Histamine (10 μg/kg) was administered i.v. prior to each resistance measurement. o: saline (n = 3); •: free drug (n = 3); Δ: liposomal drug (n = 4). (Courtesy of Liposome Technology, Inc., Menlo Park, California.)

laden liposomes may be more effective for treatment of intraperitoneally located tumors. Besides, as liposomes are removed from the peritoneal cavity via the lymphatic system, intraperitoneal injection of liposomes may be useful in the therapy of lymphatic tumors.

Intraperitoneal administration of chemotherapeutic agents has been used for many years as a way of increasing the delivery of drugs to tumors (e.g., ovarian carcinoma) located in the peritoneal cavity (Markman, 1986; Howell and Zimm, 1988). Cisplatin (Casper et al., 1983; Markman et al., 1985), cytosine arabinoside (Ara-C) (King et al., 1984; Markman et al., 1985, 1986), and bleomycin (Markman et al., 1986) are examples of intraperitoneally administered drugs which were already successfully applied in clinical settings. In comparison with the intravenous route of administration the potential advantages of intraperitoneal therapy are the avoidance of high toxic drug plasma levels and an increased (local) exposure of tumors (cells) to anticancer drugs. Whether this increased exposure

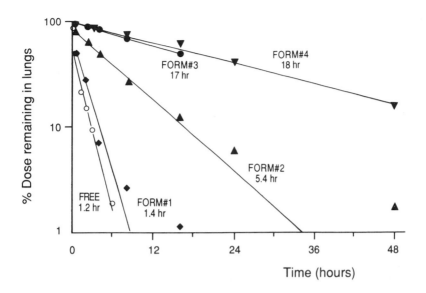

FIGURE 12   Effect of lipid composition on absorption rate of a
liposome-encapsulated bronchodilator.   Mean % of dose remaining
unabsorbed in the lungs after instillation of terbutaline (1 μmol/kg),
free or in liposomes, to guinea pigs.   Formulation 1:   saturated
short-chain acyl chains, no cholesterol.   Formulation 2:   unsaturated
mixed-length acyl chains, with cholesterol.   Formulation 3:   saturated
short-chain acyl chains, with cholesterol.   Formulation 4:   saturated
long-chain acyl chains, with cholesterol.   (Courtesy of Liposome
Technology, Inc., Menlo Park, California.)

will result in an improved therapeutic effect depends on free-surface
diffusion into the tumor.   In addition, after intraperitoneal administration, a substantial fraction of the drug will finally reach the
blood, which enables the drug to affect the tumor via the capillary
route.

    Like other high molecular weight and particulate materials such
as india ink (Tsilibary and Wissig, 1977), erythrocytes (Flessner
et al., 1983), latex particles (Bettendorf, 1979), or colloidal gold
(Langhammer et al., 1973), liposomes are removed from the peritoneal cavity via the lymphatics of the diaphragm (Parker et al.,
1981a; Ellens et al., 1983; Poste, 1983).   Regardless of where particles (liposomes) are placed within the peritoneal cavity, they are
swept upward by normal respiratory movements and enter the lymphatics of the diaphragm, presumably by leaving the cavity through
small pores (stomata) (Allen, 1936, 1956; Courticé and Simmonds,

1954; Tsilibary and Wissig, 1977; Bettendorf, 1979). Abnormal
respiration (as a result of anesthesia) affects the retention time of
intraperitoneally injected particles. Anesthesia which stimulates
respiration results in a decrease of retention time while respiration-
suppressing anesthesia results in the opposite effect (Courtice and
Simmonds, 1954).

1. Pharmacokinetic Profile After Intraperitoneal
   Injection of Liposomes

Figure 13 presents a schematic diagram for drug absorption from the
peritoneal cavity. As mentioned above, particles (e.g., erythrocytes,
bacteria, colloidal gold, and liposomes) which are not able to pass
capillary membranes are removed from the peritoneal cavity via the
lymphatic system (Fig. 13, I and II). Relatively low molecular weight
compounds (e.g., drugs) are exclusively absorbed via splenic blood
capillaries into the portal vein (Fig. 13, III).

Several studies have been performed in order to investigate the
effect of liposomal size (Hirano and Hunt, 1985), lipid composition
(Senior and Gregoriadis, 1982; Hirano et al., 1985), and lipid dose
(Ellens et al., 1983, Kim et al., 1987) on the fate of liposomes after
intraperitoneal administration. In the size range studied (0.048–
0.72 μm), no size-dependent absorption could be expected (Hirano
and Hunt, 1985). Particles larger than 22.5 μm are not expected
to enter the lymphatic capillaries (Allen, 1956). After intraperitoneal administration of multivesicular liposomes (19 ± 7 μm), Kim and
Howell (1987a) and Kim et al. (1987) showed that liposomal entrapment of Ara-C prolongs the half-life of the drug in the peritoneal

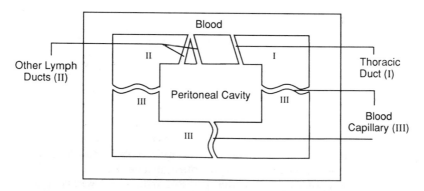

FIGURE 13 Schematic diagram for drug absorption from the peritoneal cavity. I and II represent the lymphatic system and III represents splenic blood capillaries. (Adapted from Hirano and Hunt,
1985.)

cavity by 79-fold. In the latter study (Kim et al., 1987) they demonstrated that the peritoneal half-life of these large vesicles could be further increased (619-fold) by preinjecting blank liposomes. In general, positively charged liposomes disappear from the cavity more slowly than liposomes with a negative charge. Negatively charged cell surfaces have been proposed to be responsible for the retarded absorption of positive liposomes (Hirano et al., 1985).

In comparison to intravenous administration of MLV, which usually results in a rapid and almost quantitative uptake into liver and spleen, the fraction taken up into these organs is lower after intraperitoneal injection of these large liposomes. The reason might be that liposomes are trapped in lymph nodes and degradation of the liposomes in the peritoneal cavity can occur (Ellens et al., 1981; Parker et al., 1982); besides, several types of liposomes are degraded more quickly in lymphatic fluid than in plasma (Parker et al., 1981a,b).

2. Potential Implications for Cancer Therapy

Lymph Node Metastases Tumor cells often metastasize through lymphatic channels (Abe and Taneichi, 1972; Knapp and Friedman, 1974; Musumeci et al., 1978). Therefore, enhanced localization of cytostatic drugs in the lymphatics may be therapeutically beneficial. Liposome encapsulation limits the direct passage of a drug from the peritoneal administration site to the blood circulation, while it enhances the lymphatic transport from the peritoneal cavity (Fig. 13). Hirano and Hunt (1985) showed that in contrast to the smallest liposomes tested (0.048 μm), which were hardly retained in the lymph nodes, the largest liposomes tested (0.72 μm) showed a marked retention in the lymph nodes. Therefore, encapsulation of drugs in this 0.72-μm type of liposome should be considered when increased drug availability in lymph and lymph nodes is desired. Comparable results were found by Hirano et al. (1985) and Parker et al. (1981a) who used [14C]sucrose and carboxyfluorescein, respectively, as encapsulated marker. Parker et al. (1981b) demonstrated that after intraperitoneal injection adriamycin-containing liposomes (extruded through 0.6-μm filters) were retained by certain lymph nodes; they showed similar results for liposome-encapsulated Ara-C (Parker et al., 1982). Whether these results can be applied to more effective killing of tumor cells located in the lymphatics is still an open question.

Intraperitoneally Localized Tumors Although animal models with intraperitoneal ascitic tumors are generally considered to be a poor model for human cancer, being little better than treating cancer cells in a test tube (Poste, 1983; Poste and Kirsch, 1983), several interesting studies were performed with promising therapeutic potentials using drug-containing liposomes. Kim and Howell (1987a) and

Kim et al. (1987) showed that the prolonged retention time of Ara-C
in the peritoneal cavity after intraperitoneal administration of the
drug in liposomal form as discussed above resulted in better therapeutic effects on intraperitoneally inoculated L1210 cells, as compared
to the free drug. The activity of intraperitoneally administered
cDDP on Ehrlich ascites carcinoma in mice was increased after encapsulation in neutral liposomes (Sur et al., 1983). The in vivo studies
revealed improved antitumor activity and a lower toxicity after administration of cDDP liposomes compared to free drug.

   In order to improve the therapeutic benefits further, site-
specific drug delivery into this anatomically isolated cavity by using
immunoliposomes (liposomes to which tumor-antigen-specific antibodies
are covalently linked) might be favorable. Hashimoto and colleagues
(1983) injected mammary tumor cells in mice intraperitoneally. Twenty-
four hours later actinomycin D-containing immunoliposomes were injected intraperitoneally; free drug and "naked" liposomes mixed with
the free antitumor drug were used as controls. At the dose level
used 100% survival was observed after injection of drug-containing
immunoliposomes; all control animals died rapidly. Another interesting study has been performed by Wang and Huang (1987) who demonstrated target cell-specific delivery of a liposome-encapsulated
plasmid to intraperitoneally located RDM4 lymphoma cells after intraperitoneal administration. In a recent study of Straubinger and coworkers (1988) the possibility of a liposome-based therapy of human
ovarian cancer with specific immunoliposomes is discussed. They
demonstrated that after intraperitoneal administration immunoliposomes
bonded specifically to masses of tumor cells for periods of days while
control liposomes showed a very low interaction with the tumor cells.
Whether the encapsulated cytostatic drug will exert its therapeutic
effect depends on the endocytotic capacity of the tumor cells and/or
the integrity of the immunoliposomes used. Temporarily high concentrations of drug released from bound immunoliposomes in close
proximity of the tumor cells might be very effective.

F. Intramuscular or Subcutaneous Injection of
Liposomes

Delivery of peptides and proteins via the gastrointestinal tract has
not been successful because of poor penetration through the intestinal epithelium and high levels of proteolytic activity in the gastrointestinal tract. Liposomal encapsulation of proteins and peptides
will not improve the efficiency and capacity of this absorption pathway considerably (e.g., Ryman et al., 1982; Machy and Leserman,
1987; Weiner and Chia-Ming Chiang, 1988). These difficulties in
delivery via the oral route caused the parenteral route to remain
the preferred route for the administration of therapeutic peptides

and proteins. Liposomes might have a potential to control the availability and release kinetics of the encapsulated proteins and peptides
from the site of injection in a way superior to conventional techniques
for the delivery of these drugs. A number of articles have been
published on the distribution of liposome-encapsulated compounds
over the body after intramuscular, interstitial or subcutaneous injection. Site of injection, charge, liposome size, charge, and composition were critical parameters for controlling the fate of the encapsulated compound (Patel and Ryman, 1981; Senior, 1987). After intramuscular injection of liposome dispersions differing in size in the
thigh muscle of mice "large" liposomes (0.3–2.0 μm) were cleared
more slowly from the injection site than dispersions of "small" liposomes (0.15–0.7 μm); the fraction of liposomes taken up by the
lymphatics was higher for the small liposomes than for the large
liposomes. The marker molecule (inulin) was released over a prolonged period of time; release kinetics were dependent on liposome
preparation and dose (Jackson, 1981). A prolonged release of
liposome-encapsulated inulin and a higher lymphatic uptake compared
to free inulin was also observed by other groups after intramuscular
injection (Ohsawa et al., 1985). The retention of a hydrophilic peptide from the thigh muscle of rats is shown in Fig. 14, from which
it is clear that the marker is retained in the muscle for days.

Subcutaneous injection of insulin encapsulated in liposomes in
rats resulted in prolonged hypoglycemic effects compared to a solution of free insulin; this study also indicated that a substantial
fraction of hand-shaken multilamellar vesicles could enter the circulation in intact form after subcutaneous injection (Stevenson et al.,
1982). The neutral liposomes used in this study were cleared more
slowly from the injection site than the negatively charged liposomes.
In rabbits a zero-order clearance of the bilayer label encapsulated
in multilamellar vesicles with a diameter of 6–7 μm was observed
after intramuscular injection; the half-life was 8.5 days (Arrowsmith
et al., 1984). Finally, Eppstein and Felgner (1988) reported on the
prolonged retention at the site of injection of human recombinant
β-interferon encapsulated in liposomes after intramuscular injection
in mice. The drug gradually disappeared from the injection site;
after 6 days 15% of the label remained. Kim and Howell (1987b)
reported first-order release kinetics of cytarabine, a hydrophilic
cytostatic, from the injection site after subcutaneous administration
to mice; 50% of Ara-C disappeared in 5 days. They used large
multivesicular vesicles (about 20 μm). Charge and particle size
also played a role in lymph localization of labeled liposomes after
interstitial administration in the footpads of rats: neutral and positively charged sonicated liposomes localized to a higher degree in
regional lymph nodes than negatively charged sonicated liposomes
(Osborne et al., 1979).

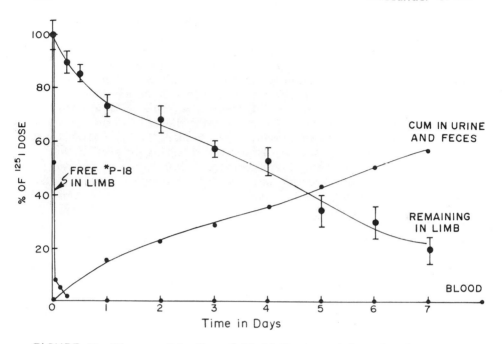

FIGURE 14   Pharmacokinetics of *P-18-liposome injected subcutaneously.   Liposomes containing ^{125}I-*P-18 (a peptide hormone with
$M_W < 5000$) at 2×10^5 cpm per injection were injected subcutaneously
into both hind limbs of rats (body weight = 100 g).   The rats were
placed in metabolic cages for urine and feces collection.   At each
time point, two rats were bled and the whole blood was counted for
radioactivity.   The rats were then killed and the hind limbs were
dissected and counted for remaining radioactivity.   The clearance of
free ^{125}I-*P-18 from the limbs is shown for comparison.   (Adapted
from Yau-Young et al., 1986.)

     In conclusion, intramuscular or subcutaneous injection of lipo-
somes can prolong the retention of the encapsulated drug.   The
clearance rate strongly depends, for instance, on liposome size and
charge.   Small vesicles tend to localize in regional lymph nodes.   In
particular, if prolonged release of high molecular weight molecules,
which can easily be denatured in nonaqueous media or dry form
(like certain proteins), is required, intramuscular or subcutaneous
delivery via liposomes is an interesting option.

G. Liposomes as Carriers of Anitgens

An increased immune response has been described after administration of liposome-associated antigens when compared with free antigen administration (reviewed by Alving, 1987). This effect may be utilized to produce liposome-based vaccines. The increased immunogenicity might be based on the strong interaction between liposomes and macrophages. Macrophages play an important role in the induction of an immune response, e.g., they act as antigen-presenting cells to T lymphocytes, which can trigger both a humoral and a cellular immune response. Macrophages have a strong tendency to phagocytose liposomes. This liposome-mediated delivery of antigens to macrophages may be one of the major factors causing liposome-associated antigens to be more immunogenic than the free antigen. No general rules can be derived from the studies performed until now to predict which conditions induce an optimum response (information is available in, for example, Scherphof et al., 1981; Machy and Leserman, 1987; a cluster of chapters in Gregoriadis, 1988c). Parameters which have been reported to influence the immune response are (1) the location of the antigen (Is it encapsulated in the aqueous phase? Does it adhere to the bilayer or is part of it taken up into the bilayer?); (2) ratio of antigen to lipid in the liposomes; (3) bilayer rigidity; (4) route of administration; (5) surface characteristics (e.g., the presence of charge-inducing agents or ligands for receptors on macrophages, like the mannose receptor).

In many experimental situations the immunogenicity of the antigen-liposome combination was still relatively low. Then adjuvants were applied to stimulate the immune response. Examples of adjuvants to antigen-liposome combinations are lipid A, muramyl dipeptide, or conjugates of these compounds. In most studies the humoral response was monitored after a primer as well as (a) booster injection(s). A limited number of reports have been published on the cellular response to antigen-liposome combinations. Liposome encapsulation showed beneficial effects on the cellular response to a soluble tumor antigen extract (Schroit and Key, 1983); after challenging with tumor cells, protection was observed. Torchilin and coworkers reported on an improved protection against influenza virus in mice when liposome-associated antigens were administered (Torchilin et al., 1988). Other promising results concerning protection of animals after immunization with liposome-based vaccines and a subsequent challenge with the relevant pathogen were reported by Desiderio and Campbell (1985) and Rhalem et al. (1988).

Thus, liposomes—with or without adjuvants—have a potential as antigen delivery systems. No clear insights exist on how to prepare liposome-based vaccines with optimum immunological properties by rationale instead of by trial and error. Therefore, much basic work is needed to unravel the mechanisms involved.

Apart from liposomes, other vehicles for delivery of antigens to the immune system, such as iscoms, emulsions, and micellar structures, are presently under investigation (Jiskoot et al., 1986b; Allison and Byars, 1986; Kersten et al., 1988a,b).

H. Topical Delivery of Drugs with Liposomes

In this section different forms of topical delivery (dermal and ocular) will be briefly discussed.

1. Dermal Delivery

In a number of studies the dermal delivery of liposome-encapsulated drug was compared with alternative delivery systems. An increase of the delivery of corticosteroids to the epidermis and dermis in combination with a decrease of the (unwanted) percutaneous delivery of the drug was demonstrated for liposomes (Mezei, 1988). Sustained release effects were also observed for liposome-encapsulated methotrexate (Patel, 1985). Methotrexate in liposomes was longer retained in the skin than administered in free form. The underlying mechanism for these effects are not fully understood yet. Improved partitioning of lipophilic compounds with skin material, a direct effect on the barrier function of the skin, reduction of skin metabolism, and prolonged localization of the drug-laden liposome at the site of application were mentioned as possible reasons for the observed effects. In vitro, less penetration of liposome-encapsulated hydrophilic compounds compared to aqueous solutions of these compounds through the hairless mouse skin was observed (Ganesan et al., 1984). Knepp et al. (1988) demonstrated a zero-order release of encapsulated material from liposomes upon administration to the skin.

Liposomes are now used extensively in cosmetics; unfortunately, very few publications with solid experimental data on the effect of liposomal material on the skin are available (Lautenschläger et al., 1988).

In conclusion, until now insufficient data have been available to define general rules to predict for which drugs liposome encapsulation will be therapeutically beneficial when applied to the skin. It is too early to draw definite conclusions about the advantage of the use of liposomes over alternative dermal delivery systems in the therapy for drugs designed to exert a pharmacological effect either locally or systemically.

2. Ocular Delivery

Liposomes have been studied to improve the ocular availability of drugs after application to the eye or after intravitreal injection. Besides, a liposomal eye drop formulation for the treatment of dry eye symptoms was developed and entered the clinical phase II stage (Guo et al., 1988).

The aim of the use of drug-laden liposomes in ocular formulations is to increase the availability of the encapsulated drug to the eye in terms of both uptake and residence time compared to conventional formulations. Liposomes and drug tend to be washed away rapidly from the cornea by the tear fluid and by blinking. An important issue is therefore the prolongation of the residence time of the liposomes by increasing the affinity of the liposomes for the corneal surface. Several approaches were developed to achieve increased affinity. The surface of the epithelial cell membrane is slightly negatively charged. Therefore, positively charged liposomes interact intensively with the surface. Inconclusive results were reported concerning the effect of particle size on this interaction (Schaeffer and Krohn, 1982; Singh and Mezei, 1984). An alternative approach is the use of liposomes exposing lectins on their surface. The lectins anchored in the liposomes bind to carbohydrate moieties on the cell surface. It was demonstrated that the liposome-cell interaction was intensified (Megaw et al., 1981).

After ocular administration, for lipophilic drugs encapsulated in liposomes drug levels in the eye were higher than for control preparations. Absorption of hydrophilic substances is significantly impaired by liposome encapsulation (Stratford et al., 1983; Singh and Mezei, 1983). Prolonged presence of liposome-encapsulated drug in the eye was observed for both triamcinolonacetonide and cytarabine compared to control preparations (Singh and Mezei, 1983; Assil and Weinreb, 1987).

The mechanism responsible for improved delivery of lipophilic drugs has not yet been clarified. Absorption of liposomes by cells is unlikely. Adsorption to cells followed by slow release of the drug from the liposome, either via diffusion through the thin aqueous tear film or via direct partitioning from the membrane of the vesicle to the membrane of the cell, was proposed as a possible pathway.

Prolonged presence of the drug at the site of injection is the aim of liposome encapsulation of drugs, which are injected in the vitreous body. Both amphotericin and gentamicin in liposome formulations were cleared from the injection site significantly more slowly than the free drug; residence times depended on liposome size and, in some cases, on bilayer composition (Tremblay et al., 1985; Barza et al., 1985, 1987; Fishman et al., 1986).

In conclusion, delivery of liposome-encapsulated drugs in eye drops can improve the extent of uptake and the residence time compared to the free drug. In particular, lipophilic substances seem to benefit from this approach. The exact mechanism behind the improved biopharmaceutical behavior still has to be unraveled. Intravitreal injection of drug-containing liposomes increases the residence times of both hydrophilic and lipophilic drugs.

VII. ADVERSE EFFECTS OF LIPOSOMES

Presently, several clinical trials with liposome-encapsulated agents
are under way and more are planned (Zonneveld and Crommelin, 1988;
Klausner, 1988). During the last 5 years, key issues related to the
pharmaceutical manufacturing of liposomes such as stability, sterilization, upscaling, and reproducibility have been successfully addressed.
Although it is generally believed that a proper selection of the bilayer components can minimize the occurrence of toxic effects due to
the use of natural body constituents, the issue of liposome-related
toxicity is not a trivial one and should be carefully studied.
     A number of aspects related to the potential toxicity of liposomes
has to be considered. They include effects due to (1) their particulate nature, (2) the individual components, and (3) a changed distribution of the encapsulated drug.
     As described before, in general, after intravenous administration
the majority of liposomes are rapidly removed from the systemic circulation by Kupffer cells in the liver and fixed macrophages in the
spleen. Single doses of biodegradable particles such as liposomes
are generally well tolerated by many animals. High doses can affect
the mononuclear phagocyte system (MPS) clearance function temporarily. The MPS recovery after single doses has already been reviewed (Allen, 1988; Storm et al., 1990a). Multiple injections of
liposomes can lead to sustained saturation of the MPS, which may
result in a prolonged impaired functioning of the MPS (reviewed by
Allen, 1988); this implies, for example, increased susceptibility to
infections, systemic endotoxemia, or increased metastatic spread of
tumors (Poste et al., 1982).
     Particle size, dose, dosage frequency, carrier composition, and
the presence of lipid peroxides are factors affecting the toxicity to
the MPS. The phospholipids usually employed in liposomes are biodegradable and similar to those found in commercial preparations for
parenteral nutrition (e.g., Intralipid and Liposyn), which can be
injected in large quantities. At modest doses, egg phosphatidylcholine (PC)-cholesterol (chol) and phosphatidic acid (PA)-chol liposomes did not induce adverse effects (Storm et al., 1990a). Depending on dose and frequency of dosing, some charged amphiphilic
lipids [e.g., phosphatidylserine (PS), dicetylphosphate (DCP), and
stearlyamine (SA)] may cause adverse effects in vivo (Allen et al.,
1981; Adams et al., 1977). Administration of liposomes composed
of PS can cause significant toxicity, probably as a result of the formation of the metabolic product of PS, lyso-PS (Bigon et al., 1979).
Rigid liposomes [i.e., those containing a high percentage of sphingomyelin, cholesterol, distearoyl-PC (DSPC), or dipalmitoyl-PC (DPPC)]
have a tendency to cause greater liver and spleen pathological reaction than those primarily composed of egg PC (Weereratne et al.,
1983; Allen et al., 1984; Allen and Smuckler, 1985).

Changing the distribution of a drug can lead to toxic effects not described before. It is possible that after liposomal delivery high concentrations of drugs (e.g., cytotoxic drugs) inside macrophages affect these cells detrimentally (Poste and Kirsch, 1983). This results in toxic effects in liver, spleen, and bone marrow which were not previously associated with the use of these drugs.

Peroxidation of lipids is another factor which must be considered in the safety evaluation of liposome administration. Smith and co-workers (1983) demonstrated that lipid peroxides can play an important role in liver toxicity. Allen et al. (1984) showed that liposomes protected by an antioxidant caused less MPS impairment than liposomes subjected to mild oxidizing conditions. From the study of Kunimoto et al. (1981) it can be concluded that the level of peroxidation in freshly prepared liposome preparations and those on storage strongly depends both on the phospholipid fatty acid composition and on the head group of the phospholipid. Addition of appropriate antioxidants to liposomes composed of lipids which are liable to peroxidation and designed for use in human studies is therefore necessary.

The potential antigenicity and immunogenicity of liposomes and liposomal components (e.g., incorporated polysaccharides and proteins) must be considered in plans for repeated, chronic administration to patients. It is generally accepted that the immunogenicity of liposomes as such is very low. Repeated injections of liposomes did not produce antiliposome antibodies in rabbits (Alving, 1984). However, in the same study (Alving, 1984) it was shown that naturally occurring antibodies against some of the most fundamental membrane phospholipids were widespread in normal human serum. These natural antibodies appeared to be similar to the monoclonal antiliposome antibodies experimentally induced by liposomes containing lipid A. Several other studies (Gregoriadis, 1980; Scherphof et al., 1981; Maneta-Peyret et al., 1988) demonstrated that only when an adjuvant was used in the liposome preparations could antiliposome antibodies be detected after repeated injections. More attention should be paid to the immune response after encapsulation or attachment of glycolipids, polysaccharides, and proteins (e.g., as homing devices in the case of drug targeting) against the liposomes themselves and especially against the encapsulated or attached agent (Alving, 1987; Van Rooyen, 1988).

Other potential adverse effects resulting from liposome administration which need to be studied are undesired complement activation (Cunningham et al., 1979), blood clotting, and pharmacological effects of the lipid components as well as physical obstruction of small capillaries by large particles.

Thus far no specific articles have been published concerning the tolerance of "pure" liposomes in humans. In all reported clinical studies (reviewed by Zonneveld and Crommelin, 1988) on parenteral administration of drug-containing liposomes, toxic effects were rare and, if they occurred, mild. Moreover, it was difficult to discriminate between toxic effects caused by the drug and carrier-induced toxicity. In a recent pilot study with an antifungal agent incorporated in sonicated liposomes administered to 15 cancer patients with fungal infections (Sculier et al., 1988), transient and slight somnolence—which might have been related to the liposome administration—was reported.

In view of the fact that large volumes have been administered to man without causing serious side effects (Zonneveld and Crommelin, 1988), it may be concluded that although liposomes can damage in vitro-cultured human cell lines (Nuzzo et al., 1985; Mayhew et al., 1987b), adverse effects observed in vivo are expected to be minimal. One should realize that liposomes reduce the toxicity of the encapsulated drug. Thus, the toxicity of liposomes may be of minor importance compared to the advantages of administration of certain chemotherapeutics encapsulated in liposomes.

VIII. LIPOSOME PRODUCTION IN THE INDUSTRIAL SETTING

The upscaling of liposome production methods from the lab bench scale to an industrial level raises a number of problems. The issues which have to be addressed involve

1. The quality of and access to raw materials, in particular the lipid sources used
2. Sterility (for injectables)
3. Pyrogen control (for injectables)
4. Reproducibility of the production process in terms of, e.g., size distribution, encapsulation efficiency, and release pattern
5. Absence of unwanted side products (e.g., residues of organic solvents in the bilayer)
6. Chemical and physical stability (degradation, aggregation or fusion, reduction of drug retention)
7. Cost

The selected procedure has to comply with good manufactoring practice (GMP) rules and must be fully validated.

Relatively few articles have been published on the industrial manufacturing of liposomes (Fildes, 1981; Rao, 1984; Ostro, 1988; Van Hoogevest and Fankhauser, 1989; Martin, 1989). A large number of patents describing procedures for large-scale production of liposomes have been filed, but for most liposome formulations now under clinical investigation details of the "integrated" protocols (all steps from raw materials to final product) are still proprietary knowledge within the different companies. It is very unlikely that one generally applicable procedure can be developed for all liposome-drug combinations. Therefore, for every new type of liposome a choice from the various procedures available for the different steps in the production process has to be made. Van Hoogevest and Frankhauser (1989) described the preparation procedure, reproducibility, and quality control protocols for muramyltripeptide-phosphatidylethanolamine (a bilayer-interacting compound) containing liposomes presently under investigation in clinical trials for their immunomodulating potential. For the commercial production of injectable liposomes containing hydrophilic drugs—which is considered a greater challenge—no such detailed information on production and quality control is available.

IX. REGULATORY AFFAIRS

Up to now the regulatory requirements concerning liposomal delivery systems have not been defined in detail. Fildes (1981) and, more recently, Ostro (1988) discussed this issue. Although regulations for approval differ between countries, the common approach to (parenteral) liposomal drug delivery systems is expected to be that these liposomal products are considered as new drug entities. The limited experience with liposomes in the clinical setting (Zonneveld and Crommelin, 1988) has not yet permitted the establishment of detailed guidelines. Testing requirements will depend on the type of disease to be treated: Is the disease life threatening? Is the medication used chronically? Are the liposomes administered parenterally? Are the liposomes administered locally and is the drug supposed to act locally?

Drug delivery via liposomes is not just the delivery of the encapsulated drug via "another" dosage form. The pharmacokinetic profile of the liposome-encapsulated drug is very different from the profile of the "free" drug. Even if the liposomes contain an "old" drug which is considered to be safe, preclinical safety studies of the liposome formulation—paying extra attention to the histopathology of organs rich in MPS cells and any new target organs—and clincial safety and efficacy studies are required. This means that a lot of preclinical and clinical work still has to be done. Therefore, liposomes designed for parenteral drug delivery are not expected to be launched on the market before the mid-1990s.

X. CONCLUSIONS AND PROSPECTS

In the early 1970s, liposomes were welcomed as drug delivery systems
with an enormous potential, in particular in the field of drug targeting. In this chapter the current state of affairs concerning liposomes
as drug and antigen delivery systems was discussed. The experience
gained in the last decade has given us a much more realistic view on
the usefulness of liposomes in the delivery of drugs and antigens.
A notable property of liposomes, which has not been appreciated
enough, is the presence of water inside liposomes. This makes them
an excellent delivery system for biotechnologically engineered proteins with tertiary and quanternary structures which are sensitive
to irreversible damage induced by dehydration, as often occurs with
alternative, particulate carrier systems.

For a number of liposome preparations—both injectables and locally administered products—the therapeutic advantages over existing
formulations have been proven in animal models; clinical trials with
liposome preparations are now under way. So far, clinical studies
showed no significant toxic effects which could be ascribed to the
lipid components of the liposomes used.

Presently, efforts are being made to solve pharmaceutical problems concerning the large-scale production of liposomes according to
GMP rules, the availability of high-quality lipid material, and the
sufficient long-term stability of the fomulation. Good progress has
been made in the last few years and large-scale production protocols
have been developed. We expect that these experiences will facilitate
the introduction of new liposome products by reducing time and finan-
cial investments.

ABBREVIATIONS

AIDS   acquired immunodeficiency syndrome
AMB   amphotericin B
Ara-C   cytosine arabinoside
AUC   area under the blood/plasma/serum concentration-time
            curve
BSA   bovine serum albumin
cDDP   cis-diamminedichloroplatinum(II), cisplatin
CF   carboxyfluorescein
chol   cholesterol
DMPC   dimyristoylphosphatidylcholine
DOPA   dioleoylphosphatidic acid
DPPC   dipalmitoylphosphatidylcholine
DPPG   dipalmitoylphosphatidylglycerol
DSC   differential scanning calorimetry
DSPC   distearoylphosphatidylcholine

DXR	doxorubicin
GMP	Good Manufacturing Practice
GUV	giant unilamellar vesicle(s)
HDL	high-density lipoproteins
HPLC	high-performance liquid chromatography
IgG	immunoglobulin G
LUV	large unilamellar vesicle(s)
MHC	major histocompatibility complex
MLV	multilamellar vesicle(s)
MPS	mononuclear phagocyte system
MTX	methotrexate
NMR	nuclear magnetic resonance
PA	phosphatidic acid
PC	phosphatidylcholine
PE	phosphatidylethanolamine
PG	phosphatidylglycerol
PL	phospholipid(s)
PS	phosphatidylserine
REV	reverse phase evaporation vesicle(s)
SUV	small unilamellar vesicle(s)
Tc	transition temperature
TLC	thin-layer chromatography

ACKNOWLEDGMENT

This work is supported by Liposome Technology, Inc., Menlo Park, California.

REFERENCES

Abe, R., and Taneichi, N. (1972). Lymphatic metastasis in experimental cecal cancer: Effectiveness of lymph nodes as barriers to the spread of tumor cells, Arch. Surg., 104, 95-98.

Abra, R. M., and Hunt, C. A. (1981). Liposome disposition in vivo. III. Dose and vesicle size effects, Biochim. Biophys. Acta, 666, 493-503.

Abra, R. M., Hunt, C. A., and Lau, D. T. (1984). Liposome disposition in vivo. VI. Delivery to the lung, J. Pharm. Sci., 73, 203-206.

Adams, D. H., Joyce, G., Richardson, V. J., Ryman, B. E., and Wisnieuwski, H. M. (1977). Liposome toxicity in the mouse central nervous system, J. Neurol. Sci., 31, 173-179.

Agrawal, A. K., Singhal, A., and Gupta, C. M. (1987). Functional targeting to erythrocytes in vivo using antibody bearing liposomes as drug vehicles, Biochem. Biophys. Res. Commun., 148, 357-361.

Ahrens, J., Graybill, J., Craven, P., and Taylor, R. (1984).
Treatment of experimental murine candidiasis with liposome-
associated amphotericin B, Sabouraudia, 22, 263-265.

Allen, L. (1936). The peritoneal stomata, Anat. Rec., 67, 89-103.

Allen, L. (1956). On the penetrability of the lymphatics of the diaphragm, Anat. Rec., 124, 639-657.

Allen, T. M., Romans, A. Y., Kercret, H., and Segrest, J. P.
(1980). Detergent removal during membrane reconstitution,
Biochim. Biophys. Acta, 601, 328-342.

Allen, T. M., McAllister, L., Mausolf, S., and Gyorffy, E. (1981).
Liposome-cell interactions: A study of the interactions of lipo-
some containing entrapped anti-cancer drugs with EMT6, S49 and
AE1 cell lines, Biochim. Biophys. Acta, 643, 346-362.

Allen, T. M., and Everest, J. M. (1983). Effect of liposome size
and drug release properties on pharmacokinetics of liposome-
encapsulated drugs in rats, J. Pharmacol. Exp. Ther., 226,
539-544.

Allen, T. M., Murray, L., MacKeigan, S., and Shah, M. (1984).
Chronic liposome administration in mice: Effects on reticuloendothelial function and tissue distribution, J. Pharmacol. Exp. Ther.,
229, 267-275.

Allen, T. M., and Smuckler, E. A. (1985). Liver pathology accompanying chronic liposome administration in mouse, Res. Commun.
Chem. Pathol. Pharmacol., 50, 281-290.

Allen, T. M., and Chonn, A. (1987). Large unilamellar liposomes
with low uptake into the reticuloendothelial system, FEBS Lett.,
223, 42-46.

Allen, T. M. (1988). Toxicity of drug carriers to the mononuclear
phagocyte system, Adv. Drug Deliv. Rev., 2, 55-67.

Allen, T. M. (1989). Stealth™ liposomes: Avoiding reticuloendothelial
uptake, in Liposomes in the Therapy of Infectious Diseases and
Cancer (G. Lopez-Berestein and I. J. Fidler, eds.), Alan R.
Liss, New York, pp. 405-415.

Allison, A. C., and Byars, N. E. (1986). An adjuvant formulation
that selectively elicits the formation of antibodies of protective
isotypes and of cell-mediated immunity, J. Immunol. Meth., 95,
157-168.

Alving, C. R., Steck, E. A., Chapman, Jr., W. L., Waits, V. B.,
Hendrick, L. D., Swartz, Jr., G. M., and Hanson, W. L.
(1978). Therapy of leishmaniasis: Superior efficacies of
liposome-encapsulated drugs, Proc. Natl. Acad. Sci. USA, 75,
2959-2963.

Alving, C. R. (1984). Natural antibodies against phospholipids and
liposomes in humans, Biochem. Soc. Trans., 12, 342-344.

Alving, C. R. (1987). Liposomes as carriers for vaccines, in Liposomes: From Biophysics to Therapeutics (M. J. Ostro, ed.),
Marcel Dekker, New York, pp. 195-218.

Alving, C. R. (1988). Macrophages as targets for delivery of liposome-encapsulated antimicrobial agents, Adv. Drug Del. Rev., 2, 107-128.

Antanavage, J., Cien, T. F., Ching, Y. D., Dunlap, L., Mueller, P., and Rudy, M. (1978). Formation and properties of cell-size single bilayer vesicles, Biophys. J., 21, A122.

Aragnol, D., and Leserman, L. D. (1986). Immune clearance of liposomes inhibited by an anti-Fc receptor antibody in vivo, Proc. Natl. Acad. Sci. USA, 83, 2699-2703.

Arrowsmith, M., Hadgraft, J., and Kellaway, I. W. (1984). The in vivo release of cortisone esters from liposomes and the intramuscular clearance of liposomes, Int. J. Pharm., 20, 347-362.

Assil, K. M., and Weinreb, R. N. (1987). Multivesicular liposomes. Sustained release of the antimetabolite cytarabine in the eye, Arch. Ophthalmol., 105, 400-403.

Aurora, T. S., Li, W., Cummins, H. Z., and Haines T. H. (1985). Preparation and characterization of monodisperse unilamellar phospholipid vesicles with selected diameters of from 300-600 nm, Biochim. Biophys. Acta, 820, 250-258.

Bangham, A. D., Standish, M. M., and Watkins, J. C. (1965). Diffusion of univalent ions across the lamellae of swollen phospholipids, J. Mol. Biol., 13, 238-252.

Barenholz, Y., Amselem, S., and Lichtenberg, D. (1979). A new method for preparation of phospholipid vesicles (liposomes). French press, FEBS Lett., 99, 210-214.

Barrat, G. M., Ryman, B. E., Begent, R. H. J., Keep, P. A., Searle, F., Boden, J. A., and Bagshawe, K. D. (1983). Improved radioimmunodetection of tumors using liposome-entrapped antibody, Biochim. Biophys. Acta, 762, 154-164.

Barza, M., Baum, J., Tremblay, C., Szoka, F., and D'Amico, D. J. (1985). Ocular toxicity of intravitreally injected liposomal amphotericin B in rhesus monkeys, Am. J. Ophthalmol., 100, 259-263.

Barza, M., Stuart, M., and Szoka, F. (1987). Effect of size and lipid composition on the pharmacokinetics of intravitreal liposomes, Invest. Ophthalmol. Visual Sci., 28, 893-900.

Batzri, S., and Korn, E. D. (1973). Single bilayer liposomes prepared without sonication, Biochim. Biophys. Acta, 298, 1015-1019.

Becker, S. (1988). Functions of the human mononuclear phagocyte system: A condensed review, Adv. Drug Deliv. Rev., 3, 1-100.

Begent, R. H. J., Green, A. J., Bagshawe, K. D., Jones, B. E., Keep, P. A., Searle, F., Jewkes, R. F., Barrat, G. M., and Ryman, B. E. (1982). Liposomally entrapped second antibody improves tumor imaging with radiolabelled (first) antitumor antibody, Lancet, 2, 739-742.

Bettendorf, U. (1979). Electronmicroscopic studies on the peritoneal resorption of intraperitoneally injected latex particles via the diaphragmatic lymphatics, Lymphology, 12, 66-70.

Bigon, E., Boarato, E., Bruni, A., Leon, A., and Toffano, G. (1979). Pharmacological effects of phosphatidylserine liposomes: The role of lysophosphatidylserine, Br. J. Pharmacol., 67, 611-616.

Black, C. D. V., Watson, G. J., and Ward, R. J. (1977). The use of Pentostam liposomes in the chemotherapy of experimental leishmaniasis, Trans. Roy. Soc. Trop. Med. Hyg., 71, 550-552.

Boden, N., and Sixtl, F. (1986). Forces between phospholipid bilayers, Faraday Disc. Chem. Soc., 81, 1-9.

Bodor, N., and Brewster, M. E. (1986). Problems of delivery of drugs to the brain, in Methods of Drug Delivery (G. M. Ihler, ed.), Pergamonn Press, Oxford, pp. 153-202.

Bradfield, J. W. B. (1984). The reticuloendothelial system and blood clearance, in Microspheres and Drug Therapy: Pharmaceutical, Immunological and Medical Aspects (S. S. Davis, L. Illum, J. G. McVie, and E. Tomlinson, eds.), Elsevier, Amsterdam, pp. 25-37.

Browning, J. L. (1981). NMR studies of the structural and motional properties of phospholipids in membranes, in Liposomes: From Physical Structure to Therapeutic Applications (C. G. Knight, ed.), Elsevier, Amsterdam, pp. 189-242.

Brunner, J., Skrabal, P., and Hauser, H. (1976). Single bilayer vesicles prepared without sonication. Physico-chemical properties, Biochim. Biophys. Acta, 455, 322-331.

Casper, E. S., Kelsen, D. P., Alcock, N. W., and Lewis, Jr., L. (1983). Ip cisplatin in patients with malignant ascites: Pharmacokinetic evaluation and comparison with the iv route, Cancer Treatm. Rep., 67, 235-238.

Connor, J., Sullivan, S., and Huang, L. (1985). Monoclonal antibody and liposomes, Pharmacol. Ther., 28, 341-365.

Courtice, F. C., and Simmonds, W. J. (1954). Physiological significance of lymph drainage of the serous cavities and lungs, Physiol. Rev., 34, 419-448.

Crommelin, D. J. A. (1984). Influence of lipid composition and ionic strength on the physical stability of liposomes, J. Pharm. Sci., 73, 1559-1563.

Crommelin, D. J. A., and Van Bommel, E. M. G. (1984). Stability of liposomes on storage: Freeze dried, frozen and as an aqueous dispersion, Pharmaceut. Res., 1, 159-163.

Crommelin, D. J. A., Fransen, G. J., and Salemink, P. J. M. (1986). Stability of liposomes on storage, in Targeting of Drugs with Synthetic Systems (G. Gregoriadis, J. Senior, and G. Poste, eds.), Plenum Press, New York, pp. 277-286.

Crommelin, D. J. A., and Storm, G. (1987). Pharmaceutical aspects of liposomes: Preparation, characterization, and stability, in Controlled Drug Delivery (B. W. Muller, ed.), Wissenschaftliche Verlagsgescellschaft mbH, Stuttgart, pp. 80-91.

Crommelin, D. J. A., and Storm, G. (1990a). Drug Targeting, in Comprehensive Medicinal Chemistry, Vol. 5, (P. G. Sammes and J. B. Taylor, eds.), Pergamon Press, Oxford, pp. 661-701.

Crommelin, D. J. A., Nässander, U. K., Peeter, P. A. M., Steerenberg, P. A., de Jong, W. H., Eling, W. M. C., and Storm, G. (1990b). Drug laden liposomes in antitumor therapy and in the treatment of parasitic diseases, J. Control. Rel., 11, 233-243.

Crowe, J. H., Crowe, L. M., and Mouradian, R. (1983). Stabilization of biological membranes at low water activities, Cryobiology, 20, 346-356.

Cullis, P. R., Hope, M. J., Bally, M. B., Madden, Th. D., Mayer, L. D., and Janoff, A. S. (1987). Liposomes as pharmaceuticals, in Liposomes: From Biophysics to Therapeutics (M. J. Ostro, ed.), Marcel Dekker, New York, pp. 39-72.

Cunningham, G. M., Kingzette, M., Richard, R. L., Alving, C. R., Lint, T. F., and Gewurz, H. (1979). Activation of human complement by liposomes: A model for membrane activation of the alternative pathway, J. Immunol., 122, 1237-1242.

Daoud, S. S., and Juliano, R. L. (1986). Reduced toxicity and enhanced antitumor effects in mice of the ionophoric drug valinomycin when incorporated in liposomes, Cancer Res., 46, 5518-5523.

Davis, S. S., and Illum, L. (1986). Colloidal delivery systems: opportunities and challenges, in Site-Specific Drug Delivery (E. Tomlinson and S. S. Davis, eds.), John Wiley and Sons, Chichester, pp. 93-110.

Deamer, D., and Bangham, A. D. (1976). Large volume liposomes by an ether vaporization method, Biochim. Biophys. Acta, 443, 629-634.

Demel, R. A., and De Kruyff, B. (1976). The function of sterols in membranes, Biochim. Biophys. Acta, 457, 109-132.

Derksen, J. T. P. (1987). Immunoglobulin-coupled liposomes and their interaction with rat liver macrophages. Ph.D. Thesis. University of Groningen, The Netherlands.

Desiderio, J. V., and Campbell, S. G. (1985). Immunization against experimental murine salmonellosis with liposome-associated O-antigen, Inf. Immun., 48, 658-663.

Ellens, H., Morselt, H., and Scherphof, G. (1981). In vivo fate of large unilamellar sphingomyelin-cholesterol liposomes after intraperitoneal and intravenous injection into rats, Biochim. Biophys. Acta, 674, 10-18.

Ellens, H., Morselt, H. W. M., Dontje, B. H. J., Kalicharan, D.,
   Hulstaert, C. E., and Scherphof, G. L. (1983). Effects of liposome dose and the presence of lymphosarcoma cells on blood
   clearance and tissue distribution of large unilamellar liposomes in
   mice, Cancer Res., 43, 2927-2934.
Emmen, F., and Storm, G. (1987). Liposomes in treatment of infectious diseases, Pharm. Weekbl. Sci. Ed., 9, 162-171.
Eppstein, D. A., and Felgner, P. L. (1988). Applications of liposome formulations for antimicrobial/antiviral therapy, in Liposomes
   as Drug Carriers: Recent Trends and Progress (G. Gregoriadis,
   ed.), John Wiley and Sons, Chichester, pp. 311-323.
Estep, T. N., Mouncastle, D. B., Biltonen, R. L., and Thompson,
   T. E. (1978). Studies on anomalous thermotropic behavior of
   aqueous dispersions of dipalmitoyl phosphatidylcholine-cholesterol
   mixtures, Biochemistry, 17, 1984-1989.
Estep, T. N., Mouncastle, D. B., Barenholz, Y., Biltonen, L., and
   Thompson, T. E. (1979). Thermal behavior of synthetic sphingomyelin-cholesterol dispersions, Biochemistry, 18, 2112-2117.
Farr, S. J., Kellaway, I. W., Parry-Jones, D., and Woolfrey, S. G.
   (1985). The clearance of ^{99m}Tc labelled liposomes from the lungs
   of volunteers, Proc. Int. Symp. Control. Rel. Bioact. Mater.,
   12, 219-220.
Fidler, I. J., Raz, A., Fogler, W. E., Kirsh, R., Bugelski, P.,
   and Poste, G. (1980). Design of liposomes to improve delivery
   of macrophage-augmenting agents to alveolar macrophages, Cancer Res., 40, 4460-4466.
Fielding, B. (1989). Drug delivery to the lungs and systemic circulation by inhalation of liposome based formulations, in Proc.
   32nd Annual Meeting Western Pharmacology Society, Brecken-
   ridge, Colorado, January 1989. In Press.
Fildes, F. J. T. (1981). Liposomes: The industrial viewpoint, in
   Liposomes: From Physical Structure to Therapeutic Applications,
   (C. G. Knight, ed.), Elsevier, Amsterdam, pp. 465-485.
Fischer, T. H., and Lasic, D. D. (1984). A detergent depletion
   technique for the preparation of small vesicles, Mol. Liq.
   Crystall., 102, 141-153.
Fishman, P. G., Peyman, G. A., and Lesar, T. (1986). Intra-
   vitreal liposome-encapsulated gentamicin in a rabbit model,
   Invest. Ophthalmol. Visual Sci., 27, 1103-1106.
Flessner, M. F., Parker, P. J., and Sieber, S. M. (1983). Perito-
   neal lymphatic uptake of fibrinogen and erythrocytes in the rat,
   Am. J. Physiol., 244, H89-H96.
Fransen, G. J., Salemink, P. J. M., and Crommelin, D. J. A.
   (1986). Critical parameters in freezing of liposomes, Int. J.
   Pharm., 33, 27-35.

Frøkjaer, S., Hjorth, E. L., and Wørts, O. (1984). Stability testing of liposomes during storage, in Liposome Technology, Vol. 1, (G. Gregoriadis, ed.), CRC Press, Boca Raton, pp. 235-245.

Fujiwara, T., Maeta, H., Chida, S., Morita, T., Watabe, Y., and Abe, T. (1980). Artificial surfactant therapy in hyaline-membrane disease, Lancet, 1, 55-59.

Gabizon, A., and Papahadjopoulos, D. (1988). Liposome formulations with prolonged circulation time in blood and enhanced uptake by tumors, Proc. Natl. Acad. Sci. USA, 85, 6949-6953.

Gabizon, A. (1989). Liposomes as a drug delivery system in cancer chemotherapy, in Drug Carrier Systems: Horizons in Biochemistry and Biophysics, Vol. 9 (F. H. Roerdink and A. M. Kroon, eds.), John Wiley and Sons, Chichester, pp. 185-211.

Gabizon A., Sulkes, A., Peretz, T., Druckmann, S., Goren, D., Amselem, S., and Barenholz, Y. (1989). Liposome-associated doxorubicin: Preclinical pharmacology and exploratory clinical phase, in Liposomes in the Therapy of Infectious Diseases and Cancer, (G. Lopez-Berestein and I. J. Fidler, eds.), Alan R. Liss, New York, pp. 391-402.

Gaines, N., and Hauser, H. (1983). Characterization of small unilamellar vesicles produced in unsonicated phosphatidic acid and phosphatidylcholine-phosphatidic acid dispersions by pH adjustment, Biochim. Biophys. Acta, 731, 31-39.

Ganesan, M. G., Weiner, N. D., Flynn, G. L., and Ho, N. F. H. (1984). Influence of liposomal drug entrapment on percutaneous absorption, Int. J. Pharm., 20, 139-154.

Graybill, J. R., Craven, P. C., Taylor, R. L., Williams, D. M., and Magee, W. E. (1982). Treatment of murine cryptococcosis with liposome-associated amphotericin B, J. Infect. Dis., 145, 748-752.

Gregoriadis, G. (1980). Recent progress in liposome research, in Liposomes in Biological Systems (G. Gregoriadis and A. C. Allison, eds.), John Wiley and Sons, New York, pp. 377-398.

Gregoriadis, G. (ed.) (1984). Liposome Technology, Vol. 1-3, CRC Press, Boca Raton.

Gregoriadis, G. (1988a). Fate of injected liposomes: Observations on entrapped solute retention, vesicle clearance and tissue distribution in vivo, in Liposomes as Drug Carriers: Recent Trends and Progress (G. Gregoriadis, ed.), John Wiley and Sons, Chichester, pp. 3-18.

Gregoriadis, G. (ed.) (1988b). Liposomes as Drug Carriers: Recent Trends and Progress, John Wiley and Sons, Chichester.

Gregoriadis, G. (ed.) (1988c). Liposomes in immunomodulation/vaccines, in Liposomes as Drug Carriers: Recent Trends and Progress (G. Gregoriadis, ed.), John Wiley and Sons, Chichester, pp. 113-307.

Grit, M., De Smidt, J. H., Struijke, A., and Crommelin, D. J. A. (1989). Hydrolysis of phosphatidylcholine in aqueous liposome dispersions, Int. J. Pharm., 50, 1-6.

Groves, M. J. (1984). The application of particle characterization methods to submicron dispersions and emulsions, in Modern Methods of Particle Size Analysis (H. G. Barth, ed.), John Wiley and Sons, New York, pp. 43-91.

Grüner, S. M., Lenk, R. P., Janoff, A. S., and Ostro, M. J. (1985). Novel multilayered lipid vesicles: Comparison of physical characteristics of multilamellar liposomes and stable plurilamellar vesicles, Biochemistry, 24, 2833-2842.

Guiot, P., Baudhuin, P., and Gotfredsen, C. (1980). Morphological characterization of liposome suspensions by stereological analysis of freeze fracture replicas from spray frozen samples, J. Microsc., 120, 159-174.

Guo, L. S. S., Sarris, A. M., and Levy, M. D. (1988). A safe bioadhesive liposomal formulation for ophthalmic applications, Invest. Ophthalmol. Visual Sci. (Suppl.), 29, 439.

Hashimoto, Y., Sugawara, M., Masuko, T., and Hojo, H. (1983). Antitumor effect of actinomycin D entrapped in liposomes bearing subunits of tumor-specific monoclonal immunoglobulin M antibody, Cancer Res., 43, 5328-5334.

Hauser, H. (1971). The effect of ultrasonic irradiation on the chemical structure of egg lecithin, Biochem. Biophys. Res. Commun., 45, 1049-1055.

Hauser, H. (1987). Spontaneous vesiculation of uncharged phospholipid dispersions consisting of lecithin and lysolecithin, Chem. Phys. Lipids, 43, 283-299.

Hauser, H., and Gaines, N. (1982). Spontaneous vesiculation of phospholipids: a simple and quick method of forming unilamellar vesicles, Proc. Natl. Acad. Sci. USA, 79, 1683-1687.

Hauser, H., and Strauss, G. (1987). Stabilization of small unilamellar phospholipid vesicles during spray-drying, Biochim. Biophys. Acta, 897, 331-334.

Heath, T. D., Lopez, N. G., and Papahadjopoulos, D. (1985a). The effects of liposome size and surface charge on liposome-mediated delivery of methotrexate-gamma-aspartate to cells in vitro, Biochim. Biophys. Acta, 820, 74-84.

Heath, T. D., Lopez, N. G., Stern, W. H., and Papahadjopoulos, D. (1985b). 5-Fluoroorotate: A new liposome-dependent cytotoxic agent, FEBS Lett., 187, 73-75.

Hirano, K., and Hunt, C. A. (1985). Lymphatic transport of liposome-encapsulated agents: Effects of liposome size following intraperitoneal administration, J. Pharm. Sci., 74, 915-921.

Hirano, K., Hunt, C. A., Strubbe, A., and MacGregor, R. D. (1985). Lymphatic transport of liposome-encapsulated drugs following intraperitoneal administration: Effect of lipid composition, Pharm. Res., 6, 271-278.

Howell, S. B., and Zimm, S. (1988). Intraperitoneal chemotherapy: Application to upper gastrointestinal neoplasms, Acta Chir. Scand. Suppl., 541, 16-21.

Hsu, J. J., and Juliano, R. L. (1982). Interactions of liposomes with the reticuloendothelial system. II. Nonspecific and receptor mediated uptake of liposomes by mouse peritoneal macrophages, Biochim. Biophys. Acta, 720, 411-419.

Huang, C. H. (1969). Studies on phosphatidylcholine vesicles. Formation and physical characteristics, Biochemistry, 8, 344-352.

Hunt, C. A., Rustum, Y. M., Mayhew, E., and Papahadjopoulos, D. (1979). Retention of cytosine arabinoside in mouse lung following intravenous administration in liposomes of different size, Drug Metabol. Dispos., 7, 124-128.

Hwang, K. J. (1987). Liposome pharmacokinetics, in Liposomes: From Biophysics to Therapeutics (M. J. Ostro, ed.), Marcel Dekker, New York, pp. 109-156.

Hwang, K. J., and Beaumier, P. L. (1988). Disposition of liposomes in vivo, in Liposomes as Drug Carriers: Recent Trends and Progress (G. Gregoriadis, ed.), John Wiley and Sons, London, pp. 19-36.

Ivey, H., Roth, S., and Kattwinkel (1977). Nebulization of sonicated phospholipids (PL) for the treatment of respiratory distress syndrome (RDS) of infancy, Pediatr. Res., 11, 573.

Jackson, A. J. (1981). Intramuscular absorption and regional lymphatic uptake of liposome-entrapped inulin, Drug Metab. Dispos., 9, 535-540.

Jiskoot, W., Teerlink, T., Beuvery, E. C., and Crommelin, D. J. A. (1986a). Preparation of liposomes via detergent removal from mixed micelles by dilution. The effect of bilayer composition and process parameters on liposome characteristics, Pharm. Weekbl. Sci. Ed., 8, 259-265.

Jiskoot, W., Teerlink, T., Van Hoof, M. M. M., Bartels, K., Kanhai, V., Crommelin, D. J. A., and Beuvery, E. C. (1986b). Immunogenic activity of gonococcal protein I in mice with three different lipoidal adjuvants delivered in liposomes and in complexes, Inf. Immun., 54, 333-338.

Johnson, S. M. (1973). The effect of charge and cholesterol on the size and thickness of sonicated phospholipid vesicles, Biochim. Biophys. Acta, 307, 27-41.

Jousma, H., Talsma, H., Spies, F., Joosten, J. G. H., Junginger, H. E., and Crommelin, D. J. A., Characterization of liposomes (1987). The influence of extrusion of multilamellar vesicles through polycarbonate membranes on particle size, particle size distribution and number of bilayers, Int. J. Pharm., 35, 263-274.

Juliano, R. L., and Stamp, D. (1975). Effect of particle size and charge on the clearance rates of liposomes and liposome encapsulated drugs, Biochim. Biophys. Res. Commun., 63, 651-658.

Juliano, R. L., and Mc Cullough, H. N. (1980). Controlled delivery of an antitumor drug: Localized action of liposome encapsulated cytosine arabinoside administered via the respiratory system, J. Pharmacol. Exp. Ther., 214, 381-387.

Juliano, R. L. (1988). Factors effecting the clearance kinetics and tissue distribution of liposomes, microspheres and emulsions, Adv. Drug Deliv. Rev., 2, 31-54.

Kagawa, Y., and Racker, E. (1971). Partial resolution of the enzymes catalyzing oxidative phosphorylation, J. Biol. Chem., 246, 5477-5487.

Kao, Y. J., and Juliano, R. L. (1981). Interaction of liposomes with the reticuloendothelial system: Effects of blockade on the clearance of large unilamellar vesicles, Biochim. Biophys. Acta, 677, 453-461.

Keep, P. A., Searle, F., Begent, R. H. J., Barrat, G. M., Boden, J., Bagshawe, K. D., and Ryman, B. E. (1983). Clearance of injected radioactively labelled antibodies to tumor products by liposome-bound second antibodies, Oncodev. Biol. Med., 4, 273-280.

Kellaway, I. W., Taylor, K. G. M., Taylor, G., and Stevens, J. (1988). The pharmacokinetics of liposomal sodium cromoglycate nebulized to volunteers, Proc. Int. Symp. Control. Rel. Bioact. Mater., 15, 197-198.

Kemps, J. M. A., and Crommelin, D. J. A. (1988a). Chemische stabiliteit van fosfolipiden in farmaceutisch preparaten. I. Hydrolyse van fosfolipiden in waterig milieu, Pharm. Weekbl., 123, 355-363.

Kemps, J. M. A., and Crommelin, D. J. A. (1988b). Chemische stabiliteit van fosfolipiden in farmaceutisch preparaten. II. Peroxidatie van fosfolipiden in waterig milieu, Pharm. Weekbl., 123, 457-469.

Kersten, G. F. A., Van de Put, A., Teerlink, T., Beuvery, E. C., and Crommelin, D. J. A. (1988a). Immunogenicity of liposomes and iscoms containing the major outer membrane protein of Neisseria gonorrhoeae: Influence of protein content and liposomal bilayer composition, Inf. Immun., 56, 1661-1664.

Kersten, G. F. A., Teerlink, T., Derks, H. J. G. M., Verkleij, A. J., Van Weezel, T. L., Crommelin, D. J. A., and Beuvery, E. C. (1988b). Incorporation of the major outer membrane protein of Neisseria gonorrhoeae in saponin-lipid complexes (Iscoms): Chemical analysis, some structural features, and comparison of their immunogenicity with three other antigen delivery systems, Inf. Immun., 56, 432-438.

Kim, S., and Howell, S. B. (1987a). Multivesicular liposomes con-
    taining cytarabine entrapped in the presence of hydrochloric
    acid for intracavitary chemotherapy, Cancer Treatm. Rep., 71,
    705-711.
Kim, S., and Howell, S. B. (1987b). Multivesicular liposomes con-
    taining cytarabine for slow release s.c. administration, Cancer
    Treatm. Rep., 71, 447-450.
Kim, S., Turker, M. S., Chi, E. M., Sela, S., and Martin, G. M.
    (1983). Preparation of multivesicular liposomes, Biochim. Bophys.
    Acta, 728, 339-348.
Kim, S., Kim, D. J., and Howell, S. B. (1987). Modulation of the
    peritoneal clearance of liposomal cytosine arabinoside by blank
    liposomes, Cancer Chemother. Pharmacol., 19, 308-310.
Kimelberg, H. K., Tracy, T. F., Biddlecome, Jr., S. M., and
    Bourke, R. S. (1976). The effect of entrapment in liposomes
    on the in vivo distribution of [^{3}H]methotrexate in a primate,
    Cancer Res., 36, 2949-2957.
King, M. E., Pfeifle, C. E., and Howell, S. B. (1984). Intraperito-
    neal cytosine arabinoside therapy in ovarian carcinoma, J. Clin.
    Oncol., 2, 662-669.
Kirby, C. J., and Gregoriadis, G. (1984). A simple procedure for
    preparing liposomes capable of high encapsulation efficiency
    under mild conditions, in Liposome Technology, Vol. 1 (G.
    Gregoriadis, ed.), CRC Press, Boca Raton, pp. 19-27.
Kirkland, J. J., Yau, W. W., and Szoka, F. C. (1982). Sedimenta-
    tion field flow fractionation of liposomes, Science, 215, 296-298.
Kitagawa, T., Inoue, K., and Nojima, S. (1976). Properties of lip-
    somal membranes containing lysolecithin, J. Biochem. (Tokyo),
    79, 1123-1133.
Klausner, A. (1988). Will 1988 be "the year of the liposome?"
    Biotechnology, 6, 20.
Klein, R. A. (1970). The detection of oxidation in liposome prepa-
    rations, Biochim. Biophys. Acta, 210, 486-489.
Knapp, R. C., and Friedman, E. A. (1974). Aortic lymph node
    metastases in early ovarian cancer, Am. J. Obstet. Gynecol.,
    119, 1013-1017.
Knepp, V. M., Hinz, R. S., Szoka, F. C., and Guy, R. H. (1988).
    Controlled drug delivery from a novel liposomal system. I.
    Investigations of transdermal potential, J. Control. Rel., 5,
    211-221.
Knight, C. G. (ed.) (1981). Liposomes: From Physical Structure
    to Therapeutic Applications, Elsevier, Amsterdam.
Kobayashi, T., Kataoka, T., Tsukagoshi, S., and Sakurai, Y.
    (1977). Enhancement of anti-tumor activity of 1-β-D-arabino-
    furanosylcytosine by encapsulation, Int. J. Cancer, 20, 581-
    587.

Konings, A. W. T. (1984). In Liposome Technology, Vol. 1 (G. Gregoriadis, ed.), CRC Press, Boca Raton, pp. 139-161.

Konno, H., Suzuki, H., Tadakuma, T., Kumai, K., Yasuda, T., Kubota, T., Ohta, S., Nagaike, K., Hosokawa, S., Ishibiki, K., Abe, O., and Saito, K. (1987). Antitumor effect of adriamycin entrapped in liposomes conjugated with anti-human alfafetoprotein monoclonal antibody, Cancer Res., 47, 4471-4477.

Kosloski, M. J., Rosen, F., Milholland, R. J., and Papahadjopoulos, D. (1978). Effect of lipid vesicle (liposome) encapsulation of methotrexate on its chemotherapeutic efficacy in solid rodent tumors, Cancer Res., 38, 2848-2853.

Kremer, J. M., Esker, M. W., Pathmamanoharan, C., and Wiersema, P. H. (1977). Vesicles of variable diameter prepared by a modified injection method, Biochemistry, 16, 3932-3935.

Kumpati, J. (1987). Liposome-loaded phenylalanine or tryptophan as sickling inhibitor: A possible therapy for sickle cell disease, Biochem. Med. Metabol. Biol., 38, 170-181.

Kunimoto, M., Inoue, E., and Nojima, S. (1981). Effect of ferrous ion and ascorbated induced lipid peroxidation on liposomal membranes, Biochim. Biophys. Acta, 646, 169-178.

Langhammer, H., Büll, U., Pfeiffer, K. J., Hör, G., and Pabst, H. W. (1973). Experimental studies on lymphatic drainage of the peritoneal cavity using [198]Au-colloid, Lymphology, 6, 149-157.

Lasic, D. D. (1988). The spontaneous formation of unilamellar vesicles, J. Colloid Interf. Sci., 124, 428-435.

Lasic, D. D., Belic, A., and Valentincic, T. A. (1988). A new method for the instant preparation of large unilamellar vesicles, J. Am. Chem. Soc., 110, 970-971.

Lautenschläger, H., Röding, J., and Ghyczy, M. (1988). Über die Verwendung von Liposomen aus Soja-Phospholipiden in der Kosmetik, Seifen. Öle. Fette. Wachse, 14, 531-534.

Lautersztain, J., Perez-Soler, R., Khokhar, A. R., Newman, R. A., and Lopez-Berestein, G. (1986). Pharmacokinetics and tissue distribution of liposome-encapsulated cis-bis-N-decyliminodiacetato-1,2-diaminocyclohexane platinum(II), Cancer Chemother. Pharmacol., 18, 93-97.

Lichtenberg, D., and Barenholz, Y. (1988). Liposomes: Preparation, characterization and preservation, in Methods of Biological Analysis, Vol. 33 (D Glick, ed.), John Wiley and Sons, New York, pp. 337-461.

Lopez-Berestein, G., Metha, R., Hopfer, R. L., Mills, K., Kasi, L., Metha, K., Fainstein, V., Luna, M., Hersh, E. M., and Juliano, R. L. (1983). Treatment and prophylaxis of disseminated infection due to Candida albicans in mice with liposome-encapsulated amphotericin B, J. Infect. Dis., 147, 939-945.

Lopez-Berestein, G., Rosenblum, M. G., and Metha, R. (1984a). Altered tissue distribution of amphotericin B by liposomal encapsulation: Comparison of normal mice and mice infected with Candida albicans, Cancer Drug Deliv., 1, 199-205.

Lopez-Berestein, G., Kasi, L., Rosenblum, M. G., Haynie, T., Jahns, M., Glenn, H., Mavligit, G. M., and Hersh, E. M. (1984b). Distribution of ^{99m}technetium labeled liposomes in patients with cancer, Cancer Res., 44, 375-378.

Lopez-Berestein, G., Hopfer, R. L., Metha, R., Metha, K., Hersh, E. M., and Juliano, R. L. (1984c). Liposome-encapsulated amphotericin B for treatment of disseminated candidiasis in neutropenic mice, J. Infect. Dis., 150, 278-283.

Lopez-Berestein, G., Fainstein, V., Hopfer, R., Mehta, K., Sullivan, M. P., Keating, M., Rosenblum, M. G., Mehta, R., Luna, M., Hersh, E. M., Reuben, J., Juliano, R. L., and Bodey, G. P. (1985). Liposomal amphotericin B for the treatment of systemic fungal infections in patients with cancer: A preliminary study, J. Infect. Dis., 151, 704-710.

Lopez-Berestein, G. (1988). Liposomal amphotericin B in antimicrobial therapy, in Liposomes as Drug Carriers: Recent Trends and Progress (G. Gregoriadis, ed.), John Wiley and Sons, Chichester, pp. 345-352.

Lopez-Berestein, G. (1989). Treatment of systemic fungal infections with liposomal-amphotericin B, in Liposomes in the Therapy of Infectious Diseases and Cancer (G. Lopez-Berestein and I. J. Fidler, eds.), Alan R. Liss, New York, pp. 317-327.

Mabrey-Gaud, S. (1981). Differential scanning calorimetry of liposomes, in Liposomes: From Physical Structure to Therapeutic Applications (C. G. Knight, ed.), Elsevier, Amsterdam, pp. 105-138.

Machy, P., and Leserman, L. (eds.) (1987). Liposomes in Cell Biology and Pharmacology, John Libbey Eurotext Ltd/INSERM, Great Britain.

Maneta-Peyret, L., Bessoule, J. J., Geffard, M., and Cassagne, C. (1988). Demonstration of high specificity antibodies against phosphatidylserine, J. Immunol. Meth., 108, 123-127.

Markman, M., Cleary, S., Lucas, W. E., and Howell, S. B. (1985). Intraperitoneal chemotherapy with high-dose cisplatin and cytosine arabinoside for refractory ovarian carcinoma and other malignancies principally involving the peritoneal cavity, J. Clin. Oncol., 3, 925-931.

Markman, M. (1986). Intraperitoneal antineoplastic agents for tumors principally confined to the peritoneal cavity, Cancer Treatm. Rev., 13, 219-242.

Markman, M., Cleary, S., Lucas, W., Weiss, R., and Howell, S. B. (1986). Ip chemotherapy employing a regimen of cisplatin, cytarabine, and bleomycin, Cancer Treatm. Rep., 70, 755-760.

Martin, F. J. (1989). Pharmaceutical manufacturing of liposomes, in
Specialized Drug Delivery Systems: Manufacturing and Production
Technology (P. Tyle, ed.), Marcel Dekker, New York. In Press.

Mason, J. T., and Huang, C. (1978). Hydrodynamic analysis of egg
phosphatidylcholine vesicles, in Liposomes and Their Uses in
Biology and Medicine (D. Papahadjopoulos, ed.), Ann. N.Y.
Acad. Sci. Vol. 308, 29-49.

Mayer, L. D., Hope, M. J., Cullis, P. R., and Janoff, A. S. (1985a).
Solute distributions and trapping efficiencies observed in freeze-
thawed multilamellar vesicles, Biochim. Biophys. Acta, 817, 193-
196.

Mayer, L. D., Bally, M. B., Hope, M. J., and Cullis, P. R. (1985b).
Uptake of antineoplastic agents into large unilamellar vesicles in
response to a membrane potential, Biochim. Biophys. Acta, 816,
294-302.

Mayer, L. D., Hope, M. J., and Cullis, P. R. (1986a). Vesicles
of variable sizes produced by a rapid extrusion procedure, Biochim. Biophys. Acta, 858, 161-168.

Mayer, L. D., Hope, M. J., and Cullis, P. R. (1986b). Uptake of
adriamycin into large unilamellar vesicles in response to a pH
gradient, Biochim. Biophys. Acta, 857, 123-126.

Mayhew, E., Papahadjopoulos, D., Rustum, Y. M., and Dave, C.
(1976). Inhibition of tumor cell growth in vitro and in vivo by
ARA-C entrapped within phospholipid vesicles, Cancer Res., 36,
4406-4411.

Mayhew, E., Rustum, Y. M., and Szoka, F. (1982). Therapeutic
efficacy of cytosine arabinoside trapped in liposomes, in Targeting of Drugs (G. Gregoriadis, J. Senior, and A. Trouet, eds.),
Plenum Press, New York, pp. 249-260.

Mayhew, E., and Rustum, Y. (1983). Effect of liposome entrapped
chemotherapeutic agents on mouse primary and metastatic tumors,
Biol. Cell, 47, 81-85.

Mayhew, E., Conroy, S., King, J., Lazo, R., Nikolopoulos, G.,
Siciliano, A., and Vail, W. J. (1987a). High-pressure continuous-
flow system for drug entrapment in liposomes. In Methods in
Enzymology, Vol. 149, Drug and Enzyme Targeting, Part B
(R. Green and K. J. Widder, eds.), Academic Press, San Diego,
pp. 64-77.

Mayhew, E., Ito, M., and Lazo, R. (1987b). Toxicity of non-drug
containing liposomes for cultured human cells, Exp. Cell Res.,
171, 195-202.

McCalden, T. A., Abra, R. M., and Mihalko, P. J. (1989). Bronchodilator efficacy of liposome formulations of metaproterenol
sulfate in the anesthesized guinea pig, J. Liposome Res., 1,
211-222.

Megaw, J. M., Takei, Y., and Lerman, S. (1981). Lectin-mediated binding of liposomes to the ocular lens, Exp. Eye Res., 32, 395-405.

Mehta, R., Lopez-Berestein, G., Hopfer, R., Mills, K., and Juliano, R. L. (1984). Liposomal amphotericin B is toxic to fungal cells but not to mammalian cells. Biochim. Biophys. Acta, 770, 230-234.

Mezei, M. (1988). Liposomes in the topical application of drugs: a review. In Liposomes as Drug Carriers: Recent Trends and Progress (G. Gregoriadis, ed.), John Wiley and Sons, Chichester, pp. 663-677.

Mihalko, P. J., Schreier, H., and Abra, R. M. (1988). Liposomes: A pulmonary perspective, in Liposomes as Drug Carriers: Recent Trends and Progress (G. Gregoriadis, ed.), John Wiley and Sons, New York, pp. 679-694.

Miller, R. P., and Bates, J. H. (1969). Amphotericin B toxicity: A follow up report of 53 patients, Ann. Intern. Med., 71, 1089-1096.

Morgan, J. R., Williams, L. A., and Howard, C. B. (1985). Technetium-labeled liposome imaging for deap-seated infection, Br. J. Radiol., 58, 35-39.

Moro, L. C. V., Neri, G., and Rigamonti, A. (1980). Verfahren zum Reinigen von Liposomensuspensionen, German Patent A61K9/00.

Musumeci, R., DePalo, G. M., Mangioni, C., Bolis, G., and Ratti, E. (1978). The lymphatic spread of ovarian germinal and stromal tumors, Lymphology, 11, 22-26.

Müller, R. H., Davis, S. S., Illum, L., and Mak, E. (1986). Particle charge and surface hydrophobicity of colloidal drug carriers, in Targeting of Drugs with Synthetic Systems (G. Gregoriadis, J. Senior, and G. Poste, eds.), Plenum Press, New York, pp. 239-263.

New, R. R. C., Chance, M. L., Thomas, S. C., and Peters, W. (1978). Antileishmanial activity of antimonials entrapped in liposomes, Nature, 272, 55-56.

Newman, M. S., Guo, L., McCalden, T. A., Levy, M., Porter, J., Wong, A., and Fielding, R. M. (1989). A colloidal dispersion of amphotericin B with reduced lethality and non lethal toxicity, Proc. AAPS Western Regional Meet., February 26—March 1, 1989.

Nichols, J. W., and Deamer, D. W. (1976). Cathecholamine uptake and concentration by liposomes maintaining pH gradients, Biochim. Biophys. Acta, 455, 269-271.

Nir, S., Bentz, J., Wilschut, J., and Düzgünes, N. (1983). Aggregation and fusion of phospholipid vesicles, Prog. Surface Sci., 13, 1-124.

Nordlund, J. R., Schmidt, C. F., Dicken, S. N., and Thompson, T. E. (1981). Transbilayer distribution of phosphatidylethanolamine in large and small unilamellar vesicles, Biochemistry, 20, 3237-3241.

Nuzzo, F., Sala, F., Biondi, D., Casati, A., Cestaro, B., and de Carli, L. (1985). Liposomes induce chromosome aberrations in human cultured cells, Exp. Cell Res., 157, 397-408.

Ohsawa, T., Matsukawa, Y., Takakura, Y., Hashida, M., and Sezaki, H. (1985). Fate of lipid and encapsulated drug after intramuscular administration of liposomes prepared by the freeze-thawing method in rats, Chem. Pharm. Bull., 33, 5013-5022.

Oku, N., and MacDonald, R. C. (1983a). Differential effects of alkali metal chlorides on formation of giant liposomes by freezing and thawing and dialysis, Biochemistry, 22, 855-863.

Oku, N., and MacDonald, R. C. (1983b). Formation of giant liposomes from lipids in chaotropic ion solutions, Biochim. Biophys. Acta, 734, 54-61.

Olson, F., Hunt, C. A., Szoka, F., Vail, W. J., and Papahadjopoulos, D. (1979). Preparation of liposomes of defined size distribution by extrusion through polycarbonate membranes, Biochim. Biophys. Acta, 557, 9-23.

Osborne, M. P., Richardson, V. J., Jeyasingh, K., and Ryman, B. E. (1979). Radionuclide-labelled liposomes—A new lymph node imaging agent, Int. J. Nucl. Med. Biol., 6, 75-83.

Ostro, M. J. (1987). Liposomes, Sci. Am., 256, 91-99.

Ostro, M. J. (1988). Industrial application of liposomes: What does that mean? in Liposomes as Drug Carriers: Recent trends and Progress (G. Gregoriadis, ed.), John Wiley and Sons, Chichester, pp. 855-862.

Özer, Y., Talsma, H., Crommelin, D. J. A., and Hincal, A. A. (1988). Influence of freezing and freeze drying on the stability of liposomes dispersed in aqueous media. Acta Pharmaceut. Technol., 34, 129-139.

Pagano, R. E., and Weinstein, J. N. (1978). Interaction of liposomes with mammalian cells, Ann. Rev. Biophys. Bioeng., 7, 435-468.

Papahadjopoulos, D., and Watkins, J. C. (1967). Phospholipid model membranes. II. Permeability properties of hydrated liquid crystals, Biochim. Biophys. Acta, 135, 639-652.

Papahadjopoulos, D., Vail, W. J., Jacobson, K., and Poste, G. (1975). Cholate lipid cylinders: Formation by fusion of unilamellar lipid vesicles, Biochim. Biophys. Acta, 394, 483-491.

Papahadjopoulos, D., and Vail, W. J. (1978). Incorporation of macromolecules within large unilamellar vesicles (LUV), Ann. N.Y. Acad. Sci., 308, 259-267.

Papahadjopoulos, D., Heath, T., Bragman, K., and Matthay, K.
    (1985). New methodology for liposome targeting to specific cells,
    Ann. N.Y. Acad. Sci., 446, 341-348.
Parker, R. J., Sieber, S. M., and Weinstein, J. N. (1981a). Effect
    of liposome encapsulation of a fluorescent dye on its uptake by
    the lymphatics of the rat, Pharmacology, 23, 128-136.
Parker, R. J., Hartman, K. D., and Sieber, S. M. (1981b). Lym-
    phatic absorbtion and tissue disposition of liposome-entrapped
    [14C]adriamycin following intraperitoneal administration, Cancer
    Res., 41, 1311-1317.
Parker, R. J., Priester, E. R., and Sieber, S. M. (1982). Com-
    parison of lymphatic uptake metabolism, excretion, and biodistri-
    bution of free and liposome-entrapped [14C]cytosine beta-D-
    arabinofuranoside following intraperitoneal administration in rats,
    Drug. Metab. Dispos., 10, 40-46.
Parsegian, V. A. (1973). Long range physical forces in the biologi-
    cal milieu, Ann. Rev. Biophys. Bioeng., 2, 221-253.
Parsegian, V. A., Fuller, N., and Rand, R. P. (1979). Measured
    work of deformation and repulsion of lecithin bilayers, Proc.
    Natl. Acad. Sci. USA, 76, 2750-2754.
Patel, H. M., and Ryman, B. E. (1981). Systemic and oral admin-
    istration of liposomes, in Liposomes: From Physical Structure
    to Therapeutic Applications (C. G. Knight, ed.), Elsevier,
    Amsterdam, pp. 409-441.
Patel, H. M. (1985). Liposomes as a controlled release system,
    Biochem. Soc. Trans., 13, 513-516.
Peeters, P. A. M., Storm, G., and Crommelin, D. J. A. (1987).
    Immunoliposomes in vivo: State of the art, Adv. Drug Del.
    Rev., 1, 249-266.
Peeters, P. A. M., Oussoren, C., Eling, W. M. C., and Crommelin,
    D. J. A. (1988). Immunospecific targeting of immunoliposomes
    F(ab')2 and IgG to red blood cells in vivo, Biochim. Biophys.
    Acta, 943, 137-147.
Peeters, P. A. M., Brunink, B. G., Eling, W. M. C., and Crom-
    melin, D. J. A. (1989). Therapeutic effect of chloroquine (CQ)
    containing immunoliposomes in rats infected with Plasmodium
    berghei parasitized mouse red blood cells; comparison with com-
    binations of antibodies and CQ or liposomal CQ, Biochim. Bio-
    phys Acta, 981, 269-276.
Perez-Soler, R., Lopez-Berestein, G., Kasi, L. P., Cabanillas, F.,
    Jahns, M., Glenn, H., Hersh, E. M., and Haynie, T. (1985).
    Distribution of 99mtechnetium labeled multilamellar liposomes in
    patients with Hodgkin's disease, J. Nucl. Med., 26, 743-749.
Perez-Soler, R., Khokhar, A. R., Hacker, M. P., and Lopez-
    Berestein, G. (1986). Toxicity and antitumor activity of cis-
    bis-cyclopentenecarboxylato-1,2-diaminocyclohexane platinum(II)
    encapsulated in multilamellar vesicles, Cancer Res., 46, 6269-6273.

Popescu, M. C., Swenson, C. E., and Ginsberg, R. S. (1987). Liposome-mediated treatment of viral, bacterial, and protozoal infections, in Liposomes: From Biophysics to Therapeutics (M. J. Ostro, ed.), Marcel Dekker, New York, pp. 219-251.

Poste, G., Bucane, C., Raz, A., Bugelski, P., Kirsh, R., and Fidler, I. J. (1982). Analysis of the fate of systemically administered liposomes and implications for their use in drug delivery, Cancer Res., 42, 1412-1422.

Poste, G. (1983). Liposome targeting in vivo: Problems and opportunities, Biol. Cell, 47, 19-38.

Poste, G., and Kirsh, R. (1983). Site-specific (targeted) drug delivery in cancer chemotherapy, Biotechnology, 1, 869-878.

Poste, G. (1984). Drug targeting in cancer therapy, in Receptor-Mediated Targeting of Drugs (G. Gregoriadis, G. Poste, J. Senior, and A. Trouet, eds.), Plenum Press, New York, pp. 427-474.

Poznansky, M. J., and Juliano, R. L. (1984). Biological approaches to the controlled delivery of drugs: A critical review, Pharmacol Rev., 36, 277-336.

Presant, C. A., Proffitt, R. T., Turner, A. F., Williams, L. E., Winsor, D., Werner, J. L., Kennedy, P., Wiseman, C., Gala, K., McKenna, R. J., Smith, J. D., Bouzaglou, S. A., Callahan, R. A., Baldeschwieler, J., and Crossley, R. J. (1988). Successful imaging of human cancer with indium-111-labled phospholipid vesicles, Cancer, 62, 905-911.

Proffitt, R. T., Williams, L. E., Presant, C. A., Tin, G. W., Uliana, J. A., Gamble, R. C., and Baldeschwieler, J. D. (1983). Liposomal blockade of the reticuloendothelial system: Improved tumor imaging with small unilamellar vesicles, Science, 220, 502-505.

Rahman, Y. E. (1988). Use of liposomes in metal poisonings and metal storage diseases, in Liposomes as Drug Carriers: Recent Trends and Progress (G. Gregoriadis, ed.), John Wiley and Sons, Chichester, pp. 485-495.

Rand, R. P. (1981). Interacting phospholipid bilayers: Measured forces and induced structural changes, Ann. Rev. Biophys. Bioeng., 10, 277-314.

Rao, L. S. (1984). Preparation of liposomes on the industrial scale: Problems and perspectives, in Liposome Technology, Vol. 1 (G. Gregoriadis, ed.), CRC Press, Boca Raton, pp. 247-257.

Reeves, J. P., and Dowben, R. M. (1969). Formation and properties of thin-walled phospholipid vesicles, J. Cell. Physiol., 73, 49-60.

Rhalem, A., Bourdieu, C., Luffau, G., and Pery, P. (1988). Vaccination of mice with liposome-entrapped adult antigens of Nippostrongylus brasiliensis, Ann. Inst. Pasteur/Immunol., 139, 157-166.

Rhoden, V., and Goldin, S. M. (1979). Formation of unilamellar
  lipid vesicles of controllable dimensions by detergent dialysis,
  Biochemistry, 18, 4173-4176.
Richardson, V. J., Ryman, B. E., Jewkes, R. F., Jeyasingh, K.,
  Tattersall, M. N., Newlands, E. S., and Kaye, S. B. (1979).
  Tissue distribution and tumor localization of ^{99m}technetium labeled
  liposomes in cancer patients, Br. J. Cancer, 40, 35-43.
Roerdink, F., Dijkstra, J., Hartman, G., Bolscher, B., and Scherphof, G. L. (1981). The involvement of parenchymal, Kupffer
  and endothelial liver cells in the hepatic uptake of intravenously
  injected liposomes, Biochim. Biophys. Acta, 677, 79-89.
Rudolph, A. S., and Crowe, J. H. (1985). Membrane stabilization
  during freezing: The role of two natural cryoprotectants, trehalose and proline, Cryobiology, 22, 367-377.
Rustum, Y. M., Chandrakant, D., Mayhew, E., and Papahadjopoulos,
  D. (1979). Role of liposome type and route of administration in
  the antitumor activity of liposome-entrapped 1-β-arabinofuranosyl-
  cytosine against mouse L1210 leukemia, Cancer Res., 39, 1390-
  1395.
Ryman, B. E., Barratt, G. M., Patel, H. M., and Tüzel, N. S.
  (1982). Possible use of liposomes in drug delivery, in Optimization of Drug Delivery (H. Bundgaard, A. Bagger Hansen, and
  H. Kofod, eds.), Munksgaard, Copenhagen, pp. 351-364.
Saunders, L., Perrin, J., and Gammack, D. B. (1962). Aqueous
  dispersion of phospholipid by ultrasonic radiation, J. Pharm.
  Pharmacol., 14, 567-572.
Schaeffer, H. E., and Krohn, D. L. (1982). Liposomes in topical
  drug delivery, Invest. Ophthalmol. Visual Sci., 22, 220-227.
Scherphof, G., Damen, J., and Hoekstra, D. (1981). Interactions
  of liposomes with plasma proteins and components of the immune
  system, in Liposomes: From Physical Structure to Therapeutic
  Applications (C. G. Knight, ed.), Elsevier, Amsterdam, pp.
  299-322.
Scherphof, G. L., Roerdink, F., Dijkstra, J., Ellens, H., de
  Zanger, R., and Wisse, E. (1983). Uptake of liposomes by
  rat and mouse hepatocytes and Kupffer cells, Biol. Cell, 47,
  47-58.
Schieren, H., Rudolph, S., Finkelstein, M., Coleman, P., and
  Wessmann, G. (1978). Comparison of large unilamellar vesicles
  prepared by a petroleum ether vaporization method with multilamellar vesicles: ESR, diffusion and entrapment analyses, Biochim. Biophys. Acta, 542, 137-153.
Schroit, A. J., and Key, M. E. (1983). Induction of syngeneic
  tumor-specific immunity by liposomes reconstituted with L$_2$C
  tumor-cell antigens, Immunology, 49, 431-438.

Schurtenberger, P., Mazer, N., Waldvogel, S., and Känzig, W.
    (1984). Preparation of monodisperse vesicles with variable size
    by dilution of mixed micellar solutions of bile salt and phospha-
    tidylcholine, Biochim. Biophys. Acta, 775, 111-114.
Schwendener, R. A., Asanger, M., and Weder, H. G. (1981). n-
    Alkylglucosides as detergents for the preparation of highly homo-
    geneous bilayer liposomes of variable sizes (60-240 φ) applying
    defined rates of detergent removal by dialysis, Biochem. Biophys.
    Res. Commun., 100, 1055-1062.
Sculier, J. P., Coune, A., Brassine, C., Laduron, C., Atassi, G.,
    Ruysschaert, J. M., and Fruhling, H. (1986). Intravenous in-
    fusion of high doses of liposomes containing NSC 251635, a water-
    insoluble cytostatic agent. A pilot study with pharmacokinetic
    data, J. Clin. Oncol., 4, 789-797.
Sculier, J. P., Coune, A., Meunier, F., Brassine, C., Laduron, C.,
    Hollaert, C., Collette, N., Heymans, C., and Klastersky, J.
    (1988). Pilot study of amphotericin B entrapped in sonicated
    liposomes in cancer patients with fungal infections, Eur. J. Cancer
    Clin. Oncol., 24, 527-538.
Senior, J., and Gregoriadis, G. (1982). Stability of small unilamellar
    liposomes in serum and clearance from the circulation: The effect
    of the phospholipid and cholesterol components, Life Sci., 30,
    2123-2136.
Senior, J. (1987). Fate and behavior of liposomes in vivo: A re-
    view of controlling factors, CRC Crit. Rev. Therapeut. Drug
    Carrier Syst., 3, 123-193.
Sharma, P., Tyrrell, D. A., and Ryman, B. E. (1977). Some prop-
    erties of liposomes of different size, Biochem. Soc. Trans., 5,
    1146-1149.
Shek, P. N., Jurima-Romet, M., Barber, R. F., and Demeester, J.
    (1988). Liposomes: Potential for inhalation prophylaxis and
    therapy, J. Aerosol Med., 1, 257-258.
Shew, R. L., and Deamer, D. W. (1985). A novel method for en-
    capsulation of macromolecules in liposomes, Biochim. Biophys.
    Acta, 816, 1-8.
Singh, K., and Mezei, M. (1983). Liposomal ophthalmic drug deliv-
    ery system. I. Triamcinolon acetonide, Int. J. Pharm., 16,
    339-344.
Singh, K., and Mezei, M. (1984). Liposomal ophthalmic drug deliv-
    ery system. II. Dihydrostreptomycin sulphate, Int. J. Pharm.,
    19, 263-269.
Smith, M. T., Thor, H., and Orrenius, S. (1983). The role of
    lipid peroxidation in the toxicity of foreign compounds to liver
    cells, Biochem. Pharmacol., 32, 763-764.
Steerenberg, P. A., Storm, G., de Groot, G., Claessen, A., Bergers,
    J. J., Franken, M. A. M., van Hoesel, Q. G. C. M., Wubs, K. L.,
    and de Jong, W. H. (1988). Liposomes as drug carrier system

for cis-diamminedichloroplatinum(II): Antitumor activity in vivo, induction of drug resistance, nephrotoxicity and Pt distribution. Cancer Chemother. Pharmacol., 21, 299-307.

Stevenson, R. W., Patel, H. M., Parsons, J. A., and Ryman, B. (1982). Prolonged hypoglycemic effect in diabetic dogs due to subcutaneous administration of insulin liposomes, Diabetes, 31, 506-511.

Storm, G., Van Bloois, L., Brouwer, M., and Crommelin, D. J. A. (1985). The interaction of cytostatics with adsorbents in aqueous media. The potential implications for liposome preparation, Biochim. Biophys. Acta, 818, 343-351.

Storm, G. (1987). In Liposomes as Delivery System for Doxorubicin in Cancer Chemotherapy. (Ph.D. Thesis.) University of Utrecht, The Netherlands.

Storm, G., Roerdink, F. H., Steerenberg, P. A., De jong, W. H., and Crommelin, D. J. A. (1987). Influence of lipid composition on the antitumor activity exerted by doxorubicin-containing liposomes in a rat solid tumor model, Cancer Res., 47, 3366-3372.

Storm, G., Nässander, U. K., Roerdink, F. H., Steerenberg, P. A., de Jong, W. H., and Crommelin, D. J. A. (1989a). Studies on the mode of action of doxorubicin-liposomes, in Liposomes in the Therapy of Infectious Diseases and Cancer (G. Lopez-Berestein and I. J. Fidler, eds.), Alan R. Liss, New York, pp. 105-116.

Storm, G., van Hoesel, Q. G. C. M., De Groot, G., Kop, W., Kruizinga, W., and Hillen, F. C. (1989b). A comparative study on antitumor effect, cardiotoxicity and nephrotoxicity of doxorubicin administered as bolus, continuous infusion or entrapped in liposomes in the Lou/M Wsl rat, Cancer Chemother. Pharmacol. 24, 341-348.

Storm, G., Oussoren, C., and Peeters, P. A. M. (1990a). Safety of liposome administration, in Membrane Lipid Oxidation, Vol. 3 (C. Vigo-Pelfrey, ed.), CRC Press, Boca Raton, in press.

Storm, G. Nässander, U., de Jong, W. H., and Crommelin, D. J. A. (1990b). Effect of lipid composition on the in vivo integrity of doxorubicin-containing liposomes. Two pathways for sustained release of doxorubicin. Submitted.

Stratford, R. E., Yang, D. C., Redell, M. A., and Lee, V. H. L. (1983). Effects of topically applied liposomes on disposition of epinephrine and inulin in the albino rabbit eye, Int. J. Pharm., 13, 263-272.

Straubinger, R. M., Lopez, N. G., Debs, R. J., Hong, K., and Papahadjopoulos, D. (1988). Liposome-based therapy of human ovarian cancer: Parameters determining potency of negatively charged and antibody-targeted liposomes, Cancer Res., 48, 5237-5245.

Strauss, G. (1984). Freezing and thawing of liposome suspensions, in Liposome Technology, Vol. 1 (G. Gregoriadis, ed.), CRC Press, Boca Raton, pp. 197-219.

Sur, B., Ray, R. R., Sur, P., and Roy, D. K. (1983). Effect of liposomal encapsulation of cis-platinum diamminodichloride in the treatment of Ehrlich ascites carcinoma, Oncology, 40, 372-376.

Szoka, F., and Papahadjopoulos, D. (1978). Procedure for preparation of liposomes with large aqueous space and high capture by reverse-phase evaporation, Proc. Natl. Acad. Sci. USA, 75, 4194-4198.

Szoka, F., and Papahadjopoulos, D. (1980). Comparative properties and methods of preparation of lipid vesicles (liposomes), Ann. Rev. Biophys. Bioeng., 9, 467-508.

Szoka, F., and Papahadjopoulos, D. (1981). Liposomes: Preparation and characterization, in Liposomes: From Physical Structure to Therapeutic Applications (C. G. Knight, ed.), Elsevier, Amsterdam, pp. 51-82.

Talsma, H., Özer, A. Y., van Bloois, L., and Crommelin, D. J. A. (1989). The size reduction of liposomes with a high pressure homogenizer (Microfluidizer). Characterization of prepared dispersions and comparison of conventional methods, Drug Dev. Indust. Pharm., 15, 197-207.

Taylor, R. L., Williams, D. M., Craven, P. C., Graybill, J. R., Drutz, D. J., and Magee, W. E. (1982). Amphotericin B in liposomes: A novel therapy for histoplasmosis, Am. Rev. Resp. Dis., 125, 610-611.

Tomlinson, E. (1987). Theory and practice of site-specific drug delivery, Adv. Drug Del. Rev., 1, 87-198.

Toonen, P. A. H. M., and Crommelin, D. J. A. (1983). Immunoglobulins as targeting agents for liposome-encapsulated drugs, Pharm. Weekbl. Sci. Ed., 5, 269-280.

Torchilin, V. P., Burkhanov, S. A., Mazhul, L. A., and Ageeva, O. N. (1988). In Liposomes as Drug Carriers: Recent Trends and Progress (G. Gregoriadis, ed.), John Wiley and Sons, Chichester, pp. 229-233.

Treat, J., Greenspan, A. R., and Rahman, A. (1989). Liposome encapsulated doxorubicin preliminary results of phase I and phase II trials, in Liposomes in the Therapy of Infectious Diseases and Cancer, (G. Lopez-Berestein and I. J. Fidler, eds.), Alan R. Liss, New York, pp. 353-365.

Tremblay, C., Barza, M., Flore, C., and Szoka, F. (1984). Efficacy of liposome-intercalated Amphotericin B in the treatment of systemic candidiasis in mice, Antimicrob. Ag. Chemother., 26, 170-173.

Tremblay, C., Barza, M., Szoka, F., Lahav, M., and Baum, J. (1985). Reduced toxicity of liposome-associated amphotericin B injected intravitreally in rabbits, Invest. Ophthalmol. Visual Sci., 26, 711-16.

Tsilibary, E. C., and Wissig, S. L. (1977). Absorption from the peritoneal cavity: SEM study of the mesothelium covering the peritoneal surface of the muscular portion of the diaphragm, Am. J. Anat., 149, 127-133.

Turner, A. F., Presant, C. A., Proffitt, R. T., Williams, L. E., Winsor, D. W., and Werner, J. L. (1988). [111]In labeled liposomes: Dosimetry and tumor depiction, Radiology, 166, 761-765.

Udayachander, M., Meenakshi, A., Muthiah, R., and Sivanandham, M. (1987). Tumor targeting potential of liposomes encapsulating Ga-67 and antibody to Dalton's lymphoma associated antigen (anti-DLAA), Int. J. Rad. Oncol. Biol. Phys., 13, 1713-1719.

Ueno, M., Tanford, C., and Reynolds, J. A. (1984). Phospholipid vesicle formation using nonionic detergents with low monomer solubility. Kinetic factors determine vesicle size and polydispersity, Biochemistry, 23, 3070-3076.

Van Bloois, L., Dekker, D. D., and Crommelin, D. J. A. (1987). Solubilization of lipophilic drugs by amphiphiles: Improvement of the apparent solubility of almitrine bismesylate by liposomes, mixed micelles and O/W emulsions, Acta Pharmaceut. Technol., 33, 136-139.

Van Bommel, E. M. G., and Crommelin, D. J. A. (1984). Stability of doxorubicin-liposomes on storage: As an aqueous dispersion, frozen or freeze dried, Int. J. Pharm., 22, 299-310.

Van Dalen, F., Kersten, G., Teerlink, T., Beuvery, E. C., and Crommelin, D. J. A. (1988). Preparation and characterization of liposomes with incorporated Neisseria gonorrhoeae protein IB and amphophilic adjuvants, J. Control. Rel., 7, 123-132.

Van Hoesel, Q. G. C. M., Steerenberg, P. A., Crommelin, D. J. A., van Dijk, A., van Oort, W., Klein, S., Douze, J. M. C., de Wildt, D. J., and Hillen, F. C. (1984). Reduced cardiotoxicity and nephrotoxicity of doxorubicin entrapped in stable liposomes in the Lou/Wsl rat, Cancer Res., 44, 3698-3705.

Van Hoogevest, P., and Fankhauser, P. (1989). An industrial liposomal dosage form for muramyl-tripeptide-phosphatidylethanolamine (MTP-PE), in Liposomes in the Therapy of Infectious Diseases and Cancer (G. Lopez-Berestein and I. J. Fidler, eds.), Alan R. Liss, New York, pp. 453-466.

Van Rooyen, N. (1988). Liposomes as immunological adjuvants: Recent developments, in Liposomes as Drug Carriers: Recent Trends and Progress (G. Gregoriadis, ed.), John Wiley and Sons, New York, pp. 159-167.

Wang, C., and Huang, L. (1987). pH-sensitive immunoliposomes mediate target-cell-specific delivery and controlled expression of a foreign gene in mouse, Proc. Natl. Acad. Sci. USA, 84, 7851-7855.

Weder, H. G., and Zumbuehl, O. (1984). The preparation of variably sized homogeneous liposomes for laboratory, clinical and

industrial use by controlled detergent dialysis, in Liposome
Technology, Vol. 1 (G. Gregoriadis, ed.), CRC Press, Boca
Raton, pp. 79-107.

Weereratne, E. A. H., Gregoriadis, G., and Crow, J. (1983). Toxicity of sphingomyelin-containing liposomes after chronic administration in mice, Br. J. Exp. Pathol., 64, 670-676.

Weiner, A. L. (1987). Lamellar systems for drug solubilization, in
Liposomes: From Biophysics to Therapeutics (M. J. Ostro, ed.),
Marcel Dekker, New York, pp. 339-369.

Weiner, N., and Chia-Ming Chiang (1988). Gastrointestinal uptake
of liposomes, in Liposomes as Drug Carriers: Recent Trends
and Progress (G. Gregoriadis, ed.), John Wiley and Sons, Chichester, pp. 599-607.

Weinstein, J. N., and Leserman, L. D. (1984). Liposomes as drug
carriers in cancer chemotherapy, Pharmacol. Ther., 24, 207-233.

Wiebe, V. J., and Degregorio, M. W. (1988). Liposome-encapsulated
amphotericin B: A promising new treatment for disseminated
fungal infections, Rev. Infect. Dis., 10, 1097-1101.

Williams, B. D., O'Sullivan, M. M., Saggu, G. S., Williams, K. E.,
Williams, L. A., and Morgan, J. R. (1986). Imaging in rheuma-
toid arthritis using liposomes labeled with technetium, Br. Med.
J. (Clin. Res.), 293, 1143-1144.

Williams, B. D., O'Sullivan, M. M., Saggu, G. S., Williams, K. E.,
Williams, L. A., and Morgan, J. R. (1987). Synovial accumula-
tion of technetium-labeled liposomes in rhematoid arthritis, Ann.
Rheum. Dis., 46, 314-318.

Williams, L. E., Proffitt, R. T., and Lovisatti, L. (1984). Possible
applications of phospholipid vesicles (liposomes) in diagnostic
radiology, J. Nucl. Med. Allied Sci., 28, 35-45.

Woolfrey, S. G., Taylor, G., Kellaway, I. W., and Smith, A. (1988).
Pulmonary absorption of liposome-encapsulated 6-carboxyfluores-
cein, J. Control. Rel., 5, 203-209.

Yatvin, M. B., Mühlensiepen, H., Porschen, W., Weinstein, J. N.,
and Feinendegen, L. E. (1981). Selective delivery of liposome-
associated cis-dichlorodiammineplatinum(II) by heat and its influence on tumor drug uptake and growth, Cancer Res., 41,
1602-1607.

Yau-Yong, A., Chow, J., and Law, M. (1986). Liposome delivery
of a biologically active peptide, 133rd Ann. Meeting Am. Pharm.
Assoc., 16, 105.

Zonneveld, G. M., and Crommelin, D. J. A. (1988). Liposomes:
Parenteral administration to man, in Liposomes as Drug Carriers:
Recent Trends and Progress (G. Gregoriadis, ed.), John Wiley
and Sons, Chichester, pp. 795-817.

Index

Abate, 90
Acetonitrile, 205
n-Acetylhydroxyproline, 203
α-Acetylmethadol, 87
Aclarubicin, 21, 250
Actinomycin D, 297, 304
Adriamycin, 21-22, 233, 243-
    250, 303
Alanine, 211
Albumin, 232, 240, 262, 282
Albuterol, 299
Alkaline phosphatase, 59, 61
Alkylcyanoacrylate, 262
Alzheimer's disease, 55
Ames mutagenicity, 110
Amino acid ester, 174-176
Aminolipids, 276
2-Aminopicoline, 179
Amino-4-picoline, 177
Amphotericin B, 285, 309
Ampicillin, 22
Anesthetics, 15
Angiogenesis, 55, 57
Antibacterial agents, 177
Antibiotics, 15, 22, 233

Antibody, 27
Anticancer agents, 15
Antigens, 25, 27-28, 307
Antimony trifluoride, 2
Ascorbic acid, 280
Aseptic processing, 12-13
Aspartic acid, 212
Aspergillus, 286
Atropine, 298
Autocatalytic degradation, 97
Autoradiography, 52-54

Baboons, 17
BCNU, 50, 53, 55, 59, 66
Benzocaine, 179
Benzyl penicillin, 298
Bethanechol, 55-56, 58
Biocompatibility, 65
Biodegradation, 14, 81, 85, 97
Biomedical membranes, 182
Biscarboxyphenoxy propane, 44
Bisphenol A, 223
Bleomycin, 239, 249-250, 300

Block copolymers, 84
Blood, 8
Bone, 239
Bovine growth hormone, 30, 60, 63
Brain, 53-55, 66
Breast cancer, 245
p-Bromoaniline, 65
Brucellosis, 287
Bulk erosion, 5, 97
Butyl p-aminobenzoate, 179
Butylated hydroxytoluene, 280
Butyl hydroxyanisole, 280
γ-Butyrolactone, 122

Cadmium acetate, 44
Calcitonin, 25
Candida, 286
Candidiasis, 285
Capronor, 71, 79, 94, 110-112
n-Carboxyanhydrides, 196, 205
Carboxyfluorescein, 275, 277, 298, 303
p-Carboxyphenoxyalkanoic acid, 49
Carcinoma, 57
Carmustine (see BCNU)
Cartilage, 61, 239
Casein microspheres, 248
Catecholamine, 172-173
Cattle synchronization, 14
CCNU, 21
Cellulose, 232-233
Cellulose acetate butyrate, 84-85
Cellulose propionate, 84
Cervical cap, 238
Chemoembolism, 244-245
Chemoembolization, 21
Chitin, 233
Chloramphenicol, 22
Chlorine dioxide, 24
Chloroform, 64
Chloroprocaine, 179
Chloropromazine, 90-91

Chloroquine, 297
Cholate, 269
Cholesterol, 264-265, 277, 279-280, 285, 288, 301, 310
Cholesterolhemisuccinate, 261
Chondrogenic stimulating proteins, 56, 61
Cisplatin, 185, 290, 300
Clinical trials, 17-18
Codeine, 87
Collagen, 232-234
  dimensional stability, 234
  mechanical strength, 234
  tissue irritation, 234
Complement, 282
Compression molding, 10, 50, 52, 63, 87, 208
Conjunctivitis, 233
Contact angles, 167-168, 183
Contact lenses, 184
Continuous-flow reactors, 172
Contraceptive hormones, 2
Contraceptives, 6, 9, 11, 28
Copper, 63
Cornea, 55, 65
Corneal shields, 236
Corticosteroids, 242, 299, 308
Cortisone, 56
Cortisone acetate, 54
Cromoglycate, 299
Cryopreservation, 278
Cryoprotectants, 277
Cyanogen bromide, 219
Cyanogen halide, 218
Cyclazocine, 18
Cyclobenzaprine hydrochloride, 145, 149, 153, 155
Cyclophosphamide, 285
Cytosine arabinoside, 90, 298, 300

Dental, 2
Deoxycholate, 269, 285
Desaminotyrosine, 220
Devonorgestrel, 143

Dexamethasone, 237
Dexon, 13
Dextran, 232-233
Dextran acetate, 7
Diabetes, 57, 60
Dibucaine, 24
Dicetylphosphate, 310
Dideoxykanamycin, 23
Differential scanning colorim-
    etry (DSC), 3
Diffusion, 17, 26, 88
Diffusion coefficients, 86
Diketene acetal, 122-123, 127
4-Dimethylaminopyridine, 45
Dimiristoylphosphatidylcholine,
    286
Dimiristoylphosphatidylglycerol,
    286
Dioleoylphosphatidic acid, 266
Diphenol, 218
1,6-Diphenyl-1,3,5-hexatriene,
    276
Diphosgene, 45
Diquat, 12
Distearoylphosphatidylcholine,
    288
Dopamine, 172-173, 177
Doxorubicin, 233, 243, 247,
    250, 272, 290-294
DuPont, 2

Earth metal oxides, 44
Ehrlich ascites carcinoma, 244,
    250
ELISA, 30
Endothal, 12
Endothelial cells, 65
Enzymatic degradation, 105
Enzymes, 56
Epinephrine, 172
Epirubicin, 244
Epoxyhydrodienes, 279
Erythrocytes, 301
Erythromycin, 235
Erythromycin estolate, 235

Estradiol, 238
Estradiol benzoate, 16
Estrogen, 14, 27
Estrus, 26
Ethinyl estradiol, 16, 87
Ethylene oxide, 13
Ethylene vinyl acetate, 65, 88
Ethyl oleate, 94
Extrusion, 87

Fabrication, 87
Fibers, 11, 87, 232
Fibrinogen, 282
Fibrinogen microspheres, 250
Fibroblasts, 103
Fibronectin, 282
Films, 232, 234
Fluidized bed coating, 8-9
5-Fluorouracil, 90, 238-239, 243,
    249-250
Flurbiprofen, 237
Fluridone, 12, 90
Formulations, 45
    compression molding, 10, 45,
        50, 52, 63, 87, 208
    "hot melt" microencapsulation,
        46
    injection molding, 10, 11, 45
    microsphere preparation by
        solvent removal, 46
Fracture, 2
FSH, 14
Fumaric acid, 49
Fur, 24

β-Galactosidase, 60, 62
Gamma irradiation, 13
Ganglioside, 282
Gelatin, 232, 240
Gelatin microspheres, 248-250
Gelfoam, 66
Gel permeation chromatography,
    44, 151

Gentamicin, 22, 234-235, 237,
    246, 309
Giant cells, 103
Glass transition temperature, 83
Glaucoma, 233, 235
Glioblastoma, 66
Glucose, 58
Glucose-6-phosphate dehydro-
    genase, 172-173
Glutaric dialdehyde, 172
Glutathione, 298
GnRH, 27
Good Manufacturing Practices,
    12, 31
Growth hormone, 30
Growth promoter, 14

β-HCG, 28
Heparin, 168-170
Hepatic cancer, 245
Hepatocytes, 282
Herbicides, 12, 88, 90
Herniorrhaphy, 2
1,6-Hexanediol, 126, 129, 144,
    152-153, 155-156
Histamine, 300
Horse, 17
Humidity, 3
Hyaluronic acid, 232-233
Hydrilla, 90
Hydrocortisone, 24
Hydrocortisone acetate, 24
Hydrogels, 183, 187-188, 232-
    233
Hydroperoxides, 279
ω-Hydroxybutyric acid, 122
20-Hydroxyecdysone, 97
Hydroxyphosphazenes, 175
Hydroxyproline, 202, 205, 208

Immunization, 27
Immunoglobulins, 282
Implants, 10

India ink, 301
Indomethacin, 233
Inflammation, 65-66
Infrared spectroscopy, 63
Inherent viscosity, 13
Injection molding, 10-11, 50, 65,
    151-152
Insect hormones, 24
Insecticide, 90
Insulin, 29, 56-60, 242
Interferon, 30, 93
Intraepithelial cervical dysplasia,
    238
Intraocular lenses, 184
Inulin, 232-233, 305

Juvenile hormone, 24

Kidney, 185
Kinetics, 47
Krebs cycle, 5

Latex, 301
Lauryl alcohol, 2
Leishmaniasis, 282
Levonorgestrel, 11, 16, 88, 94, 96
    110-112, 142, 144, 146
Lewis lung carcinoma, 21
LHRH analog, 97, 100
[D-Trp6, des-Gly10] LHRH di-
    ethylamide, 97
LHRH polypeptide, 14, 25-27
Lidocaine, 24
Ligament reconstruction, 2
Light-scattering, 274
Lipoproteins, 282, 288
Listeriosis, 287
Local anesthetics, 179
Lypressin, 25
L-lysine, 196
Lysine, 211

Lysophospholipids, 272

Macrophages, 8, 103
Magnesium hydroxide, 140
Magnetic albumin microspheres,
    246-247
Magnetite, 246
Mammary adenocarcinoma, 247
MDP-B30, 29
Mechanical strength, 11
Medisorb, 2
Medroxyprogesterone, 234
Meglumine antimoniate, 284
Melatonin, 24-25
Melt rheometers, 11
Melt-spinning, 87
Meperidine, 87
Meshes, 232
Metaproterenol, 299
L-Methadone, 18-20, 87, 92-93
Methoprene, 24
Methotrexate, 290, 308
Methylene chloride, 46-47, 50,
    213
Methylprednisolone, 24
Microcapsules, 6, 8-9, 85, 87,
    90, 92-93, 98-99
Microfluidized bed coating, 19
Microspheres, 6, 8-10, 16-17,
    19-25, 28-29, 87, 91, 93,
    232, 240-249
Milk, 28
Misonidazole, 22
Mitomycin C, 21, 243-245, 249
Molecular weight, 44-45, 47
Monkeys, 99
Monocytes, 8
Monosialoganglioside, 288
Morphine, 19, 97
Muramyl dipeptide, 28-29
Muramyltripeptide, 313
Mutagenic, 65
Mycobacterium, 287

Nafarelin acetate, 26
Naloxone, 18
Naltrexone, 18-19, 87, 93-94,
    97-99, 122
Nanoparticles, 240
Naproxen, 175
Narcotic antagonists, 2, 15, 18,
    92
Narcotics, 86
p-Nitroacetanilide, 135
p-Nitroaniline, 49, 65, 207, 208
Nitrofurantoin, 91
Noregestimate, 16
Norepinephrine, 172
Norethindrone, 87, 141-142
Norethisterone, 6, 9-10, 14-18,
    122
Norgestrel, 15, 87, 242

Ocular inserts, 232, 234-235
Opsonins, 288
Oxytocin, 298

n-Palimtoylhydroxyproline, 206
Papain, 240
Papaverine hydrochloride, 233
PCPP:SA, 59
Pentaerythritol, 123, 128-129
Periodontal disease, 14, 23
Permeability, 86, 103
Peroxidation, 279
Petroleum ether, 46-47
Phagocytosis, 102
Pharmacokinetic studies, 94
Phase separation, 8-9, 22
Phatidylcholine, 264
L-Phenylalanine mustard (mel-
    phalan), 175
p-Phenylenediamine, 65
Phosgene, 45
Phosphatidylcholine, 261, 266,
    270, 275, 277, 279

Phosphatidylglycerol, 261, 279
Phosphatidylinositol, 288
Phosphatidylserine, 261, 310
Phospholipids, 279, 310
Phthalic acid, 154
Phthalic anhydride, 133, 144-
    145, 148, 153-155
Pilocarpine, 234-235, 237
n-Pivalohydroxyproline, 203
Plasma extenders, 187
Platinum antitumor agents, 185
Pluronic F68, 88
Poly(acryl) starch, 232
Poly(n-acylhydroxyproline
    ester), 202, 204, 207,
    209
Poly[bis(methoxyethoxyethoxy)-
    phosphazene] "MEEP",
    184
Poly[bis(methylamino)phospha-
    zene], 186
Poly[bis-(p-methylphenoxy)-
    phosphazene], 169
Poly(bisphenol A carbonate),
    213
Poly(bisphenol A iminocar-
    bonate), 213, 224
Poly[bis(phenoxy)phosphazene],
    170
Polycaprolactone, 12
Poly-ε-Caprolactone:
  anionic polymerization, 73
  biodegradable packaging, 71
  biodegradation, 81, 97
  blends, 84
  cationic polymerization, 75
  coordination polymerization,
    77
  crystallinity, 81, 102-103, 105
  glass transition temperature,
    81
  heat of fusion, 81
  melting point, 81
  monomer synthesis, 72
  permeability, 86
  physical properties, 84
  radical polymerization, 80

[Poly-ε-Caprolactone]
  solubility, 82
  synthesis, 72
  tensile strength, 84
  toxicology, 110
Poly(carboxyphenoxy acetic
    acid), 51
Poly[(p-carboxyphenoxy) alkane],
    47
Poly(carboxyphenoxy) hexane,
    62, 64
Poly(carboxyphenoxy) methane,
    62, 64
Poly(carboxyphenoxyoctanoic
    acid), 51
Poly(carboxyphenoxy) propane,
    62
Poly(carboxyphenoxypropane-
    sebacic acid), 47, 54, 64-
    66
Poly(carboxyphenoxyvaleric acid),
    51
Polycytidylic acid, 30
Poly(diphenoxyphosphazene),
    169
Polyesters, 71
Polyester-urethanes, 84
Polyether ether ketone, 210-211
Polyglactin, 8
Poly(glutamic acid), 196
Poly(hydroxyethyl methacrylate),
    65
Poly(hydroxyproline esters), 208
Poly(iminocarbonates), 212-215,
    218-219, 221-223, 225
Polyisoprene, 4
Polyisosinic acid, 30
Polylactic/glycolic acid, 85
Polylactide/polyglycolide:
  biodegradation, 3, 5-7, 13
    enhancement, 5
    enzymes, 5
  crystallinity, 3, 6
  degradation, 7
  density, 4
  fibers, 11
  glass transition temperature, 3

[Polylactide/polyglycolide]
  hydrolysis, 5
  hydrophobicity, 3
  implants, 10
  in vivo, 6
  melting point, 3-4
  microcapsules, 8-9
  microspheres, 6, 8-10
  molecular weight, 2-3, 13
  monomer acidity, 3
  monomer purity, 3
  monomer stereochemistry, 3
  polymer characteristics, 3
  solubility, 4
    dioxane, 4
    ethyl acetate, 4
    halogenated hydrocarbons,
    4
    hexafluoroisopropanol, 4
    tetrahydrofuran (THF), 4
  synthesis, 2
  toxicity, 2
  water uptake, 3, 5
Polymer hydrolysis, 128
Polymer processing, 150
Polymer synthesis, 121
Poly(γ-methyl-D-glutamate), 222
Polymyxin B, 22
Polyols, 123
Poly(n-palmitoylhydroxyproline),
  210
Poly(n-palmitoylhydroxyproline
  ester), 204-206
Poly(phenylenedipropionic
  acid), 62
Polyphosphazenes:
  biocompatibility, 166-177
  hydrophilic, 168
  hydrophobic, 167
  insoluble, 168
  irritation, 175
  synthesis, 164
  toxicity, 175
Poly(Schiff's base), 201
Poly(sebacic acid), 62, 64
Poly(L-serine), 200
Polyserine, 197

Poly(serine ester), 197
Poly(serine imine), 201
Polystyrene, 4
Poly(terephthalic acid), 54
Poly(terephthalic acid: sebacic
  acid), 54
Poly(n-tetradecanolyhydroxy-
  proline ester), 204
Poly(L-tyrosine), 212, 223-224
Polyvinylalcohol, 7
Poly-(4-vinylpyridine), 45
Pore size, 88
Prednisolone, 242
Procaine, 179
Procaine penicillin, 235
Progesterone, 15-16, 86-89, 91,
  177-178, 238
Progestins, 27
Prostate cancer, 14
Protein A, 247
Proteins, 10, 15
2,3-Pyridine dicarboxylic an-
  hydride), 133
Pyrimethamine, 21

Quinazoline, 20

γ-Radiation, 183-184
Regulatory approval, 2
Release characteristics, 49
Release kinetics, 11
Reservoir devices, 93, 95-96,
  98-99
Rheumatoid arthritis, 242
RIA, 30
Right Valley fever, 30
RNA, 25, 30

Safety, 66
Salmonellosis, 287
Screw extrusion, 10

Sebacic acid, 44
Sebacoyl chloride, 45
L-serine, 198, 200
Serine-β-lactones, 200
Shipping fever, 30
Silastic, 79
Silicone oil, 9, 46, 47
Smooth muscle cells, 65
Sodium cromoglycate, 245
Soft-tissue prostheses, 184
Solvent evaporation, 8, 16, 22,
    24, 29
Solvent film casting, 29, 87
Somatostatin, 25
Sphingomyelin, 282, 288
Spleen, 185
Sponges, 232
Stability, 62
Stannous chloride, 2
Stannous octoate, 2
Starch, 30, 232-233
Stealth liposomes, 289
Stearylamine, 261, 310
Stereoisomers, 2
Sterilization, 12
Steroids, 6, 9, 11, 15, 86, 175-
    176
Streptozotocin, 57
Sulfadiazine, 20, 177, 242
Surgical dressings, 2
Surgicel, 66
Suture, 2, 6, 11, 233
Synthesis, 44

Terbutaline, 301
Testosterone, 16-17, 19-20,
    87, 94-95
Testosterone propionate, 16
Tetracycline, 14, 22-23, 88,
    238-239
Thermogravimetric analysis, 221
Thermoplastic, 154
Thioridazine, 5
Timolol, 233

Tobramycin, 236-237
α-Tocopherol, 280
p-Toluene sulfonic acid, 2, 44
Toxic, 65
Trachea, 2
Trans-cyclohexanedimethanol,
    126, 144-145, 152-153,
    156
Trans-retinoic acid, 238
Triamcinolone, 242
Tricaproin, 94
Triglycerides, 267
Trituration, 50-53
Trypsin, 172, 240
Tyramine, 220
L-tyrosine, 211
Tyrosine, 197, 201, 210, 212,
    214-215, 220, 222-223,
    225

Unsaturated polyanhydrides, 48

Vaccine, 14, 27-28
Valinomycin, 272
Vancomycin, 237
Vinblastine, 272
Viscosity, 44
Von Willebrand factor, 282

Walker 256 carcinoma, 244
Wool, 24
Wound dressings, 233

Xylocaine, 24

Young's modulus, 222

Zero-order degradation, 48
Zero-order release, 176
Zinc, 63
$ZnEt_2$-H_2O, 44